D0876266

Leslie Kish

Selected Papers

WILEY SERIES IN SURVEY METHODOLOGY
Established in part by WALTER A. SHEWHART AND SAMUEL S. WILKS

Editors: *Robert M. Groves, Graham Kalton, J. N. K. Rao, Norbert Schwarz, Christopher Skinner*

A complete list of the titles in this series appears at the end of this volume.

Leslie Kish

Selected Papers

Edited by

STEVEN HEERINGA
Institute for Social Research
University of Michigan
Ann Arbor, Michigan

GRAHAM KALTON
WESTAT

A JOHN WILEY & SONS, INC., PUBLICATION

This text is printed on acid-free paper. ♾

Published simultaneously in Canada.

For ordering and customer service, call 1-800-CALL WILEY.

Library of Congress Cataloging-in-Publication Data:

Kish, Leslie, 1910–
[Selections, 2003]
Leslie Kish : selected papers / [compiled and edited by] Steven Heeringa and Graham Kalton.
p. cm — (Wiley series in survey methodology)
A selection of his key research papers, chosen for their relevance for survey research, published between 1949 and 1999, and augmented by commentaries of colleagues.
Includes bibliographical references and index.
ISBN 0-471-26661-2 (acid-free paper)
1. Social surveys—Methodology. 2. Sampling (Statistics) 3. Social sciences—Statistical methods. I. Kalton, Graham. II. Heeringa, Steven, 1953– III. Title. IV. Series.

HM538.K572 2003
300'.7'23—dc21 2002192251

Printed in the United States of America.

10 9 8 7 6 5 4 3 2 1

Leslie Kish

Contributors

Martin Frankel
14 Patricia Ln
Cos Cob CT 06807-1734

Robert Groves
Survey Research Center
Institute for Social Research
University of Michigan
Ann Arbor MI 48106

Steven Heeringa
Survey Research Center
University of Michigan
Ann Arbor MI 48106

Graham Kalton
Westat
1650 Research Blvd
Rockville MD 20850

James Lepkowski
Joint Program in Survey Methodology
Institute for Social Research
University of Michigan
Ann Arbor MI 48106

Colm O'Muirheartaigh
University of Chicago
1155 East 60th St
Chicago IL 60637-2745

Keith Rust
Westat
1650 Research Blvd
Rockville MD 20850

Contents

FOREWORD

XAVIER CHAROY

President, International Association of Survey Statisticians (IASS)

I am both proud and honored to write these few lines as a foreword to this volume of a selection of Leslie Kish's papers that is co-published by John Wiley and Sons and the International Association of Survey Statisticians (IASS).

Leslie Kish must be considered as one of the "founding fathers" of the IASS, created in 1975 as a section of the International Statistical Institute (ISI). As a very active member of the small committee that had been set up to establish the IASS, he was one of the very few who played an instrumental role in its creation. He then became Vice President of the Association (1977–79) and President (1983–85). He initiated the principle of workshops held in conjunction with the ISI biennial sessions: short half-day meetings at first (1973–85), then well structured 2 to 3 day workshops (1987 and after). As an ordinary member of the Association, he remained active until 1994 by taking responsibility for the "Questions/ Answers" column he had initiated in the review of the Association, the *Survey Statistician*.

This deep and long-lasting involvement of Leslie Kish in such an endeavour may be considered as a summary of his life, entirely devoted to teaching, training, and disseminating sampling techniques throughout the whole world.

Leslie Kish was born in 1910 in Poprad, then in the Austro-Hungarian empire (and now in Slovakia). His family moved to the United States in 1926, and due to the early death of his father, he had to work while studying in high school and college. In 1937, he interrupted his studies and joined the inter-national brigades which fought in the Spanish Civil War. In 1939, he completed his undergraduate degree in mathematics and joined the U.S. Bureau of the Census and then the Department of Agriculture. When the United States entered the Second World War, he volunteered to serve in the Army Air Corps. When the war was over, he spent another two years in the Department of Agriculture. He was, in 1947, one of the founders of the Institute for Social Research (ISR) at the University of Michigan. All his career was then to take place in ISR and the University. Even after his retirement in 1981 as Emeritus Professor, he continued being involved in statistical work through articles, speeches, and consultation.

Teaching was Leslie Kish's passion. Besides his regular activity as a professor at the University of Michigan, he established the summer Sampling Program for

Foreign Statisticians in 1961. About 500 statisticians from over 100 countries have been trained there in the last 40 years. Many of them now have key positions in their countries' statistical agencies and universities. In order to enable statisticians from developing countries to attend these courses, the "Leslie Kish International Fellows Fund" was established in 2000. His activities within IASS mentioned above also show his passion for teaching.

In addition to a considerable number of articles, Leslie Kish produced three major books. *Survey Sampling* (1961), now a great classic, covers all aspects of the subject. *Statistical Design for Research* (1987, published by John Wiley and Sons) presents many innovations based on a wide practical experience. *Sampling Methods for Agricultural Surveys* (1989) was written for the attention of statisticians working in the area of agriculture, especially in developing countries. His articles and other short contributions cannot be counted. I will only mention the book printed by IASS in 1995, a collection of the 42 most interesting "Questions/Answers" he had produced over 17 years in the *Survey Statistician* (1995a).

But teaching and writing were not all of Leslie Kish's life. He also engaged in considerable international activity, traveling all over the world— particularly in developing countries for which he always showed great concern—to help design sample surveys, often together with former trainees of his courses in the University of Michigan. He thus not only maintained an extensive network of correspondents and friends but constantly kept theory in line with practice. I will also mention here the great interest he had in the culture—history, politics, arts, philosophy, cuisine, etc.—of the countries he visited. All that made him a citizen of the world.

Publishing this book has been made possible thanks to its editors, Graham Kalton and Steven Heeringa, both former colleagues and friends of Leslie Kish. With the help of Colm O'Muircheartaigh and a number of Leslie's other former students, colleagues, and friends, they have made a selection of his most interesting and significant articles in order to make a complete and consistent homage to the great Professor. Mrs. Rhea Kish, Leslie's widow, provided editorial support, thus actively contributing to the tribute paid to her late husband.

As co-publisher of the book, the International Association of Survey Statisticians financially contributes to the publication. However, the cost would have exceeded its possibilities if the Kish family had not donated funds to the IASS through a fund at the Institute for Social Research of the University of Michigan.

In my personal name, in that of the IASS and its members, in that of all of Kish's former colleagues and students, I want to extend my warm gratitude to the Kish family, to the editors, and to all those who contributed in some way to have this book printed.

PREFACE

During the more than 50 years of our marriage Leslie was always: thinking about ideas for a new article on a sampling problem; drafting and discussing it with colleagues; getting it published; and adding a copy to his reference file, available to students and fellow samplers. He believed that journal articles had important advantages over books: they could tackle timely, specific problems; they would be widely and easily available; they could serve as continuing education for students and colleagues who had little access to fellow statisticians/samplers—even with those in adjacent countries. Moreover, the different emphases of various journals meant they could cover his broad interests in theory and practice.

In the summer of 2000 Leslie was engrossed in plans for an entirely new version of his 1965 book, *Survey Sampling*, which he was to write jointly with Colm O'Muircheartaigh. So when he was asked once again by a colleague to publish a volume of collected works, he put the idea aside for a later time. After his death in October, it seemed appropriate that the time for a volume of his selected works had arrived. So in 2001 I approached a number of his colleagues (many of them former students) with the idea of putting together a representative collection of his important and still useful articles. They all responded enthusiastically. Graham Kalton and Steven Heeringa agreed to take on the task of editors, with assistance from Colm O'Muircheartaigh. John Wiley and Sons showed interest in publishing the book, but financial support was needed.

I wanted to be sure that the book would be available at no cost to practicing samplers—particularly reaching those Leslie had taught in over 100 countries. Graham and Colm approached the Executive Committee of the International Association of Survey Statisticians (IASS), who generously agreed to support one-half of the printing costs and then to provide a free copy to each of its members. My daughters and I gratefully agreed to support the other half of the costs. In addition, David Featherman, Director of the Institute for Social Research, and Robert Groves, Director of the Survey Research Center, contributed organizational funds to ensure that future students in the Summer Program in Survey Sampling would also receive free copies.

I'd like to describe, briefly, the process used to select the articles included here. Graham, Steve, and Colm carefully put together a list of 82 unduplicated articles (from a larger total) and sent this to 10 additional colleagues. All 13 then ranked their choices on a scale of 1 to 5. They were asked to consider Leslie's broad range of contributions to sampling—theoretical, practical, and philosophical—when making their choices. Everyone responded, their choices were ranked, and this book contains their 17 principal selections. The editors formed the articles into related groups, and it was decided that a commentary about each group would be useful. So colleagues have augmented the collection in each group with brief comments on the historic, current, or continuing relevance of the articles and with other pertinent references. Two of the articles also represent recognition of Leslie's active involvement in various professional organizations: his Presidential Address to the American Statistical Society and his address in honor of the 100th anniversary of the Italian Statistical Society.

All are the original articles as printed; they have been retyped with a few minor corrections, in an attempt at consistent notation in the formulas. I have received permission from all the journals in which they originally appeared, and am grateful for their prompt approvals. Permission from each journal is appropriately noted for each article.

This book could never have come to fruition without the remarkable help of many people. I thank Leslie's colleagues who enthusiastically agreed to the idea and who cheerfully participated in our small survey to rank their choices. I also appreciate the cooperation of those who contributed to the commentaries. I am especially grateful that we were able to include the comments of Charles Alexander before his untimely death. He and Leslie had forged a warm personal and professional relationship as they jointly pursued the further development of rolling samples.

Most of all I owe a great debt to the editors who provided unfailing good humor, encouraging advice, and constructive attention to detail. They spent much energy and a great deal of time from their very busy schedules to deal with the IASS, ISR, SRC, and the publisher, and to review all details about formulas, typography, style, arrangement of articles, etc.

However, this volume could never have been produced without the extraordinary contribution of Gail Arnold at ISR. She faithfully retyped every single article into a Word document and reformatted every one into camera-ready copy. She spent hours and hours at these tasks out of what I can only attribute to her devotion to Leslie and the goodness of her heart. There is no way to thank her completely; I only hope she knows how much Carla Kish, Andrea Kish, and I appreciate her work.

One last thought. I would like to share something I experienced while reading Leslie's works. I do not understand statistics, but I found that his personality, his fundamental character, including humor, creativity, critical—yet kind—comments, and optimistic expectations for the future of the subject he loved are clearly apparent throughout. This has been a bittersweet project for me, and I am most grateful to have been part of it.

Rhea Kish
January 2003

ROLE OF PROBABILITY SAMPLING IN SCIENTIFIC RESEARCH

GRAHAM KALTON AND STEVEN G. HEERINGA[1]

Professor Leslie Kish, one of the pioneers of survey sampling, died on October 7, 2000, at the age of 90. In the Foreword to this volume, Xavier Charoy, President of the International Association of Survey Statisticians for 2001–03, provides some details of Kish's illustrious career of more than 50 years, nearly all of it spent at the Survey Research Center at the University of Michigan.[2] Kish has had a major impact on the development of survey statistics, achieved both through his impressive research contributions and through his extremely successful promotion of the use of scientific probability sampling methods throughout the world. His wide-ranging research always focused on issues of practical importance, and his innovations facilitated the use of effective probability sampling in diverse areas. He promoted the practice of sound probability sampling methods through his expository writings (particularly for sociologists and demographers), through his numerous consultancies and advisory services, and through his training of survey statisticians, particularly those from developing countries.

This volume contains a selection of Kish's key research papers, chosen for their enduring relevance for survey research. They span his career, from his famous first published paper on a method for the objective selection of one person in a sampled household (Kish 1949), using what is now known as the Kish selection grid, to a paper on his central research interest at the time of his death, namely cumulating and combining population surveys (Kish 1999).

Kish's research was always driven by important practical problems. Much of his writing is concerned with aspects of sample design and analysis, but he also always recognized the significance of other aspects of survey quality in addition to sampling. His wide-ranging contributions to survey research thus extend into research on sources of nonsampling errors. Furthermore he made important contributions to the literature on general quantitative research design in social

[1] Graham Kalton, Westat, 1650 Research Blvd., Rockville, MD 20850. grahamkalton@westat.com. Steven Heeringa, Survey Research Center, 426 Thoompson St., University of Michigan, Ann Arbor, MI 48106. sheering@isr.umich.edu.

[2] For further details of Kish's full life and career, see the interview that Frankel and King (1996) conducted with him, Fellegi (2000) and Kalton (2002).

research. To provide a structure for this volume, the selected papers have been separated into sections with common themes: the role of probability sampling in scientific research, applied sampling methods, issues of inference from survey data; nonsampling errors, domain estimation, and professional leadership and training. Each section starts with a short introduction by two of Kish's sometime colleagues and/or students that aims to place the papers selected for the section in context. The introductions also provide references to Kish's other papers in the subject area and to other, more recent, work.

This first section contains two papers. The first, which reviews the history of survey sampling in the twentieth century, was an invited talk that Kish gave at a Centenarian on Representative Sampling in Rome in 1995. It is entitled "The Hundred Years' War of Survey Sampling." In general, reviewing history is valuable since many of the controversies of the past recur, although sometimes in a different form, and that holds true for survey sampling. The first half of the twentieth century saw the laying of firm foundations for probability sampling for surveys, with an explosion in applied sampling thereafter: as Kish points out in this paper, "The developments after 1945 were sudden, dramatic, and far reaching." Since he was a major participant in, and thoughtful contributor to, the development of probability sampling methodology in the second half of the century, the saying "The only good histories are those written by those who had command in the events they describe" seems particularly apposite for this paper.

The second paper in this section contains a broad assessment of quantitative designs for social research. This widely cited and influential paper on "Some Statistical Problems in Research Design," published in 1959, was his first on this topic. Throughout his career Kish maintained an interest in the kinds of designs suited to different analytic objectives. In particular, the designs included surveys, experiments, and other investigations (also known as observational studies, controlled observations, and quasi-experimental designs). Other publications of his in this area include Kish (1975), Kish (1985), and his 1987 book. It is interesting to note that Kish shares with Cochran, another famous pioneer in survey sampling whom Kish greatly admired, the publication of major texts in both survey sampling and observational studies—Kish's (1965) *Survey Sampling* and his (1987) *Statistical Design for Research* and Cochran's (1953) *Sampling Techniques* (and subsequent editions in 1963 and 1977) and his (1983) *Planning and Analysis of Observational Studies*.

The literature on observational studies extends across several disciplines, including the social sciences, epidemiology, and evaluation studies. Shadish, Cook, and Campbell (2002) provide a comprehensive treatment from a social science perspective. Rosenbaum (2002) gives a detailed treatment of analysis issues, with examples mainly from the health sciences. What Kish was able to bring to this subject was the survey sampling perspective and his appreciation of the similarities and differences among surveys, experiments, and observational studies. He was always keenly aware of the importance of well-chosen terminology to catch attention for key concepts. In the area of research design, he adapted the term "the three Rs" to denote realism, randomization, and representativeness as desirable properties for studies of causal relationships. In his writings, he used this term effectively to focus on the differing relative strengths that different modes of study have on these dimensions.

1

THE HUNDRED YEARS' WARS OF SURVEY SAMPLING

LESLIE KISH

1. INTRODUCTION: CONCEPTION

This will be neither a full history nor a detailed study of just one of its many aspects. Rather I will describe the development of survey sampling with a series of controversies, or battles, as the title implies. But unlike military battles, our scientific controversies are seldom settled once and for all. Instead they tend to continue, although much changed in form and content. I shall talk of battles past, present, and future, more or less in chronological order.

The 1895 paper of A. N. Kiaer (1895a, b) can well serve for an official birth date for survey sampling, though surveys were done even earlier, by LaPlace and Lavoisier among others. But first I mention briefly the necessary "conception" of "statistics" that must precede the birthday of survey sampling. That may be placed in 1820 when Quetelet founded "statistics" on the concept of "population." I propose that POPULATION is the most basic or central concept of statistics, followed by REPLICATION and much later by RANDOM REPLICATION and RANDOM VARIABLES. T. M. Porter (1987) shows in *The Rise of Statistical Thinking: 1820-1900* that the name and field of "statistics" arose when Quetelet showed that the laws of probability could be applied to *real* populations of *diverse* individuals. Later Galton took those ideas to biostatistics and Maxwell to physics.

These conceptions of statistics based on populations of diverse elements were necessary bases for the birth of sampling 75 years later.

2. THE TWO HALVES OF A CENTURY OF SAMPLING

The century since Kiaer's 1895 paper (1895a, b) splits neatly into two equal halves, when at the end of World War II in 1945 survey sampling really "took off." However,

---

Reprinted with permission from *Statistics in Transition* (1995), 2(5), 813–830.
Presented at the Centenarian on Representative Sampling arranged by the Italian Statistical Society in Rome on 31 May 1995.

there were some important developments in the first half also, of which I shall now merely mention a few.

First, early activities of the International Statistical Institute are memorialized in the two papers of 1926 (Section 4).

Second, came developments in Russia, Kiev, and the early Soviet Union (see Zarkovich 1956 and 1962), with A. A. Tchuprow (1923), who is remembered in the Tchuprow-Neyman allocation, and A. G. Kovalevsky's *Basic Theory of Sampling Methods* (1924) was the first book ever on sampling. But it was in Russian and it disappeared together with its author.

Then came the writings of Neyman, *An Outline of the Theory and Practice of the Representative Method* (1933) in Warsaw in Polish with English summary, referred to and used in his majestic foundation paper for sampling (1934): *On the Two Different Aspects of the Representative Method: The Method of Stratified Sampling and the Method of Purposive Selection.* Later he gave his 1938 conferences in Washington, but he left sampling when he founded his statistical department and laboratory at Berkeley.

Third, in England, A. L. Bowley conducted and wrote about surveys (1913), also about random sampling (1926). At Rothamsted, R. A. Fisher appreciated sampling as "quite the most important question in practical statistics" (Yates 1946) and the most influential samplers came from there: Mahalanobis, Yates, Cochran and Snedecor all worked with Fisher at Rothamsted and Cambridge. But today the parallel and related developments in experimental design at Rothamsted are better known.

Fourth, in the United States the most important developments were the unemployment surveys by the Works Projects Administration, the WPA in 1934-1940 (Frankel and Stock 1942; Stephan 1948), which in 1943 became the famous Labor Force Surveys of the Census Bureau.

The developments after 1945 were sudden, dramatic, and far-reaching. Five textbooks appeared in quick succession in English, all of which have become classics: Yates 1949; Deming 1950; Cochran 1953; Hansen, Hurwitz, and Madow 1953; and Sukhatme 1954. There were a few courses in sampling since 1939, but they were based on articles, reports, and notes. But after these five books many courses appeared in the USA, in the U.K., in India and elsewhere. The samples of the U.S. Census Bureau became the Current Population Surveys of 357 primary sampling areas by 1963 (U.S. Bureau of the Census 1967; Kish 1965a, 10.4). The technical staff, directed by M. H. Hansen, became a national and international resource on sample surveys.

Another center of world wide influence was the Indian Statistical Institute organized by Mahalanobis (1944; 1954), which trained hundreds of statisticians in all the states of India and many of the "less developed countries" (LDCs). The Indian National Sample Surveys covered the states of the subcontinent and gained worldwide fame.

Survey sampling by now has spread to most of the national statistical offices, though not evenly, either in time or in quality. Still more uneven is the history of sample surveys at universities (Section 9). Here I single out two centers for their international influences: in the USA and the world. At Iowa State University the premier center of sampling was organized by Snedecor, Wallace, Sarle, and by

Cochran, who gave the first sampling course in 1939. Then in 1948 a sampling program began at our Survey Research Center at the University of Michigan.

The international influences of the Statistical Divisions of the FAO (Rome) and the UN (New York) began early (1950) and these have spread over time the practice of sample surveys. Even in this brief note the role of *multinational* survey designs must be noted, especially the World Fertility Surveys and the Demographic Health Surveys (Kish 1994); more in Section 10.

Now I turn to a "chronological list of major conflicts of survey sampling," with apologies for all those terms. First, chronology of those conflicts is uncertain because their origins are vague, and because most of them are still with us in some form. Second, what defines a "major" conflict depends on personal choice. Third, the very existence and certainly the magnitude of these conflicts also represent my personal judgments. Yours may differ from mine.

3. FROM 1895 TO 1926: CENSUSES AND MONOGRAPHS VERSUS SAMPLING

The battles led by A.N. Kiaer at the biennial meetings of the ISI were conducted chiefly against the forces of complete enumeration entrenched in the official statistical offices. In addition to complete enumerations on large, national scales, sampling also was opposed by the method of "monography." "Monographs" covered with accurate, detailed, and complete enumeration and description some single geographical unit. That unit (or a few units) would be chosen with purpose and care to be "typical" on some carefully chosen variables. Faith in complete censuses survived entirely the first half century in statistical offices, and still does in many aspects. And selecting one or a few sites may still be imposed by force of circumstances or by tradition (Section 6).

Kiaer's approach was based on four principles:

(a) "representativeness" of the sample, though this was to be better defined in 1926 and later;
(b) objectivity of selection by the enumerators in the field and by systematic selection in the office;
(c) assessment of reliability by built-in replicates of some kind; and
(d) adequate description of the process of selection in the reports.

A. L. Bowley conducted stratified element samples of households, measured their reliability, and used simple statistical formulas to describe his results (1913). He became a leader for "random" sampling, which was defined as equal selection probabilities. Cluster sampling had to be used in practice but was not recognized in theory. In general, theory lagged behind practice then and still does. The best description of this period is by O'Muircheartaigh and Wong (1981); also Hansen, Dalenius, and Tepping (1985) are helpful; also Kruskal and Mosteller (1979a,b,c and 1980).

4. PURPOSIVE SELECTION VERSUS "RANDOM SELECTION"

Jensen (1926, p. 59; 1928) reported that: "When the International Institute of Statistics discussed the matter twenty-two years ago it was a question of the recognition of the 'representative' method in principle that claimed most interest. Nowadays there is hardly one statistician, who in principle will contest the legitimacy of the representative method."

In May 1924 in Rome, the ISI appointed a commission for the purpose of studying the application of the representative method in statistics (Bowley 1926). Its report stated that:

> Considering that it is necessary in many cases to draw general conclusions based upon practical investigations...
>
> I. With reference to the Resolution passed at the Session at Berlin in 1903, again calls attention to the very considerable advantages which can be obtained by applying the Representative Method under the following conditions:
> The results of a partial investigation should only be generalized provided that the sample used is in its nature sufficiently representative of the totality. In such respects the sample may be selected in different ways; the following two main cases, however, are to be distinguished:
> (A) *Random Selection.* A number of units are selected in such a way that exact equality of chance of inclusion is the dominant rule. Then precision is related to the number included which should be large enough to render insignificant accidental deviations;
> (B) *Purposive Selection.* A number of groups of units are selected which together yield nearly the same characteristics as the totality. In order to have any knowledge of the precision of the estimate it is necessary that sufficient groups should be included to allow the variation between the characteristics of the groups to be measured. But since the precision often depends to a great extent on the discretion used in making the selection, the following controls are recommended:
> 1. The selection on the same principle should be made twice or more; after their comparison, the samples can be merged. (This recommendation is also applicable to the Random Selection);
> 2. In repeated observations, the relation between the part and the whole should from time to time be examined more minutely;
>
> II. Recommends that the investigation should be so arranged wherever possible, as to allow of a mathematical statement of the precision of the results, and that with these results should be given an indication of the extent of the error to which they are liable;
>
> III. Repeats the wish expressed in the Resolution of 1903, that in the report on the results of every representative investigation an explicit account in detail of the method of selecting the sample adopted should be given.

This was quoted by Yates (1946) who then remarks:

> Two main features will strike the present-day reader: first, the considerable prominence given to the method of purposive selection, and second, the lack of any very clear conception of the possibility, except by the selection of units wholly at

> random, or by the inadequate procedure of sub-dividing the sample into two or more parts, of so designing sampling enquiries that the sampling errors should be capable of exact estimation from the results of the enquiry itself.

In 1995 purposive selection has largely disappeared from articles and courses of sampling methodology, also from the design of large and national surveys. However, purposive or judgment selections survive as alternatives to probability sampling in different forms, mentioned later: selection of a single site or a few sites; model sampling; balanced sampling; controlled selection and multiple or deep stratification; quota sampling; focus groups.

5. BALANCED VERSUS RANDOM SELECTION

Neyman's 1934 paper (of 68 pages) marks a turning point for survey sampling and it formed the basis of sampling courses, such as I took in 1941. He brought ideas of stratified clustered probability selection and of estimation (Markov, Tchuprow, Neyman) from Russia and Poland to the London meeting of the Royal Statistical Society, with discussants Bowley, Egon Pearson and Fisher.

It was based on a 1929 paper of Gini and Galvani (1929) who took a balanced sample of 29 from the 214 districts (circondari) of Italy to save about 13.5 percent of the 1921 population census of Italy. Neyman writes: "Having thus fixed the units of selection, the authors proceed to the choice of the principle of sampling; should it be random sampling or purposive selection?" Thus, we view the clash of theories by leaders of three national schools. (Aside: theory and theatre have common roots in the Greek word for "view.")

> The Italian statisticians... (to their credit)... did not find their results satisfactory. The comparison of the sample and the whole country showed, in fact, that though the average values of seven controls used are in satisfactory agreement, the agreement of average values of other characters, which were not used as controls, is often poor. The agreement of other statistics besides the means, such as frequency distributions, etc., is still worse. This applies also to the characters used as controls.

They understood that sample/population equality for control variables does not imply such equality for survey variables—better than some of our contemporaries.

Neyman then changed population and principles together with the sizes and the numbers of sampling units. He used a stratified random selection of 1,621 from the total of 123,383 "statistical districts" of Poland. These contained an average of 250 persons, but not equal. He also specified limits of variability for the results. He also exhibits a similar sample of farming conditions from the 5,000 villages of Bulgaria by Oskar Anderson.

Neyman's paper centers on sampling design and establishes the triumph of probability sampling of many small, unequal clusters, stratified for better representation. These ideas provide the bases for the theory, the practice, and the teaching of survey sampling. Purposive selection for representative sampling has been pushed out of the mainstream of theory and teaching, but it exists in practice, as we discuss together with balanced sampling in Section 6.

For representative sampling, Neyman's greatest effect was on sample design, but for most statisticians his introduction of new methods of estimation is most salient. These have been discussed often, and especially well by Smith (1976). Confidence intervals, Markov's Theorem, and related ideas were presented by Neyman on an international stage, and widely developed. They resulted in immediate stimulation and conflict. The relatively mild conflict at this meeting grew into the battles around Neyman-Pearson theory among academic statisticians. We now enter areas of current conflicts.

6. THE CHALLENGES OF FEW SITES AND MULTIPLE STRATIFICATION

The virtues of the designs favored by Neyman and by modern literature depend on the asymptotic properties of large samples. Also these are assumed (explicitly or implicitly) to serve a single statistic ($\bar{y}$ or $\hat{Y}$) based on an entire large (national) sample. However, the demands on samples differ greatly from that simple model and tend to force research back often toward the purposive and balanced selections that we would prefer to avoid.

(a) *Few sites.* Small and even medium-sized studies may need to be restricted to a single site or to a few sites, perhaps 4 or 10. Such restrictions cause conflicts, because the unbiased random selections can easily result in selections that are far from "representative" of the target population. For example, for pilot studies I have advocated and used four sites (districts) chosen as 2×2×2×2 Graeco-Latin squares (Kish 1987a, 3.1). For samples of 10 schools or hospitals in most situations we would prefer multiple controls, to be based on the many auxiliary variables available. These lead many researchers to choose purposive selections over probability sampling, but a few others to a "controlled selection" probability design.

(b) *Regions and design subclasses.* Even large national samples based on many PSUs (primary sampling units) generally suffer from too few PSUs (districts, counties) for provinces, for states, and for other geographic (or design) subclasses. These differ from "crossclasses" like age-sex, occupation, social "crossclasses," which are spread into and based on all the PSUs. The stability of regional statistics depends greatly on the number of PSUs on which they rest and may lead to some form of "pooling" of data. Stability for subclasses has been the original motivator and the continuing need for "controlled selections" (multiple stratification) of large samples (Kish 1965a, 12.8).

(c) *National samples of 20 to 60 PSUs.* Even national samples may have to be restricted to a few dozen PSUs because of costs and resources, and sometimes by the need for local depth within PSUs. Then "controlled selections" may provide adequate stratification, as needed. (I wonder if the problem of Gini and Galvani could be solved by a controlled selection of 29 provinces, and what would Neyman say? A daring proposal, I admit.)

What are the available alternatives for those faced with a practical problem of too few PSUs?

1. More large units (like districts); but this is expensive.
2. Change to many smaller units (communes). This is what Neyman did, but he changed to Poland, and he and we don't know if it was feasible in Italy in 1926.
3. Selection of only a few large units, as stratified paired selections, with much random variation left.
4. Select a purposive sample, like Gini and Galvani, or some modern "balanced sample" version.
5. Use "controlled probability sampling" (multiple stratification, deep stratification).

One method can satisfy the requirement of *known, positive probability* of selection for all elements. But this method may be difficult for some resources. Furthermore, the computed sampling errors overestimate the achieved variances, hence the method lacks strict measurability (Section 8) (Kish 1965a, 12.8; Groves and Hess 1975).

7. MODELS VERSUS REPRESENTATIVE (POPULATION) SAMPLING

Neyman agreed with Gini and Galvani that it is not "possible to give any *precise* sense to *a generally representative sample*...(but) it is possible to define what should be termed a *representative method of sampling and a consistent method of estimation...irrespectively of the unknown properties of the population* studied" (Neyman 1934).

"Representative method" is not a term often used today, but for Neyman and in most modern books and courses it would mean *probability* sampling. (A fine Italian example is Cicchitelli, Herzel, and Montanari 1992.) This means that *all* elements in the *population frame* receive *known* positive probabilities of selection, which are operationally defined, and not necessarily equal. Sometimes this may be called "random" sampling. Sometimes it may include "measurability" for assessing sampling variability, and followed by some reference to the need for large samples. Most national statistical offices today probably use probability samples for some purposes, and none would resort to purposive samples.

On the other hand, purposive samples of various kinds, but that are far from probability sampling, such as "quota sampling," are common in market research, in political polls, in medical research, and in a great deal of academic research in psychology, anthropology, economics, etc.

In four articles on "Representative Sampling," Kruskal and Mosteller (1979a,b,c and 1980) listed six categories of meanings found in nonstatistical publications today.

1. General acclaim and approbation for data.
2. Absence of selective forces.
3. Mirror or miniature of the population, with similar distributions.
4. Typical or ideal group or case.
5. Coverage of the population to reflect variation among domains.
6. Probability sampling, with procedures to give every probability element a known positive probability of selection.

We may disregard 1, 2 and 4; and probability sampling (6) denotes the only feasible method recognized by statisticians to achieve the aims stated in (3), and less clearly in (5).

The stated aim of representing a "framed" population is an important concept and it is still in conflict with theoretical models accepted in econometrics and mathematical statistics (Section 8). This conflict has been discussed often (especially by Hansen, Dalenius, and Tepping 1985). Here is a list of pairs of diverse names used for the two opposing views:

model-dependent/population-bound;
model-based/design-based;
modeling/survey sampling;
population-free/representation;
theoretical/randomization;
mathematical/physical-empirical;
model-dependent/mode-based (Kish 1987a, 1.8).

In my view, *representative sampling* is a term that can be avoided and it is disappearing from the technical vocabulary. At different times it has been used for *random* sampling, for proportionate sampling, and for *purposive* selection too. In general, it often denotes *aims* of representing a population well with a sample; and in this sense it denotes *population* sampling and *survey* sampling (Kish 1965a, 1.6).

8. CONFLICTS AROUND MEASURABILITY

It is virtuous and easy to be in favor of the principle of measurability: the term coined by Fisher for the principle that the data themselves must provide the feasibility of computing measures of its variability, and this principle gained general acceptance. Kiaer included in 1895 a vague form among his four principles. In reality, however, even Neyman failed to compute standard errors from his data for his 1934 publication. This great discrepancy between principle and practice exists even in 1995, into the third human generation since the arrival of good formulas, programs and computers. Although probability sampling has made great (but far from universal) conquest, even for most probability samples valid sampling errors are still not computed today. Allow me to avoid here the question whether sampling errors should and can be computed for purposive samples also. This is a titillating but mute question, because I know of only one special case when it was done (Moser and Stuart 1953).

Here follows a list of some of the most important and debatable (and debated) problems of measurability, and my brief opinions on them without supporting evidence.

1. Measurability should be a distinct criterion, rather than one to be included automatically in the definition of probability sampling.
2. Measurability may be second in importance to probability selection, a higher position than this criterion gets in practice. Nevertheless, I respect probability samples even when they are not measurable, or not "strictly" measurable.
3. On data tapes, for descriptive statistics (such as $\bar{y}, s^2$, or r_{xy}) identification and weights (if any) suffice, but for measurability of complex samples, the codes for ultimate clusters and for strata are also needed.
4. Surveys are often vastly multipurpose and their samples complex. This and 3 above are the chief reasons sampling errors are seldom computed and presented adequately (Kish 1995b).
5. Are sampling errors necessary and sufficient, when other survey errors, such as measurement, are often potentially larger but unknown? Yes they are necessary, though not sufficient. And they become *relatively* more important for small subclasses and for comparisons and other analytical statistics (Kish 1987a, 2.3, 7.1).
6. Few PSUs obstruct the stability of sampling errors even more than of the main descriptive statistics. Although two replicates can provide "unbiased" estimates of variances in theory, practical needs require much more precision (Kish 1995a, 18.1).

We finish this crowded section with several examples of point 2 above: probability designs that are often used, although they may fail to yield strictly unbiased measures of stability: systematic selection of PSUs, controlled probability selections, and balanced repeated replications (BRR) with only 16 replicates. Interpenetrating samples of only 4 replicates; even 10 replicates for jackknife design lack desirable stability. Other samples could be added, which should not be used, such as a single large cluster selected either with known probability, or by purposive judgment.

9. MATHEMATICAL STATISTICS VERSUS SURVEY SAMPLING

There is a wide, deep, and lamentable chasm between survey sampling and mathematical statistics (MS), which preoccupies the Departments of Statistics. There is an equally wide gap between survey sampling and "sampling theory," or its "foundations," as these try to approach asymptotically statistical mathematics. The deplorable separation prevails in most universities and in all countries, also in statistical publications, and textbooks. There exists a similar neglect of experimental design, of statistical design in general: statistical design and the collection of data are neglected in universities and publications, which are occupied only by statistical analysis and mathematics.

John Nelder (1994) complains that "our statistics students...cannot distinguish an experimental design from an ANOVA table..." and "In many departments in the USA the subject has almost disappeared from view..." Jeffers (in the same issue) complains that "statistical methods seem currently to be focused on analysis...we should give more attention to the design of data collection...Proper attention to the fundamental principles of sampling and experimental design before any data are collected would save a great deal of anguish later..." And these come from England, and Rothamsted, the ancestral home of experimental design and sampling and of the union of statistical practice with theory.

My central complaint is that over 95 percent of statistical attention in academia, textbooks, and publications is devoted to mathematical statistical analysis and only 2 percent to design. This neglect has deep and wide effects because statistical design is a statistical subject and is best taught in connection with other statistics, with mathematical statistics and analysis. The consequences of that neglect are too often poor designs by nonstatisticians (engineers, economists, etc.). In turn the consequences of poor designs are poor data that too often cannot be corrected, not even by the best statistical analysis! As somebody said: "Garbage in, garbage out; GIGO."

Why does statistical theory and analysis drive out design? Because theory and analysis can be made mathematical, so that with mathematics it is easier to teach and to set exams; also to write publishable articles. It is easier, that is, for a faculty fluent in mathematics, and mathematics is a necessary condition to join the faculty of statistics departments. Necessary OK, but, not sufficient, I say. Some of the faculty should know enough design to be able to teach well some of it. However, it is difficult to make design mathematical, hence difficult to teach and to set exams.

Please don't get me wrong. I feel that mathematics may be the most beautiful creation of the human mind. Because it is so beautiful, it is most seductive. I also admire theory. But theory is not synonymous with mathematics, and statistical theory is more than statistical mathematics. Statistical curricula should have more statistical theory and inference and philosophy of science too. Mathematics, theory, and methodology are all related but not synonymous.

There are several theoretical differences between mainline, classical, mathematical statistics (MS), and survey sampling (SS).

(a) MS assumes a universe of IID, "identically and independently distributed random variables," whereas the populations and samples of survey sampling are stratified, clustered, and often not "epsem," hence weighted. Real worlds—whether physical, biological, or social—are never IID, and it is seldom convenient or efficient to select a simple random sample.

(b) There is no *clean* boundary between small samples and large samples, and MS asks us to disregard that boundary (ever since R. A. Fisher and likelihood). But SS says that we can't forget it because we need the Central Limit Theorem, and that n of 1 or 2 is too small for probability samples, but n of 100 or 1000 is large enough, often n of 30 or even 12.

(c) It is not true, though often said or assumed, that: "The object of most surveys is to produce one sample mean or sample total." Samples are typically multipurpose, often multipopulation, etc.

(d) Surveys are also subject to errors of observation, often correlated, and to errors of nonobservation, and to other plagues.

10. ALTERNATIVES TO SURVEY SAMPLING

> Comparisons of survey samples with experiments and with controlled observations are discussed in this section. These three methods are justified as preferred strategic choices for the three criteria of representation, randomized treatments, and realism. Another kind of strategic choice involves the relative advantages of sample surveys against censuses or registers. Still another comparison concerns the choice between widespread sample surveys and local studies confined to one or a few sites. Thus, sampling surveys are shown as alternatives to three distinctly different methods of data collection, and that is the only reason for their central position in Figure 1.
>
> Experiments are strong on control of the explanatory variables through the randomization of predictor variables over subjects (i.e., subjects over treatments); but they are weak on representation over defined target populations, and often also on the realism of measurements. Surveys are strong on representation, but they are weak on control of variables. Investigations are weak on control and often on representation; their great prevalence is due often to low cost and relative convenience, and at times to the need for and feasibility of realism of measurements in "natural settings." We are faced usually with conflicts between desires for randomization, for representation, and for realism. It is seldom that desires for all three criteria can be satisfied for even two of the three. More often people merely emphasize one criterion because it is least costly and most convenient, and because it may appear on theoretical grounds—convincingly or only hopefully—the most justified. But when hope or wish is the father of the thought, the offspring may be illegitimate. The criteria that must be sacrificed must be considered more thoroughly. We ask for a greater role for explicit models of the diverse sources of variation that arise in the three types of research designs. (Kish 1987a, p. 7)

The alternatives of focus groups and quota sampling belong to another category, but survey practitioners encounter these also. Now I turn to some personal views about the battles that survey sampling will face in the near *future*.

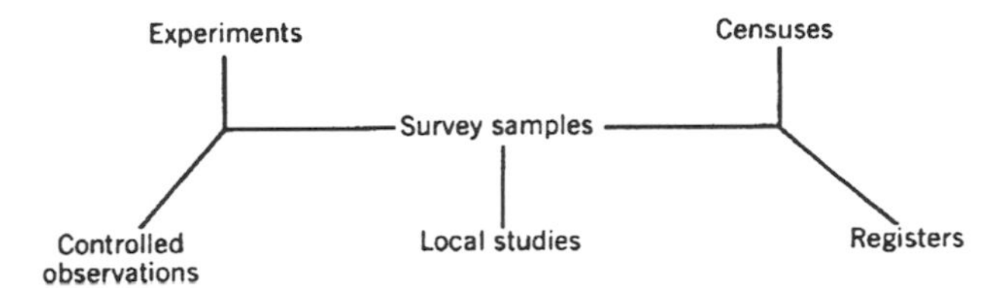

Figure 1. Three Different Comparisons of Survey Samples. (Kish 1987a)

11. PERIODIC SURVEYS, PANELS, AND ROLLING SAMPLES

Decennial censuses of the population have been the first and foremost system for survey data in most countries in the world for many decades. In addition to complete and detailed basic data on numbers, gender, and age, they also collect increasing volumes of socio-economic variables, sometimes with large "census samples" of 5 to 20 percent. The decennial censuses emphasize spatial detail at the expense of coarseness over the temporal dimension. The census may be viewed as a systematic sample of every tenth year for some variables like income; and as a systematic sample of 1 in 3650 days for some variables, like residence on the census day. In addition to providing instant photos of the population, they have also been used as periodic surveys for comparisons between and over the decades. Similarly, in the spatial dimension the censuses also serve for comparisons between provinces, districts, and other geographical detail.

However, many countries have decided that decennial data are too coarse over time, and that more temporal detail is needed for many critical variables, such as labor force participation, economic data, public health, etc., which are subject to great changes over time. Labor Force Surveys (LFS) and Current Population Surveys (CPS) are two of several names for monthly or quarterly surveys now conducted in many countries. These give up-to-date statistics for national totals, for major regions, and sometimes for other large geographical and other subclasses. However, based on periodic samples of a few thousands or tens of thousands, they cannot give spatial detail for small units, like blocks, with any precision. They serve chiefly to monitor monthly or quarterly national changes, and they are designed with overlaps of segments and dwellings between periods in order to measure those changes efficiently.

These two systems, decennial censuses and monthly surveys, serve different aims, and there is little mutual support between them. Generally two other important aims remain unfilled. One is a lack of *panels* of the same individuals (and families) that would yield data in micro (gross, individual) changes. These data often differ greatly from net changes and could shed new light on social dynamics.

The other kind of statistics that are much needed are *annual* surveys with details on geographic and other domain data. These would overcome both the lack of spatial detail in the monthly surveys, and the obsolescence of decennial censuses. However, these would require samples of 1/100 to 1/10 of the population, and experience has shown that samples that large are very expensive. A sample of 1/10 may cost half as much as a complete census.

I have advanced the idea of "rolling samples" in several papers to meet this conflict. Periodic samples of 1/F each month (or each week) would provide estimates of the whole population. They should be so designed that the cumulation of 12 samples would provide a sample of 12/F of the population and broadly spread. If F=120, the 120 months of ten years would cover the whole population. I propose "asymmetrical cumulation": one month for national data, 120 months for the smallest domains, etc.

For simplicity I assume no overlaps for the monthly samples. If overlaps are desired (though they are far from vital) they can be built into the design. I also propose a "split-panel design" to yield efficient, multipurpose overlaps, and panels at the same time (Kish 1988a; Kish 1987a, chap. 6, especially 6.5).

This proposal is now being considered in several countries. The concept of a "cumulated population" is novel in statistics, though accepted and used in some fields. The argument for the "certainty" that decennial censuses yield (every 3650 days) sounds to me like the arguments against sampling before 1895, thus coming to a full circle.

12. MULTINATIONAL AND MULTIPOPULATION DESIGNS

The following five major types of designs have each been treated in statistical and sampling literature. The five types of designs appear to be very different, and they are treated at length entirely separately in the literature of survey sampling, thus losing the similarities I shall emphasize here. Also each of these types is becoming more commonly applied, as both the need for them and the means for conducting them—financial backing, survey organizations and methods—are encountered more frequently. International surveys and periodic samples particularly are emerging as the waves of the future. I note common aspects in these five major types of designs:

(a) Multinational survey designs
(b) Multidomain survey designs
(c) Controlled observations (quasi/pseudo experiments)
(d) Periodic surveys
(e) Combinations and cumulations—rolling samples

Despite the great differences in purposes, methods, and uses of the five designs, we shall note great similarities between them in all seven major aspects of design:

1. Definition of concepts, variables, and populations
2. Survey design and methods of measurements
3. Substantive analysis
4. Weighting procedures
5. Statistical analysis
6. Sample design and selection
7. Sizes (and fractions) of samples

I emphasize a basic difference and contrast between aspects 1 to 3 on one hand, and 6 and 7 on the other. Comparative designs of all five types of designs should try for the utmost uniformity and similarity for aspects 1–3 to support comparisons, but they should allow for great flexibility and divergences in aspects 6 and 7 for the sake of feasibility and efficiency. Furthermore, this difference and contrast affect all five types of designs (a) to (e) similarly. Aspects 4 and 5 have some properties in common with the survey aspects 1–3 that need similarities, but also some properties that are functions of the sample design, like aspects 6 and 7. The sharp contrast in treatment is justified by their different appearances in the Mean Square Errors of comparisons. These are often expressed as a combination of bias and variance terms:

$$\text{MSE}(\bar{x}-\bar{y}) = [\text{Bias}(\bar{x}) - \text{Bias}(\bar{y})]^2 + \text{Var}(\bar{x}) + \text{Var}(\bar{y}) - 2\text{Cov}(\bar{x},\bar{y})$$

The deliberate *design* of multinational surveys is new, and they are increasing in scope and in numbers. The examples I could find (Kish 1994, Table 6.1) all date since 1965. These new opportunities arise not only because of new methods and techniques, but especially because of emerging finances needed for these large efforts, new demands for valid multinational comparisons, and also new statistical research centers able to implement the national surveys. We must emphasize *survey* skills, not only sampling.

Multidomain design follows naturally from multinational, and this is most obvious with the states of India, the republics of the former Yugoslavia, but also the provinces of China, Spain, etc. These domains of national surveys are always both compared and combined and national samples and populations are always combinations of domains, which are more or less diverse.

This point and the similarities to controlled observations I have discussed recently (Kish 1994 and at length in Kish 1987a, chap. 3). The connection to periodic surveys and to combinations of domains was discussed in Section 11.

13. CONCLUDING REMARKS

We may expect further developments along the lines outlined in each of the twelve sections discussed above. I am especially hopeful about the emergence of more uses and methods for periodic samples and of "rolling samples" proposed in Section 11. Also for multinational and multipopulation samples discussed in Section 12. There will be many others and I propose a few as *major* future developments. My criteria are that these should be

(a) *general*, not only of local interest in one country or one project;
(b) *important* statistically, not only a trivial improvement;
(c) *sampling* problems, rather than other matters, like measurement, that pertain to other skills; and
(d) *feasible* possibilities, not mere desires.

Multipurpose sample design is on the top of my list, meeting all four of the criteria. Sampling theory has been built around the myth of a single mean $\bar{y}$ or total $\hat{Y}$ as the sole or principal aim of samples. Actually, most sample surveys are highly multipurpose in several dimensions. The single aim is a harmful myth that can be overcome and will be, I hope (Kish 1988a).

Small domain estimation has been an active area of development for about 25 years, and I hope to see further developments. Combining *randomized experiments* with probability sampling bases is another area that should see developments.

Let me end then on a happy note. We have seen a great deal of useful development in survey sampling in the past 100 years. And there is a great deal of interesting developments left for the next three generations to accomplish in the next 100 years.

2

SOME STATISTICAL PROBLEMS IN RESEARCH DESIGN

LESLIE KISH

Several statistical problems in the design of research are discussed: (1) The use of statistical tests and the search for causation in survey research are examined; for this we suggest separating four classes of variables: explanatory, controlled, confounded, and randomized. (2) The relative advantages of experiments, surveys, and other investigations are shown to derive respectively from better control, representation, and measurement. (3) Finally, three common misuses of statistical tests are examined: "hunting with a shot-gun for significant differences," confusing statistical significance with substantive importance, and overemphasis on the primitive level of merely finding differences.

1. INTRODUCTION

Statistical inference is an important aspect of scientific inference. The statistical consultant spends much of his time in the borderland between statistics and the other aspects, philosophical and substantive, of the scientific search for explanation. This marginal life is rich both in direct experience and in discussions of fundamentals; these have stimulated my concern with the problems treated here.

I intend to touch on several problems dealing with the interplay of statistics with the more general problems of scientific inference. We can spare elaborate introductions because these problems are well known. Why then discuss them here at all? We do so because, first, they are problems about which there is a great deal of misunderstanding, evident in current research; and, second, they are *statistical* problems on which there is broad agreement among research statisticians—and on which these statisticians generally disagree with much in the current practice of research scientists.[1]

Reprinted with permission from *American Sociological Review* (1959), 24, 3, 328–338.
This research has been supported by a grant from the Ford Foundation for Development of the Behavioral Sciences. It has benefited from the suggestions and encouragement of John W. Tukey and others. But the author alone is responsible for any controversial opinions.

[1] *Cf.* Fisher (1953, pp. 1-2): "The statistician cannot evade the responsibility for understanding the processes he applies or recommends. My immediate point is that the questions involved can be disasso-

Several problems will be considered briefly, hence incompletely. The aim of this paper is not a profound analysis, but a clear elementary treatment of several related problems. The footnotes contain references to more thorough treatments. Moreover, these are not *all* the problems in this area, nor even necessarily the most important ones; the reader may find that his favorite, his most annoying problem, has been omitted. The problems selected are a group with a common core, they arise frequently, yet they are widely misunderstood.

2. STATISTICAL TESTS OF SURVEY DATA

That correlation does not prove causation is hardly news. Perhaps the wittiest statements on this point are in George Bernard Shaw's preface to *The Doctor's Dilemma*, in the sections on "Statistical Illusions," "The Surprises of Attention and Neglect," "Stealing Credit from Civilization," and "Biometrika." (These attack, alas, the practice of vaccination.) The excellent introductory textbook by Yule and Kendall (1937) deals in three separate chapters with the problems of advancing from correlation to causation. Searching for causal factors among survey data is an old, useful sport; and the attempts to separate true explanatory variables from extraneous and "spurious" correlations have taxed scientists since antiquity and will undoubtedly continue to do so. Neyman (1938/1952) and Simon (1954, 1956) show that beyond common sense, there are some technical skills involved in tracking down spurious correlations. Econometricians and geneticist have developed great interest and skill in the problems of separating the explanatory variables.[2]

The researcher designates the explanatory variables on the basis of substantive scientific theories. He recognizes the evidence of other *sources of variation* and he needs to separate these from the explanatory variables. Sorting all sources of variation into four classes seems to me a useful simplification. Furthermore, no confusion need result from talking about sorting and treating "variables," instead of "sources of variation."

I. The *explanatory* variables, sometimes called the "experimental" variables, are the objects of the research. They are the variables among which the researcher wishes to find and to measure some specified relationships. They include both the "dependent" and the "independent" variables, that

---

ciated from all that is strictly technical in the statistician's craft, and *when so detached*, are questions only of the right use of human reasoning powers, with which all intelligent people, who hope to be intelligible, are equally concerned, and on which the statistician, as such, speaks with no special authority. The statistician cannot excuse himself from the duty of getting his head clear on the principles of scientific inference, but equally no other thinking man can avoid a like obligation."

[2] See the excellent and readable article by Wold (1956). Also the two-part technical article by Kendall (1951, 1952). The interesting methods of "path coefficients" in genetics have been developed by Wright for inferring causal factors from regression coefficients. See Wright (1954) and Tukey (1954). Also Li (1956). I do not know whether these methods can be of wide service in current social science research in the presence of numerous factors, of large unexplained variances, and of doubtful directions of causation.

is, the "predictand" and "predictor" variables.[3] With respect to the aims of the research all other variables, of which there are three classes, are extraneous.

II. There are extraneous variables which are *controlled*. The control may be exercised in either or both the selection and the estimation procedures.

III. There may exist extraneous uncontrolled variables which are *confounded* with the Class I variables.

IV. There are extraneous uncontrolled variables which are treated as *randomized* errors. In "ideal" experiments (discussed below) they are actually randomized; in surveys and investigations they are only assumed to be randomized. Randomization may be regarded as a substitute for experimental control or as a form of control.

The aim of efficient design both in experiments and in surveys is to place as many of the extraneous variables as is feasible into the second class. The aim of randomization in experiments is to place all of the third class into the fourth class; in the "ideal" experiment there are no variables in the third class. And it is the aim of controls of various kinds in surveys to separate variables of the third class from those of the first class; these controls may involve the use of repeated cross-tabulations, regression, standardization, matching of units, and so on.

The function of statistical "tests of significance" is to test the effects found among the Class I variables against the effects of the variables of Class IV. An "ideal" experiment here denotes an experiment for which this can be done through randomization without any possible confusion with Class III variables. (The difficulties of reaching this "ideal" are discussed below.) In survey results, Class III variables are confounded with those of Class I; the statistical tests actually contrast the effects of the random variables of Class IV against the explanatory variables of Class I confounded with unknown effects of Class III variables. In both the ideal experiment and in surveys the statistical tests serve to separate the effects of the random errors of Class IV from the effects of other variables. These, in surveys, are a mixture of explanatory and confounded variables; their separation poses severe problems for logic and for scientific methods; statistics is only one of the tools in this endeavor. The scientist must make many decisions as to which variables are extraneous to his objectives, which should and can be controlled, and what methods of control he should use. He must decide where and how to introduce statistical tests of hypotheses into the analysis.

As a simple example, suppose that from a probability sample survey of adults of the United States we find that the level of political interest is higher in urban than in rural areas. A test of significance will show whether or not the difference in the "levels" is large enough, compared with the sampling error of the difference, to be considered "significant." Better still, the confidence interval of the difference will disclose the limits within which we can expect the "true" population value of

---

[3] Kendall points out that the latter term is preferable. See Kendall (1951, 1952), and Kendall and Buckland (1957). I have also tried to follow in IV below his distinction of "variate" for random variables from "variables" for the usual (nonrandom) variable.

the difference to lie.[4] If families had been sent to urban and rural areas respectively, after the randomization of a true experiment, then the sampling error would measure the effects of Class IV variables against the effects of urban *versus* rural residence on political interest; the difference in levels beyond sampling errors could be ascribed (with specified probability) to the effects of urban *versus* rural residence.

Actually, however, residences are not assigned at random. Hence, in survey results, Class III variables may account for some of the difference. If the test of significance rejects the null hypothesis of no difference, *several* hypotheses remain in addition to that of a simple relationship between urban *versus* rural residence and political interest. Could differences in income, in occupation, or in family life cycle account for the difference in the levels? The analyst may try to remove (for example, through cross-tabulation, regression, standardization) the effects due to such variables, which are extraneous to his expressed interest; then he computes the difference, between the urban and rural residents, of the levels of interest now free of several confounding variables. This can be followed by a proper test of significance—or, preferably, by some other form of statistical inference, such as a statement of confidence intervals.

Of course, other variables of Class III may remain to confound the measured relationship between residence and political interest. The separation of Class I from Class III variables should be determined in accord with the nature of the hypothesis with which the researcher is concerned; finding and measuring the effects of confounding variables of Class III tax the ingenuity of research scientists. But this separation is beyond the functions and capacities of the statistical tests, the tests of null hypotheses. Their function is not explanation; they cannot point to causation. Their function is to ask: "Is there anything in the data that *needs* explaining?"—and to answer this question with a certain probability.

Agreement on these ideas can eliminate certain confusion, exemplified by Selvin (1957) in a recent article:

> Statistical tests are unsatisfactory in nonexperimental research for two fundamental reasons: it is almost impossible to design studies that meet the conditions for using the tests, and the situations in which the tests are employed make it difficult to draw correct inferences. The basic difficulty in design is that sociologists are unable to randomize their uncontrolled variables, so that the difference between "experimental" and "control" groups (or their analogs in nonexperimental situations) are a mixture of the effects of the variable being studied and the uncontrolled variables or correlated biases. Since there is no way of knowing, in general, the sizes of these correlated biases and their directions, there is no point in asking for the probability that the observed differences could have been produced by random errors. The place for significance tests is after all relevant correlated biases have been controlled...In design and in interpretation,

[4] The sampling error measures the chance fluctuation in the difference of levels due to the sampling operations. The computation of the sampling error must take proper account of the actual sample design, and not blindly follow the standard simple random formulas. See Kish (1957).

> in principle and in practice, tests of statistical significance are inapplicable in nonexperimental research.[5]

Now it is true that in survey results the explanatory variables of Class I are confounded with variables of Class III; but it does not follow that tests of significance should not be used to separate the random variables of Class IV. Insofar as the effects found "are a mixture of the effects of the variable being studied and the uncontrolled variables;" insofar as "there is no way of knowing, in general, the sizes" and directions of these uncontrolled variables, Selvin's logic and advice should lead not only to the rejection of statistical tests; it should lead one to refrain altogether from using survey results for the purposes of finding explanatory variables. *In this sense*, not only tests of significance but any comparisons, any scientific inquiry based on surveys, any scientific inquiry other than an "ideal" experiment, is "inapplicable." That advice is most unrealistic. In the (unlikely) event of its being followed, it would sterilize social research—and other nonexperimental research as well.

Actually, much research—in the social, biological, and physical sciences—must be based on nonexperimental methods. In such cases the rejection of the null hypothesis leads to several alternate hypotheses that may explain the discovered relationships. It is the duty of scientists to search, with painstaking effort and with ingenuity, for bases on which to decide among these hypotheses.

As for Selvin's advice to refrain from making tests of significance until "after all relevant" uncontrolled variables have been controlled—this seems rather farfetched to scientists engaged in empirical work who consider themselves lucky if they can explain 25 or 50 per cent of the total variance. The control of all relevant variables is a goal seldom even approached in practice. To postpone to that distant goal all statistical tests illustrates that often "the perfect is the enemy of the good."[6]

---

[5] In a criticism of this article, McGinnis (1958) shows that the separation of explanatory from extraneous variables depends on the type of hypothesis at which the research is aimed.

[6] Selvin (1957) performs a service in pointing to several common mistakes: (a) The mechanical use of "significance tests" can lead to false conclusions. (b) Statistical "significance" should not be confused with substantive importance. (c) The probability levels of the common statistical tests are not appropriate to the practice of "hunting" for a few differences among a mass of results. However, Selvin gives poor advice on what to do about these mistakes; particularly when, in his central thesis, he reiterates that "tests of significance are inapplicable in nonexperimental research," and that "the tests are applicable only when all relevant variables have been controlled." I hope that the benefits of his warnings outweigh the damages of his confusion.

I noticed three misleading references in the article. (a) In the paper which Selvin appears to use as supporting him, Wold (1956, p. 39) specifically disagrees with Selvin's central thesis, stating that "The need for testing the statistical inference is no less than when dealing with experimental data, but with observational data other approaches come to the foreground." (b) In discussing problems caused by complex sample designs, Selvin writes that "Such errors are easy enough to discover and remedy" (p. 520), referring to Kish (1957). On the contrary, my article pointed out the seriousness of the problem and the difficulties in dealing with it. (c) "Correlated biases" is a poor term for the confounded uncontrolled variables and it is not true that the term is so used in literature. Specifically, the reference to Cochran is misleading, since he is dealing there only with errors of measurement which may be correlated with the "true" value. See Cochran (1953, p. 305).

3. EXPERIMENTS, SURVEYS, AND OTHER INVESTIGATIONS

Until now, the theory of sample surveys has been developed chiefly to provide descriptive statistics—especially estimates of means, proportions, and totals. On the other hand, experimental designs have been used primarily to find explanatory variables in the analytical search of data. In many fields, however, including the social sciences, survey data must be used frequently as the analytical tools in the search for explanatory variables. Furthermore, in some research situations, neither experiments nor sample surveys are practical, and other investigations are utilized.

By "experiments" I mean here "ideal" experiments in which all the extraneous variables have been randomized. By "surveys" (or "sample surveys"), I mean probability samples in which all members of a defined population have a known positive probability of selection into the sample. By "investigations" (or "other investigations"), I mean the collection of data—perhaps with care, and even with considerable control—without either the randomization of experiments or the probability sampling of surveys. The differences among experiments, surveys, and investigations are not the consequences of statistical techniques; they result from different methods for introducing the variables and for selecting the population elements (subjects). These problems are ably treated in recent articles by Wold (1956) and Campbell (1957).

In considering the larger ends of any scientific research, only part of the total means required for inference can be brought under objective and firm control; another part must be left to more or less vague and subjective—however skillful—judgment. The scientist seeks to maximize the first part, and thus to minimize the second. In assessing the ends, the costs, and the feasible means, he makes a strategic choice of methods. He is faced with the three basic problems of scientific research: measurement, representation, and control. We ignore here the important but vast problems of measurement and deal with representation and control.

Experiments are strong on control through randomization; but they are weak on representation (and sometimes on the "naturalism" of measurement). Surveys are strong on representation, but they are often weak on control. Investigations are weak on control and often on representation; their use is due frequently to convenience or low cost and sometimes to the need for measurements in "natural settings."

Experiments have three chief advantages:

1. Through randomization of extraneous variables the confounding variables (Class III) are eliminated.
2. Control over the introduction and variation of the "predictor" variables clarifies the *direction* of causation from "predictor" to "predictand" variables. In contrast, in the correlations of many surveys this direction is not clear—for example, between some behaviors and correlated attitudes.
3. The modern design of experiments allows for great flexibility, efficiency, and powerful statistical manipulation, whereas the analytical use of survey data presents special statistical problems (Kish 1957).

The advantages of the experimental method are so well known that we need not dwell on them here. It is the scientific method *par excellence*—when feasible. In many situations experiments are not feasible and this is often the case in the social sciences; but it is a mistake to use this situation to separate the social from the physical and biological sciences. Such situations also occur frequently in the physical sciences (in meteorology, astronomy, geology), the biological sciences, medicine, and elsewhere.

The experimental method also has some shortcomings. First, it is often difficult to choose the "control" variables so as to exclude *all* the confounding extraneous variables; that is, it may be difficult or impossible to design an "ideal" experiment. Consider the following examples: The problem of finding a proper control for testing the effects of the Salk polio vaccine led to the use of an adequate "placebo." The Hawthorne experiment demonstrated that the design of a proposed "treatment *versus* control" may turn out to be largely a test of *any* treatment *versus lack* of treatment (Roethlisberger and Dickson, 1939).[7] Many of the initial successes reported about mental therapy, which later turn into vain hopes, may be due to the hopeful effects of *any* new treatment in contrast with the background of neglect. Shaw, in "The Surprises of Attention and Neglect," writes: "Not until attention has been effectually substituted for neglect as a general rule, will the statistics begin to show the merits of the particular methods of attention adopted."

There is an old joke about the man who drank too much on four different occasions, respectively, of scotch and soda, bourbon and soda, rum and soda, and wine and soda. Because he suffered painful effects on all four occasions, he ascribed, with scientific logic, the common effect to the common cause: "I'll never touch soda again!" Now, to a man (say, from Outer Space) ignorant of the common alcoholic content of the four "treatments" and of the relative physiological effects of alcohol and carbonated water, the subject is not fit for joking, but for further scientific investigation.

Thus, the advantages of experiments over surveys in permitting better control are only relative, not absolute (Cornfield 1954). The design of proper experimental controls is not automatic; it is an art requiring scientific knowledge, foresight in planning the experiment, and hindsight in interpreting the results. Nevertheless, the distinction in control between experiments and surveys is real and considerable; and to emphasize this distinction we refer here to "ideal" experiments in which the control of the random variables is complete.

Second, it is generally difficult to design experiments so as to represent a specified important population. In fact, the questions of sampling, of making the experimental results representative of a specified population, have been largely ignored in experimental design until recently. Both in theory and in practice,

---

[7]Troubles with experimental controls misled even the great Pavlov into believing *temporarily* that he had proof of the inheritance of an acquired ability to learn: "In an informal statement made at the time of the Thirteenth International Physiological Congress, Boston, August, 1929, Pavlov explained that in checking up these experiments it was found that the apparent improvement in the ability to learn, on the part of successive generations of mice, was really due to an improvement in the ability to teach, on the part of the experimenter." From Greenberg (1929, p. 327).

experimental research has often neglected the basic truth that causal systems, the distributions of relations—like the distributions of characteristics—exist only within specified universes. The distributions of relationships, as of characteristics, exist only within the framework of specific populations. Probability distributions, like all mathematical models, are abstract systems; their application to the physical world must include the specification of the populations. For example, it is generally accepted that the statement of a value for mean income has meaning only with reference to a specified population; but this is not generally and clearly recognized in the case of regression of assets on income and occupation. Similarly, the *statistical* inferences derived from the experimental testing of several treatments are restricted to the population(s) included in the experimental design (McGinnis 1958, p. 412).[8] The clarification of the population sampling aspects of experiments is now being tackled vigorously by Wilk and Kempthorne (1955, 1956) and by Cornfield and Tukey (1956).

Third, for many research aims, especially in the social sciences, contriving the desired "natural setting" for the measurements is not feasible in experimental design. Hence, what social experiments give sometimes are clear answers to questions the meanings of which are vague. That is, the artificially contrived experimental variables *may* have but a tenuous relationship to the variables the researcher would like to investigate.

The second and third weaknesses of experiments point to the advantages of surveys. Not only do probability samples permit clear statistical inferences to defined populations, but the measurements can often be made in the "natural settings" of actual populations. Thus in practical research situations the experimental method, like the survey method, has its distinct problems and drawbacks as well as its advantages. In practice one generally cannot solve simultaneously all of the problems of measurement, representation, and control; rather, one must choose and compromise. In any specific situation one method may be better or more practical than the other; but there is no over-all superiority in all situations for either method. Understanding the advantages and weaknesses of both methods should lead to better choices.

In social research, in preference to both surveys and experiments, frequently some design of controlled investigation is chosen—for reasons of cost or of feasibility or to preserve the "natural setting" of the measurements. Ingenious adaptations of experimental designs have been contrived for these controlled investigations. The statistical framework and analysis of experimental designs are used, but not the randomization of true experiments. These designs are aimed to provide flexibility, efficiency, and, especially, some control over the extraneous variables. They have often been used to improve considerably research with controlled investigations.

---

[8] McGinnis points out that usually "it is not true that one can uncover 'general' relationships by examining some arbitrarily selected population...There is no such thing as a completely general relationship which is independent of population, time, and space. The extent to which a relationship is constant among different populations is an empirical question which can be resolved only by examining different populations at different times in different places."

These designs are sometimes called "natural experiments." For the sake of clarity, however, it is important to keep clear the distinctions among the methods and to reserve the word "experiment" for designs in which the uncontrolled variables are randomized. This principle is stated clearly by Fisher (1953, pp. 17–20),[9] and is accepted often in scientific research. Confusion is caused by the use of terms like "ex post facto experiments" to describe surveys or designs of controlled investigations. Sample surveys and controlled investigations have their own justifications, their own virtues; they are not just second-class experiments. I deplore the borrowing of the prestige word "experiment," when it cloaks the use of other methods.

Experiments, surveys, and investigations can all be improved by efforts to overcome their weaknesses. Because the chief weakness of surveys is their low degree of control, researchers should be alert to the collection and use of auxiliary information as controls against confounding variables. They also should take greater advantage of changes introduced into their world by measuring the effects of such changes. They should utilize more often efficient and useful statistics instead of making tabular presentation their only tool.

On the other hand, experiments and controlled investigations can often be improved by efforts to specify their populations more clearly and to make the results more representative of the population. Often more should be done to broaden the area of inference to more important populations. Thus, in many situations the deliberate attempts of the researcher to make his sample more "homogeneous" are misplaced; and if common sense will not dispel the error, reading Fisher (1953, pp. 99–100) may.[10] When he understands this, the researcher can view the population base of his research in terms of efficiency—in terms of costs and variances. He can often avoid basing his research on a comparison of one sampling unit for each "treatment." If he cannot obtain a proper sample of the entire population, frequently he can secure, say, four units for each treatment, or a score for each.[11]

---

[9] "Controlled investigation" may not be the best term for these designs. "Controlled observations" might do, but "observation" has more fundamental meanings.

[10] Fisher says: "We have seen that the factorial arrangement possesses two advantages over experiments involving only single factors: (i) Greater *efficiency*, in that these factors are evaluated with the same precision by means of only a quarter of the number of observations that would otherwise be necessary; and (ii) Greater *comprehensiveness* in that, in addition to the four effects of single factors, their 11 possible interactions are evaluated. There is a third advantage which, while less obvious than the former two, has an important bearing upon the utility of the experimental results in their practical application. This is that any conclusion, such as that it is advantageous to increase the quantity of a given ingredient, has a wider inductive basis when inferred from an experiment in which the quantities of other ingredients have been varied, than it would have from any amount of experimentation, in which these had been kept strictly constant. The exact standardisation of experimental conditions, which is often thoughtlessly advocated as a panacea, always carries with it the real disadvantage that a highly standardized experiment supplies direct information only in respect of the narrow range of conditions achieved by standardisation. Standardisation, therefore, weakens rather than strengthens our ground for inferring a like result, when, as is invariably the case in practice, these conditions are somewhat varied."

[11] For simplicity the following illustration is a simple contrast between two values of the "explanatory" variable, but the point is more general; and this aspect is similar whether for true experiments or controlled observations. Incidentally, it is poor strategy to "solve" the problem of representation by

Suppose, for example, that thorough research on one city and one rural county discloses higher levels of political interest in the former. It is presumptuous (although common practice) to present this result as evidence that urban people in *general* show a higher level. (Unfortunately, I am not beating a dead horse; this nag is pawing daily in the garden of social science.) However, very likely there is a great deal of variation in political interest among different cities, as well as among rural counties; the results of the research will depend heavily on which city and which county the researcher picked as "typical." The research would have a broader base if a city and a rural county would have been chosen in each of, say, four different situations—as different as possible (as to region, income, industry, for example); or better still in twenty different situations. A further improvement would result if the stratification and selection of sampling units followed a scientific sample design.

Using more sampling units and spreading them over the breadth of variation in the population has several advantages. First, some measure of the variability of the observed effect may be obtained. From a probability sample, statistical inference to the population can be made. Second, the base of the inference is broadened, as the effect is observed over a variety of situations. Beyond this lies the combination of results from researches over several distinct cultures and periods. Finally, with proper design, the effects of several potentially confounding factors can be tested.

These points are brought out by Keyfitz (1953) in an excellent example of controlled investigation (which also uses sampling effectively):

> Census enumeration data were used to answer for French farm families of the Province of Quebec the question: Are farm families smaller near cities than far from cities, other things being equal? The sample of 1,056 families was arranged in a 2^6 factorial design which not only controlled 15 extraneous variables (income, education, etc.) but incidentally measured the effect of 5 of these on family size. A significant effect of distance from cities was found, from which is inferred a geographical dimension for the currents of social change.

The mean numbers of children per family were found to be 9.5 near and 10.8 far from cities; the difference of 1.3 children has a standard error of 0.28.

4. SOME MISUSES OF STATISTICAL TESTS

Of the many kinds of current misuses this discussion is confined to a few of the most common. There is irony in the circumstance that these are committed usually by the more statistically inclined investigators; they are avoided in research presented in terms of qualitative statements or of simple descriptions.

---

obtaining a good sample, or complete census, of some small or artificial population. A poor sample of the United States or of Chicago *usually* has more over-all value than the best sample of freshman English classes at X University.

First, there is "hunting with a shot-gun" for significant differences. Statistical tests are designed for distinguishing results at a predetermined level of improbability (say at $P = .05$) under a specified null hypothesis of random events. A rigorous theory for dealing with individual experiments has been developed by Fisher, the Pearsons, Neyman, Wold, and others. However, the researcher often faces more complicated situations, especially in the analysis of survey results; he is often searching for interesting relationships among a vast number of data. The keen-eyed researcher hunting through the results of one thousand random tosses of perfect coins would discover and display about fifty "significant" results (at the $P = .05$ level).[12] Perhaps the problem has become more acute now that high-speed computers allow hundreds of significance tests to be made. There is no easy answer to this problem. We must be constantly aware of the nature of tests of null hypotheses in searching survey data for interesting results. After finding a result improbable under the null hypothesis the researcher must not accept blindly the hypothesis of "significance" due to a presumed cause. Among the several alternative hypotheses is that of having discovered an improbably random event through sheer diligence. Remedy can be found sometimes by a reformulation of the statistical aims of the research so as to fit the available tests. Unfortunately, the classic statistical tests give clear answers only to some simple decision problems; often these bear but faint resemblance to the complex problems faced by the scientist. In response to these needs the mathematical statisticians are beginning to provide some new statistical tests. Among the most useful are the new "multiple comparison" and "multiple range" tests of Tukey (1949), Duncan (1955), Scheffé (1953), and others. With a greater variety of statistical statements available, it will become easier to choose one without doing great violence either to them or to the research aims.

Second, statistical "significance" is often confused with and substituted for substantive significance. There are instances of research results presented in terms of probability values of "statistical significance" alone, without noting the magnitude and importance of the relationships found. These attempts to use the probability levels of significance tests as measures of the strengths of relationships are very common and very mistaken. The function of statistical tests is merely to answer: Is the variation great enough for us to place some confidence in the result; or, contrarily, may the latter be merely a happenstance of the specific sample on which the test was made? This question is interesting, but it is surely *secondary*, auxiliary, to the main question: Does the result show a relationship which is of substantive interest because of its nature and its magnitude? Better still: Is the result consistent with an assumed relationship of substantive interest?

---

[12] Sewell (1952, pp. 158–159) points to an interesting example: "On the basis of the results of this study, the general null hypothesis that the personality adjustments and traits of children who have undergone varying training experiences do not differ significantly cannot be rejected. Of the 460 chi square tests, only 18 were significant at or beyond the 5 per cent level. Of these, 11 were in the expected direction and 7 were in the opposite direction from that expected on the basis of psychoanalytic writings...Certainly, the results of this study cast serious doubts on the validity of the psychoanalytic claims regarding the importance of the infant disciplines and on the efficacy of prescriptions based on them." Note that by chance alone one would expect 23 "significant" differences at the 5 per cent level. A "hunter" would report either the 11 or the 18 and not the hundreds of "misses."

The results of statistical "tests of significance" are functions not only of the magnitude of the relationships studied but also of the numbers of sampling units used (and the efficiency of design). In small samples significant, that is, meaningful, results may fail to appear "statistically significant." But if the sample is large enough the most insignificant relationships will appear "statistically significant."

Significance should stand for meaning and refer to substantive matter. The statistical tests merely answer the question: Is there a big enough relationship here which *needs* explanation (and is not merely chance fluctuation)? The word *significance* should be attached to another question, a substantive question: Is there a relationship here *worth* explaining (because it is important and meaningful)? As a remedial step I would recommend that statisticians discard the phrase "test of significance," perhaps in favor of the somewhat longer but proper phrase "test against the null hypothesis" or the abbreviation "TANH."

Yates (1951, pp. 32–33), after praising Fisher's classic *Statistical Methods*, makes the following observations on the use of "tests of significance":

> Second, and more important, it has caused scientific research workers to pay undue attention to the results of the tests of significance they perform on their data, particularly data derived from experiments, and too little to the estimates of the magnitude of the effects they are investigating.
>
> Nevertheless the occasions, even in research work, in which quantitative data are collected solely with the object of proving or disproving a given hypothesis are relatively rare. Usually quantitative estimates and fiducial limits are required. Tests of significance are preliminary or ancillary.
>
> The emphasis on tests of significance, and the consideration of the results of each experiment in isolation, have had the unfortunate consequence that scientific workers have often regarded the execution of a test of significance on an experiment as the ultimate objective. Results are significant or not significant and this is the end of it.

For presenting research results statistical estimation is more frequently appropriate than tests of significance. The estimates should be provided with some measure of sampling variability. For this purpose confidence intervals are used most widely. In large samples, statements of the standard errors provide useful guides to action. These problems need further development by theoretical statisticians (Cox 1958, pp. 357–372).

The responsibility for the current fashions should be shared by the authors of statistical textbooks and ultimately by the mathematical statisticians. As Tukey (1954a, p. 710) puts it:

> *Statistical methods should be tailored to the real needs of the user.* In a number of cases, statisticians have led themselves astray by choosing a problem which they could solve exactly but which was far from the needs of their clients....The broadest class of such cases comes from the choice of significance procedures rather than confidence procedures. It is often much easier to be "exact" about

significance procedures than about confidence procedures. By considering only the most null "null hypothesis" many inconvenient possibilities can be avoided.[13]

Third, the tests of null hypotheses of *zero* differences, of no relationships, are frequently weak, perhaps trivial statements of the researcher's aims. In place of the test of zero difference (the nullest of null hypotheses), the researcher should often substitute, say, a test for a difference of a specific size based on some specified model. Better still, in many cases, instead of the tests of significance it would be more to the point to measure the magnitudes of the relationships, attaching proper statements of their sampling variation. The magnitudes of relationships cannot be measured in terms of levels of significance; they can be measured in terms of the difference of two means, or of the proportion of the total variance "explained," of coefficients of correlations and of regressions, of measures of association, and so on. These views are shared by many, perhaps most, consulting statisticians—although they have not published full statements of their philosophy. Savage (1957, pp. 332–333) expresses himself forcefully: "Null hypotheses of no difference are usually known to be false before the data are collected; when they are, their rejection or acceptance simply reflects the size of the sample and the power of the test, and is not a contribution to science."

Too much of social research is planned and presented in terms of the mere existence of some relationship, such as: individuals high on variate x are high on variate y. The *exploratory* stage of research may be well served by statements of this order. But these statements are relatively weak and can serve *only* in the primitive stages of research. Contrary to a common misconception, the more advanced stages of research should be phrased in terms of the quantitative aspects of the relationships. Again, to quote Tukey (1954a, pp. 712–713):

> *There are normal sequences of growth in immediate ends.* One natural sequence of immediate ends follows the sequence:
>
> 1. Description
> 2. Significance statements
> 3. Estimation
> 4. Confidence statement
> 5. Evaluation
>
> ....There are, of course, other normal sequences of immediate ends, leading mainly through various decision procedures, which are appropriate to development research and to operations research, just as the sequence we have just discussed is appropriate to basic research.

At one extreme, then, we may find that the contrast between two "treatments" of a labor force results in a difference in productivity of 5 per cent. This difference may appear "statistically significant" in a sample of, say, 1000 cases. It may also mean a difference of millions of dollars to the company. However, it "explains" only about one per cent of the total variance in productivity. At the other extreme

[13] See also Cox (1958) and Duncan (1955).

is the far-away land of completely determinate behavior, where every action and attitude is explainable, with nothing left to chance for explanation.

The aims of most basic research in the social sciences, it seems to me, should be somewhere between the two extremes; but too much of it is presented at the first extreme, at the primitive level. This is a matter of over-all strategy for an entire area of any science. It is difficult to make this judgment off-hand regarding any specific piece of research of this kind: the status of research throughout the entire area should be considered. But the superabundance of research aimed at this primitive level seems to imply that the over-all strategy of research errs in this respect. The construction of scientific theories to cover broader fields—the persistent aim of science—is based on the synthesis of the separate research results in those fields. A coherent synthesis cannot be forged from a collection of relationships of unknown strengths and magnitudes. The necessary conditions for a synthesis include an *evaluation* of the results available in the field, a coherent interrelating of the *magnitudes* found in those results, and the construction of models based on those magnitudes.

APPLIED SAMPLING METHODS

STEVEN G. HEERINGA AND JAMES M. LEPKOWSKI[1]

The application of probability sampling theory to actual research design raises many methodological and practical questions. Throughout his career as a researcher, statistician, and educator, Leslie Kish enjoyed the challenge of finding practical, efficient answers to applied sampling questions.

In Leslie's work as a sampler, first at the U.S. Department of Agriculture and, then, for over 50 years at the University of Michigan Survey Research Center (SRC), he addressed applied sampling problems on a day-to-day basis. He recognized very early in his career that these applied problems were more than just technical details to be dealt with in everyday work, but that they were important topics to be researched, and that the solutions warranted publication in the statistical literature.

A perfect example is the first paper in this section which describes the "Kish method" for the objective random selection of a single respondent from a sample household (Kish 1949). The respondent selection method described in this paper was originally designed to control the age and gender distribution of randomly selected respondents from households with multiple eligible persons. However, Kish saw that the procedure had secondary and maybe even more important features: improvements in the coverage of household surveys and improvements to the application of random selection in field settings. The method requires systematic listing of all eligible respondents in the household by gender and age before the selection. This requirement improved the coverage of household members. The preprinted "selection tables" that were part of the procedure reduced interviewer subjectivity and error in the respondent selection. Over 50 years later, despite changes in the demographic composition of U.S. households, the systematic rostering of eligible household members and predetermined selection remain strong justifications for use of objective respondent selection in household surveys.

---

[1] Steven Heeringa, Survey Research Center, University of Michigan, Ann Arbor, MI 48106–1248. sheering@isr.umich.edu.
James Lepkowski, Joint Program in Survey Methodology, University of Michigan, Ann Arbor, MI 48106–1248. jimlep@isr.umich.edu.

Leslie often cited his experience as a medical laboratory assistant at Rockefeller Institute for Medical Research as the spark for his lifelong interest in research design. While there, he read Fisher and Pearson, and later read Yates on applied statistics and experimental design. His study of experimental design early in his career, including Latin and Graeco-Latin squares, later influenced his joint work with Roe Goodman on a controlled selection method for probability samples (Goodman and Kish 1950). Controlled selection (sometimes also called "lattice sampling") is a probability sampling technique that is used for "deep" stratification of sample elements, typically for sample designs where the combinations of stratifying variables exceed the physical number of sample selections. Just as some classical experimental designs control variation in experimental measurements when the number of control groups exceeds the number of units, controlled selection reduces variances of sample estimates by highly restricting combinations of population elements that may be chosen for a given sample. Goodman and Kish showed how such controls can be achieved while adhering to probability sampling principles. Bryant et al. (1960) judged the Goodman and Kish controlled selection method to be very complex if manual solutions were required and proposed a related controlled rounding method for sampling problems involving two stratifiers. Later work led to the development of computer programs for solving controlled selection problems (Groves and Hess 1975; Lin 1992) and eventually to unique applications of transportation theory and linear programming techniques (Causey et al. 1985; Sitter and Skinner 1994) to controlled selection of samples.

By 1960, the Survey Research Center, like the U.S. Bureau of the Census and many other data collection organizations, recognized the benefits of maintaining a master sample of primary stage units (PSUs) that would be used for a decade of household survey research. Maintaining a master sample typically involves updating the stratification and population measures of size, and re-selecting PSUs following each decennial release of Census data. Master samples enable survey organizations to hire and employ permanent field interviewing staff in the PSUs. Procedures to maximize the retention of existing sampling units when new samples are selected were and are important in retaining experienced and trained staff from sample to sample. These procedures were first introduced by Keyfitz (1951). The paper by Kish and Scott (1971) that is included in this volume extends the Keyfitz method to maximize the retention of sampling units when the new or updated sample design changes both the stratification and probabilities of selecting sample units. The Kish-Scott procedure has been used repeatedly in the decennial re-selection of SRC's master National Sample of households to optimize the overlap between the master sample of PSUs. It has also been used to maximize overlap in PSUs for samples of special populations such as school children or racial and ethnic subpopulations. Kish and Anderson (1978), also included elsewhere in this volume, investigated the related problem of stratification for multi-purpose surveys, answering the practical question of how much is gained by forming more (or fewer) strata and to what extent the number of stratification variables influences sample precision. Related papers by Anderson, Kish, and Cornell (1976, 1980) discuss the gains in sample precision due to stratification and the correspondence between sample stratification and methods for grouping and matching that are more common in clinical and epidemiological research.

Leslie had a career-long interest in research design for surveys of household and nonhousehold populations. Papers on several topics of major interest, such as small area estimation, combining censuses and samples, and rolling samples, are included elsewhere in this monograph. Leslie's own 1987 book, *Statistical Design for Research*, is a compendium of ideas and specific guidance from his experience as a sampler and social science researcher. In that book, and other earlier publications, designs for longitudinal studies, including panels and repeated samples with partial overlap ("split panels"), were a favorite topic (Kish 1965). The 1950s and 1960s were fertile periods for new research and generated many interesting sample design problems. SRC's economists and organizational psychologists engaged Leslie in the design of a number of samples of schools, businesses, hospitals and other organizations. These survey populations presented special problems for design and estimation—problems that were not always intuitively understood by researchers and policy analysts. The 1965 paper, "Sampling Organizations and Groups of Unequal Size," was written for social scientist colleagues who conducted or used research on nonhousehold populations. Other investigations required sampling of rare populations (Kish 1988b) or populations that were mobile in time and location (Kish et al. 1961).

The common thread that ran through Leslie's career and was passed on to his many students was an intellectual curiosity about the theory of sampling and, more importantly, about the everyday, applied, practical problems that sampling statisticians and researchers all face. The selection of papers that follow exemplifies the breadth and depth of this curiosity.

3

A PROCEDURE FOR OBJECTIVE RESPONDENT SELECTION WITHIN THE HOUSEHOLD

LESLIE KISH

In modern survey methods growing emphasis is placed on the objective selection of the sample. For surveys of the general population, increasing use is made of area sampling to obtain probability samples of households. Heretofore, scant attention has been given to the question of how to make an objective selection among the members of the household.

A procedure for selecting objectively one member of the household is given as used in four surveys of the adult population. Demographic data as found in the sample are compared with outside sources for available factors.

1. THE PROBLEM

To obtain random samples for surveys,[1] two basic conditions must be satisfied:

1. The sampling method must provide for a known probability of inclusion, other than zero, for every element of the population.
2. The method must be translated into a procedure that can and will be applied in practice.

Area sampling is gaining acceptance as a practical and reliable procedure for obtaining samples of households with known probabilities of selection (King and Jessen 1945, Hansen and Hauser 1945, Houseman 1947, U.S. Bureau of the Census 1947). In general practice these probabilities are equal either within specified strata or throughout the sample.

Reprinted with permission from the *Journal of the American Statistical Association*, 44, 380–387.

Presented at the 107th annual meeting of the American Statistical Association, New York City, December 30, 1947.

[1]For discussion of sources of biases that may arise when personal judgment enters into the selection see Hauser and Hansen (1944).

Given a sampling procedure—such as area sampling—which uses dwellings as units of sampling: when does the question of selection within households arise? There is no need of selection:

(A) If the respondent is uniquely determined (as head of household or homemaker); or
(B) If the household is the unit of analysis and any adult member can give equally valid information.[2]

However, if the household contains more than one member of the desired population, it may be regarded as a cluster of population elements.

One may decide to include in the sample every member within the household.[3] However, this may be a statistically inefficient procedure in general (depending on the cost-variance relationship of the survey design), unless one of these three conditions holds:

(A) Information about all members can be obtained from one of them.
(B) There is seldom more than one member of the population in the household. For example, there is an average of 1.2 spending units per household.[4]
(C) If the intra-class correlation within the household of the variables measured is of negligible size, or if it is negative.

These conditions are not met generally in surveys of the attitudes of the adult population. Furthermore, multiple interviews in one household may lead to undesirable interview situations. Hence, there is need for a procedure of selection that will translate a sample of the households into a sample of the adult population. There are no great theoretical difficulties, but a practical procedure must meet the demands of efficiency and of applicability. There are several alternatives, and the choice among them depends on a number of factors: the nature and distribution of the population, the objectives and design of the survey, and the available facilities. Under the latter we may distinguish the factors of cost, and of the nature and training of the field force.

2. THE CONDITIONS

It may be useful to describe very briefly the general sampling procedures used by the Survey Research Center.[5] A stratified random selection of dwellings is obtained by means of area sampling. Small areas, segments, define the clusters of households within counties (or groups of adjacent counties) selected as primary

---

[2] This assumption may be unjustified. See Deming (1944).
[3] See, for example, Watson (1947).
[4] Based on the Consumer Finances Surveys of the Federal Reserve Board, conducted by the Survey Research Center.
[5] For a fuller description see Goodman (1947).

sampling areas. In each of these areas there are trained interviewers who are employed on a part-time basis. Detailed sampling instructions are given to them in order to insure correct understanding and execution of the different procedures required by a variety of types of surveys.

The interviews are of the fixed question, free-answer kind, requiring generally from 30 minutes to one and one-half hours. Conducting the interviews in the home is believed to contribute toward a satisfactory interview situation. The interviewer calls at pre-designated dwellings; the respondent is then selected by a fixed procedure. Return calls are made to find the not-at-homes.

For each of a series of four surveys, conducted in June, August and December, 1946 and April, 1947 respectively, it was desired to obtain samples of about 600 interviews of the adult population of the continental United States. The samples were distributed in 31 primary sampling areas: in four of the 12 largest metropolitan areas plus 27 scattered counties.

The population of the four surveys was limited to adults living in private households, excluding, because of interviewing difficulties and cost considerations, some segments of the population: armed forces; hospitals; religious, educational and penal institutions; trailer, logging and labor camps; and hotels and large rooming houses.

A procedure was devised for selecting one adult in each sample household. This procedure was favored over some alternatives (such as selecting every other adult found in sample households) for two reasons:

1. It was desired to take no more than one interview in any household, in order to obtain each interview before the respondent had a previous opportunity to discuss the questions. Furthermore, multiple interviews were thought to be statistically inefficient because of the expected correlation of attitudes within the household.
2. An interview in every sample household was desired in order to avoid making futile calls on dwellings without interviews.

With this procedure unbiased estimates may be obtained by giving each respondent as weight the number of adults in the household. Such differential weighting may in general increase the sampling error. However, in the present instance this increase is not great because of the concentration in two-adult households. About 60 per cent of the households have two adults, about 10 per cent have more than three adults and about 1 per cent have more than five. Another result of this high concentration is that, unless the variable has a high correlation with the number of adults in the households, the difference between the weighted estimate and that in which each respondent has equal weight will be small; in case of small samples this difference may be negligible compared to the sampling error. For all attitudes thus tested in these studies the difference was inconsiderable compared with sampling error. However, in the comparison given in Table 1 there appear two differences of about two percentage points.

In order that the procedure may be applied and checked without great difficulty it is desirable to have a variable for ordering the members of the household; a variable that can be obtained by the interviewer objectively and easily. The age and sex of the members of the household were used for this purpose.

3. THE PROCEDURE

A "face sheet" is assigned by the sampling section to each sample dwelling unit; on each of these there are, in addition to the address of the dwelling, a form for listing the adult occupants, and a table of selection. At the time of the first contact with the household, the interviewer lists each adult separately on one of the six lines of the form; each is identified by entering in the first column his relationship to the head of the household (wife, son, brother, roomer, etc.). In the next two columns the interviewer records the sex and (if needed) the age of each adult. Following this the interviewer assigns a serial number to each adult: first the males are numbered in order of decreasing age, followed by the females in the same order. To assign these serial numbers it is necessary to obtain the ages of all adults only in that small portion of households in which there are two adults of the same sex and not connected by parent-child relationship.

Then the interviewer consults the table of selection; this table tells him the number of the adult to be interviewed. One of the six tables (A to F) displayed in Table 1 is printed on each face sheet; each of the six tables is assigned to one-sixth of the sample addresses in a systematic manner. This procedure was applied to the sample addresses in the four surveys.

The slightly changed order of selection in the tables given in Table 2 has certain advantages over that of those in Table 1. The low numbers of selection are concentrated in tables A, B and C; therefore the procedure will yield a male respondent in a great majority of the addresses to which these tables have been assigned. Evening calls are necessary to find at home most of the male respondents; the interviewer may concentrate his evening calls at these addresses. Conversely the interviewer may use his time during the day by calling at addresses to which tables D, E and F are assigned, with an increased chance of success.

The proper fractional representation of each adult is approximated more closely with the tables in Table 2 without the necessity of printing many more forms; the chances of selection are exact for all adults in households 1, 2, 3, 4, and 6 adults. Because numbers above six are disallowed, there are one or two adults in a thousand (generally young females) who are not represented; there is a "compensation" for them in the overrepresentation of number five in the households with five adults.

It may be noted that the procedure can be modified easily to select a constant proportion, say half or one-third, of the adults. In that case, of course, each adult would have the same chance of selection, regardless of the size of the household. In some of the households more than one interview would be taken, in some no interviews at all. The tables below may be modified readily to apply the changed procedure.

TABLE 1.

A	Number of adults in D.U.	1	2	3	4	5	6 or more
	Interview adult numbered:	1	1	3	2	5	1
B	Number of adults in D.U.	1	2	3	4	5	6 or more
	Interview adult numbered:	1	2	1	3	4	2
C	Number of adults in D.U.	1	2	3	4	5	6 or more
	Interview adult numbered:	1	1	2	4	1	3
D	Number of adults in D.U.	1	2	3	4	5	6 or more
	Interview adult numbered:	1	2	3	1	2	4
E	Number of adults in D.U.	1	2	3	4	5	6 or more
	Interview adult numbered:	1	1	2	1	3	5
F	Number of adults in D.U.	1	2	3	4	5	6 or more
	Interview adult numbered:	1	2	1	4	3	6

TABLE 2.

Relative frequency of use	Table number	If no. of adults in household is					
		1	2	3	4	5	6 or more
		Select adult numbered					
1/6	A	1	1	1	1	1	1
1/12	B1	1	1	1	1	2	2
1/12	B2	1	1	1	2	2	2
1/6	C	1	1	2	2	3	3
1/6	D	1	2	2	3	4	4
1/12	E1	1	2	3	3	3	5
1/12	E2	1	2	3	4	5	5
1/6	F	1	2	3	4	5	6

4. RESULTS: CHECKS AGAINST OUTSIDE SOURCES

The distributions of the respondents were checked with outside sources whenever we could obtain or adapt reliable data for valid comparison. They are presented in Tables 3 and 4.

The check data are those deemed most nearly valid for comparison: the per cent white and the educational attainment are for April 1947 from Census Series P-20 No. 15, age is for July 1946 from Series P-S No. 19, and the per cent native born is the 1940 census figures for ages 16–64. In the data obtained from the survey results each respondent is weighted by the number of adults (from one to six)

TABLE 3. Comparison of Respondents in Sample With Check Data

	Check data	Total for 4 surveys N=2372: Weighted	Un-weighted	Data from the surveys of: June 1946 N=585	Aug. 1946 N=592	Dec. 1946 N=570	April 1947 N=625
White	90.6	89.1	89.4	89	90	88	89
Native born	90.0	90.6	90.8	90	91	91	90
Age in years							
21–29	22.8	22.7	20.6	22	24	22	22
30–44	33.9	33.0	34.1	34	33	36	30
45–59	25.6	27.3	26.7	26	27	25	32
60 and over	17.7	16.1	17.8	17	15	17	15
NA		0.9	0.8	1	1	0	1
Education							
Not finished grammar	46.1	25.7	26.2	24	25	24	30
Finished grammar		19.1	19.3	17	17	22	20
Some high	17.4	17.1	17.4	18	19	18	14
Finished high	22.9	20.4	20.0	20	21	18	22
Some college	7.1	9.9	9.4	11	11	10	7
Finished college	5.0	7.0	6.9	9	6	7	6
NA	1.5	0.8	0.8	1	1	1	1

TABLE 4. Percentages of Males

Check data: Census estimate July, 1946	All adults in sample households	Respondents in sample weighted by number in household: 4 surveys combined	June 1946	August 1946	December 1946	April 1947
48.2	47.9	46.8	45	46	45	52

in his household. However, the column "unweighted," in which each respondent has unit weight, is included for purposes of comparison.

The appropriate sampling error (i.e., two standard errors) of the estimates in Table 3 range from three to six percentage points for various items on each of the surveys, and from two to four percentage points for the four surveys combined. The estimates of color, nativity and age groups are in general agreement, with only one of the 30 estimates lying slightly beyond the given ranges.

The data of the first three surveys showed an apparent upward bias in reports of college education. This was suspected to be a response bias to the single question on this topic. For the fourth survey three more questions were inserted to get more data on persons claiming 12 years of school or more; this involved additional work on only a third of the respondents. The fourth survey shows close agreement with the check data on college education, pointing to elimination of a response bias.[6]

The results for the checks on the sex ratio are presented in Table 4. The Bureau of Census estimate is for the civilian non-institutional adult population for July, 1946 (Series P-S No. 19). The second figure is an internal check, obtained by tabulating all adults listed as living in sample households.

Males appeared to be underrepresented among the respondents of the first three surveys. Although the difference was small, its presence in three surveys pointed to possible occasional deviation from rigorous procedure in the field. It may be noted that a 3 per cent bias in the sex ratio would require an unlikely 33 per cent sex difference in an attitude studied in order to produce a 1 per cent difference in survey results; a difference within the limits of accuracy of these surveys.

Tabulations of all adults in the sample households, as listed on the face sheet, give close agreement with the check data. Continued investigation of the problem points to two sources of this small bias, both due to the fact that males are more difficult to find at home even with repeated call-backs: overrepresentation of males among the nonresponses, and an occasional substitution on the part of a few interviewers.

Two other small sources of error present in the first three surveys were eliminated by minor changes in the procedure.

5. CONCLUSION

The described procedure of selection within the household gave results that were satisfactory within the demands of the survey objectives. While there were occasional departures from correct procedure in the field, the procedure is such that extensive control over its field application, hence improvement, is feasible. It must be emphasized, however, that a practical sampling procedure is not an automatic device. For success it depends on a field force having both the training and the morale necessary for correct application.

---

[6] These statements are also borne out by results on three more national surveys in which the same procedure was used.

4

CONTROLLED SELECTION—A TECHNIQUE IN PROBABILITY SAMPLING

ROE GOODMAN AND LESLIE KISH

A sampling technique is defined as introducing control into the selection of n out of N sampling units when it increases the probabilities of selection for preferred combinations of units (and decreases the probabilities for non-preferred combinations). Methods used in the past have by no means exhausted the possibilities of controlled selection, however. Procedures are developed by which the probabilities of selection for preferred combinations are sharply increased and the theoretical basis for the methods is stated. The methods are applied to a specific problem and the procedures are described in detail. It is found that as a result the variances of estimates for several important items are reduced as compared with the corresponding variances for stratified random sampling.

1. INTRODUCTION

Among the decisions involved in designing a sample is that of the method of selecting the sampling units. Simply stated the problem is the following: Given a population consisting of N rigorously defined sampling units, how is one to select n of these units for the sample?

In probability sampling the essential condition is that each of the N sampling units shall have a specific, known probability of selection and that the probability for none of the units shall be zero. Within this limitation the possibilities are myriad.

The methods of controlled selection presented in this paper appear to have their major (although by no means their only) usefulness in the selection of first-stage sampling units in multi-stage sampling.[1] For the sampling of widely dispersed populations first-stage units commonly consist of counties, part-counties, or combinations of them (areas containing perhaps 1000 to 5000 square

Reprinted with permission from the *Journal of the American Statistical Association*, 45, 350–372.

[1] Many of the terms used in this paper are as defined in United Nations Statistical Office (1950).

miles). Although the techniques of controlled selection have wide, general applicability in all types of probability sampling, it was for the selection of first-stage units that these methods were developed and the emphasis in this paper is upon their application to this particular problem.

2. STRATIFIED SAMPLING

A selection procedure which is widely used today is called "stratified random." By this method sampling units are selected at random within each stratum independently of the selection of units within other strata. During recent years units within a stratum have frequently been assigned varying probabilities[2] of selection. When this is done the strata are usually defined in such a way that only one unit is to be selected from each.

3. STRATIFICATION VIEWED AS A METHOD OF CONTROL

It is to be noted that stratification introduces restrictions or controls in the process of selection. In simple random sampling the probability of selection of any of the $C(N,n)$ possible combinations of n out of N sampling units is a function solely of the probabilities of selection assigned each of the n units. With stratified random sampling on the other hand the selection of units is partially controlled in that the probability of selection of the combination depends on the strata with which the n units are associated. For example, if two units are to be selected from each stratum, every combination of n units consisting of two units from each stratum has a positive probability of selection, whereas all other combinations have a zero probability of selection. By means of stratification then, the number of possible sample combinations is reduced. At the same time the probabilities of selection of admitted combinations are increased.

4. CONTROLLED SELECTION—DEFINITION

As we define it, the expression "Controlled Selection" has a rather broad meaning. This expression is defined to mean any process of selection in which, while maintaining the assigned probability for each unit, the probabilities of selection for some or all preferred combinations of n out of N units are larger than in stratified random sampling (and correspondingly the probabilities of selection for at least some non-preferred combinations are smaller than in stratified random sampling).

It is to be noted that while in stratified random sampling the probabilities of selection for certain combinations are increased, the possibilities of increasing the

[2] Although the word random is generally taken to connote equal probability, its use in this paper implies merely specific, known probabilities, not necessarily equal. The notion of making a selection among units with varying assigned probabilities was introduced about six years ago. See Hansen and Hurwitz (1943, 1949).

probabilities of selection for preferred combinations are by no means exhausted by this process. That is, with controlled selection it is in general possible to go very much further in increasing the probabilities of selection of preferred combinations of units than is done by stratified sampling alone. At the same time the probability of selection for vast numbers of other combinations is reduced, generally to zero. It is with the use of controls after the possibilities of stratification have been exhausted that this paper is concerned.

5. NEED FOR CONTROLS BEYOND STRATIFICATION

In selecting first-stage units for a sample survey of a large geographical area it is usually desirable to stratify in respect to a number of classifications of the population. The need for this multiple stratification arises in part because various kinds of information may be obtained from the same survey; furthermore a sample of first-stage units may need to service over a period of time a great number of diverse surveys with a multitude of variables and objectives often unknown in advance. At the same time the number of strata one can use is restricted to the number of first-stage units which in turn is limited by considerations of cost. Moreover, the use of additional modes of stratification is possible only at the expense of reducing the control, hence possibly the efficiency, of other classifications. Even with a few classifications one may find himself with more strata cells than the number of first-stage units he can use.

The sampler may now ask this question: Is it possible to achieve controls, similar to those of additional stratification without actually increasing the number of cells? Several types of controls have been designed and used in the past; there is reference to these in the latter part of this paper. These procedures aimed at obtaining balance with respect to control factors, but usually accepted the possibility of biased estimates. In contrast, the emphasis in this paper is upon securing unbiased estimates through rigorous procedures of probability sampling, at the same time only approximating a balance with respect to the control factors.

6. OBJECTIVE OF CONTROLLED SELECTION

As for most sampling techniques the objective of controlled selection is to reduce the variances of estimates obtained at a given cost. To the extent that the choice among alternative methods of selection does not affect costs appreciably the objective is simply one of reducing the variances. As will be shown later, by extending the control beyond that of stratification it is possible either to reduce the variance of any item or to increase it depending upon the utility of the controls used.

Since it was logical to suppose that the procedures contemplated would result in some reduction in the variances of estimates a sample was drawn for the North Central States by the new method. Later additional test samples were drawn from the same population and the between-first-stage-unit component of the variance was computed for several important items. It was found that these between-components were reduced from 11% to 32% below those corresponding to stratified random sampling. The details of these tests are given later in this paper.

7. SIMPLE ILLUSTRATION OF CONTROL BEYOND STRATIFICATION

It is possible to introduce a measure of additional control in the selection process very easily. Suppose we have two strata consisting of first-stage units from each of which one unit is to be selected. Suppose further that in stratum 1 units *B*, *C*, and *F* lie adjacent to the ocean or other major waterway, that in stratum 2 unit *d* is similarly located, and that all other units are located inland. Assume the units, and the probability of selection assigned to each of them, are as shown below. The sum of probabilities for *B*, *C*, *F* and *d* is 0.65. It is considered desirable to select one inland and one coastal unit, and undesirable to select two coastal units. The procedure is as follows:

Re-arrange the units in the first stratum by listing *B*, *C*, *F* first, followed by *A*, *D*, and *E*. Re-arrange the units in the second stratum by shifting *d* to the end, that is, by placing *e* above *d*. Then draw a random number from 1 to 100 and let this one number determine the selection in both strata. Thus if the random number is 45 or less a coastal unit will be selected in stratum 1 and an inland unit in stratum 2, if the number is 46 to 80 an inland unit will be selected in both strata, and if the number is greater than 80 an inland unit will be selected in stratum 1 and a coastal unit in stratum 2.

By means of the above procedure the probability of selecting one inland and one coastal unit is .65 and the probability of selecting two inland units is .35. With stratified random sampling on the other hand the probabilities are the following: one inland and one coastal unit, .47; two inland units, .44; two coastal units, .09. If one were more interested in securing what appeared to be a balanced sample than a strictly random sample he would be inclined to make sure that he selected one inland unit and one coastal unit to represent the two strata. By a procedure of controlled selection, however, the original assigned probabilities of each unit are rigorously maintained and the probability of securing one inland and one coastal unit is made as large as possible within the limitation of probability sampling.

Stratum 1		Stratum 2	
Unit	Probability	Unit	Probability
A	.10	*a*	.15
B	.15	*b*	.30
C	.10	*c*	.10
D	.20	*d*	.20
E	.25	*e*	.25
F	.20		

8. APPLICATION TO A SPECIFIC PROBLEM

There follows now a description of the application of controlled selection to a larger problem. As will be seen, the procedure is somewhat complex, although its concurrence with the rules of probability sampling is apparent. The description may also be helpful in showing the logical basis for the use of controlled sel2ection. In giving the details of the procedure that was used there is no pretense of showing the best way of controlling the sample selection but merely one way that it may be done.

The objective was the selection of 21 primary sampling units to represent the North Central States. The Detroit, Cleveland, Chicago, and St. Louis metropolitan areas were taken as separate strata, each of which was then selected with certainty. There remained the selection of 17 primary sampling units to represent all areas in the region outside these metropolitan areas. The first-stage sampling unit consisted of a single county or a county joined with part(s) or all of an adjacent county (or counties). In forming the primary sampling units, there were three objectives as follows:

1. To increase diversity within the unit with regard to concentration of population;
2. To limit the distance from the central city in which the interviewer would live, to the periphery, to a range of 25 to 30 miles.
3. Not to increase the size of the unit beyond a single county if doing so would decrease the proportion of the population residing in the central city to less than 40% of the total for the unit. The last objective was a concession to interviewers many of whom prefer to do a substantial part of their work near their homes.

The sampling units were grouped into 17 strata of about 1.8 million each, primarily on the basis of the size of largest city, and secondarily, for the more rural counties, by major type of farming areas. The strata were numbered from 1 to 17, the ordering being from those with largest cities to those with rural populations only. Within each stratum one unit was to be selected; the probability of selection assigned each unit was proportional to its 1940 population.

It was thought possible, by the use of controls, to insure a greater geographic spread and balance than one could expect from simple stratified random sampling. Another aim was to achieve a better balance with regard to per cent urbanization in the middle 7 strata, in which there remained considerable variation within the strata with respect to this variable. The geographic balance was intended to yield approximately proportionate representation of the various states in the sample and to make sure that the larger cities, as well as the less densely populated places, were well distributed geographically.

9. SELECTION IN SUCCESSIVE STAGES—STEP ONE

In order to simplify the mechanics of the work it was found desirable to divide the procedure into two successive steps. The various schemes used to accomplish control in these two stages will be discussed with the aid of the tables.

The first step was designed to assure proper representation among four groups of states as shown in the heading of Table 1. The table shows the sums of the probabilities of selection assigned the different sampling units within each of the four groups (A, B, C, and D) for each of the 17 strata, and for various groups of

TABLE 1. Assigned Probabilities of Sampling Units Classed in Four State Groups in Each of 17 Strata, North Central States[a]

Stratum no.	Mich., Wis. A	Ind., Ohio B	Ill., Iowa, Minn. C	Mo., N.D., S.D., Neb., Kan. D	Total
1			.4946	.5054	1.0000
2	.4146	.5854			1.0000
3		1.0000			1.0000
4	.2827	.3296	.2602	.1275	1.0000
Total 1–4	.6973	1.9150	.7548	.6329	4.0000
5	.1059	.1093	.5165	.2683	1.0000
6	.4937	.3213	.1850		1.0000
7	.1034	.7412	.1554		1.0000
8	.4361		.3478	.2161	1.0000
Total 5–8	1.1391	1.1718	1.2047	.4844	4.0000
Total 1–8	1.8364	3.0868	1.9595	1.1173	8.0000
9	.3040	.4550	.1601	.0809	1.0000
10		.2222	.4737	.3041	1.0000
11	.2915	.0759	.2981	.3345	1.0000
12	.0146	.2476	.4657	.2721	1.0000
13	.2666	.2149	.1922	.3263	1.0000
Total 9–13	.8767	1.2156	1.5898	1.3179	5.0000
14	.0261	.2719	.4297	.2723	1.0000
15	.2914	.1891	.2963	.2232	1.0000
16	.0363	.0663	.3409	.5565	1.0000
17	.2164	.0299	.1570	.5967	1.0000
Total 14–17	.5702	.5572	1.2239	1.6487	4.0000
Total 9–17	1.4469	1.7728	2.8137	2.9666	9.0000
Total 1–17	3.2833	4.8596	4.7732	4.0839	17.0000

[a]Strata of approximately equal total populations, assigned probabilities adding to one. Strata order by decreasing size of cities.

strata 1 to 4, 5 to 8, etc. These probabilities are proportional to the population figures of the 1940 Census and expressed as a fraction of the total population for the stratum. The ordering of the strata is meaningful, representing decreasing size of the central city within the sampling units. For example stratum 1 consists of three densely populated units, stratum 6 of 13 units having medium sized towns, and stratum 17 is made up of 195 entirely rural units.

The data in the table may be interpreted in terms of expected number of selections, as well as probabilities of selection. For example, the total 1.1718 under B for strata 5 to 8 means that of the 4 units to be selected in these strata, on the average 1.1718 will be selected in Ohio or Indiana. By the use of controlled selections it is possible *to make sure that in any one sample either one or two of the four units selected will be drawn from this pair of states*. By this method, when a draw is made the chances are .1718 that two units will be selected from these states and .8282 that a single unit will be selected. As will be seen, this is the major type of control achieved by the first step.

In order to show how the procedure was carried through we turn to a discussion of Table 2. This will be followed by a description of one procedure for obtaining such a table.

Table 2 is set up in such a way as to conform with Table 1. The number of cells in each box has been limited to 16 in order to simplify the work. The numbers in the cells represent the number of sampling units to be drawn from each group of states. The P's are the probabilities of selection for each combination; they add to 1.0000. Thus the first combination (represented by the first box) has a probability of .1391 of selection. If this combination were selected, the drawings for the first four strata would be taken as follows: none from the A group, two from the B group, 1 from the C group and 1 from the D group. The number of units to be selected from each group is similarly shown for strata 5–8, 9–13, and 14–17.

The probabilities of all cells of Table 1 are preserved in Table 2, that is, the product in Table 2 of the entry in a cell and the probability for the combination when summed over all combinations equals the entry for the corresponding cell in Table 1. For example for group A in strata 1–4 we have: (0)(.1391) + (0)(.0033) + ... + (1)(.0940) + (1)(.0489) = (0)(.3027) + (1)(.6973). In other words: the probability is .3027 of selecting no A, and .6973 of selecting one A for strata 1–4. Again for strata 5–8 the probability is .8609 of choosing one A, and .1391 of choosing 2 A's. Again for strata 1–8 combined we have for group A: (1)(.1636)+(2)(.8364) = 1.8364. Similarly for the other cells and finally for the total of all strata 1–17 the probability is .7167 of selecting three A's, and .2833 of selecting four A's. The same restrictions hold for groups B, C and D over all the totals shown in Table 1. Also for all the totals of Table 1, the corresponding entry, or sum of entries, of Table 2 never differs by more than a fraction.

There are 13 combinations shown in Table 2; this appeared to be the least necessary to satisfy the requirements of the seven lines of totals of Table 1.

There are various devices by which a set of combinations such as those in Table 2 may be derived. One manner of arriving at this result is described below. It should be noted that the solution given in Table 2 is not unique. Rather, when one

TABLE 2. Thirteen Alternative Combinations Which Jointly Satisfy the Requirements of Table 1*

Strata	No. of strata	$P = .1391$	$P = .0033$	$P = .0835$	$P = .0334$	$P = .0434$	$P = .0850$
		A B C D	A B C D	A B C D	A B C D	A B C D	A B C D
1–4	4	0 2 1 1	0 2 1 1	0 2 1 1	0 2 1 1	0 2 1 1	1 1 1 1
5–8	4	2 1 1 0	1 2 1 0	1 2 1 0	1 1 1 1	1 1 1 1	1 2 1 0
9–13	5	1 1 2 1	1 1 2 1	1 1 1 2	1 1 2 1	1 2 1 1	0 2 2 1
14–17	4	0 1 1 2	1 0 1 2	1 0 2 1	1 1 1 1	1 0 1 2	1 0 1 2
Total	17	3 5 5 4	3 5 5 4	3 5 5 4	3 5 5 4	3 5 4 5	3 5 5 4
Strata	$P = .2047$	$P = .0022$	$P = .0383$	$P = .0838$	$P = .1404$	$P = .0940$	$P = .0489$
1–4	1 2 0 1	1 2 0 1	1 2 0 1	1 2 1 0	1 2 1 0	1 2 1 0	1 2 1 0
5–8	1 1 2 0	1 1 1 1	1 1 1 1	1 1 1 1	1 1 1 1	1 1 1 1	1 1 1 1
9–13	1 1 2 1	1 1 2 1	0 2 2 1	1 1 2 1	1 1 1 2	1 1 1 2	1 2 1 1
14–17	0 1 1 2	0 1 1 2	1 0 1 2	0 1 1 2	1 0 2 1	1 1 1 1	1 0 1 2
Total	3 5 5 4	3 5 4 5	3 5 4 5	3 5 5 4	4 4 5 4	4 5 4 4	4 5 4 4

*P = Probability of selection for the combinations; A, B, C, D denote state groups. Entries show the number of units to be selected in each cell.

faces two possible ways of completing a combination, one may select that which appears preferable. In the present example this freedom was utilized to control the number of units coming from groups A and B combined, hence also from C and D combined. Contrarily, if a desired restriction cannot be entirely satisfied, because of conflicting restrictions one may nevertheless work toward minimizing the occurrence of an undesirable combination.

10. OBTAINING CONTROL TABLE 2

We begin by writing down a desirable combination such as the one in the first box in Table 2. To do this, we place a zero on the first line under A, since from Table 1 it is found that the probability for this cell is only .6973. In other words, the number in this cell must sometimes be zero. (We could have just as well written a first combination in which this entry was 1, it doesn't matter.) When the first entry is zero the second should be 2, since the sum of the entries in Table 1 (.6973 + 1.9105) is greater than 2, hence the sum of the first two entries should never be less than 2 (since a control was attempted in respect to A+B and C+D). Having made these entries, the other two entries on the first line should each be 1, since the total must be 4 (the number of selections from the first four strata) and neither entry should be greater than 1 since the probabilities for the cells (.7548 and .6329) are each less than 1. On the second line we place a 2 under A, since the entry for the cell in Table 1 (1.1391) is greater than 1 and the total for the first 8 strata under A (1.8364) means that the sum of the first two entries under A should usually be 2. Having entered the 2 under A, the entries of 1, 1, 0 under B, C, D, respectively automatically follow since the sum of the four entries must be 4 and neither of the entries under B or C should be less than 1, since the probabilities in Table 1 (1.1718 and 1.2047) are both greater than 1. The entries on the third line are those in the corresponding cells in Table 1 rounded to whole numbers. They add to the required total of 5. The total for the last 9 strata in Table 1 (1.4469) indicates that the sum of the last two entries under A should more often be 1 than 2. A zero is accordingly entered on the last line under A. The entry under B should then be 1, since the sum of entries in Table 1 (.5702+.5572) is slightly greater than 1. The entry of 1 under C and 2 under D results in totals of 3 for the last two lines in each of these columns, which totals are in harmony with the corresponding totals of 2.8137 and 2.9666 in Table 1. It may be noted also that the totals for the entire columns in this combination (3, 5, 5, 4) are in harmony with totals in Table 1. The probability of .1391 is the largest desirable probability for this combination, since the entry on the second line under A can be a 2 only this proportion of the time, if it is never to be less than 1, and the assigned probabilities are not to be violated.

Table 3 is useful at this point in that it shows the permissive probabilities for each desired number of selections in each cell. The .1391 assigned the combination just discussed is subtracted from each permissive probability corresponding to the cell numbers of selection appearing in this particular combination. As additional combinations are set down the probabilities are successively subtracted until the remainders are all zero. Thus we are guided in assigning probabilities to the different combinations in that the restrictions imposed by Table 1 are made ironclad by means of Table 3. For example, it is obvious from the table that the maximum probability for the first combination is .1391, since the subtractions in the first column leave a zero on line A, strata 5–8. The second and third combinations were originally written as one, that is, the second combination was assigned the probability of .0868; however, it was later found convenient to split this combination and slightly alter the arrangement in the lower right corner yielding the third combination. The probability of .0868 was the maximum for a combination having a total of 4 for the first eight strata under B.

It is perhaps fairly easy to see how certain "improvements" may be made in the combinations and their probabilities as entered in Table 2. For example, the combinations could probably have been written in such a way as to harmonize with totals for strata 5 to 13 in Table 2, as well as with the other totals. Also it is readily seen that there was freedom to use a much greater number of combinations than were used, if so desired. The procedure that was used, however, was based on the belief that until it is established that techniques of this kind result in gains in sampling precision, efforts may well be concentrated on the more obviously desirable aspects.

Table 2 is intermediate to Table 4, which provides a basis for selection of specific groups of sampling units. Using the data in Tables 1 and 2, a table such as Table 4 can quite readily be written down. It will be noted that for each box in Table 2 there is a corresponding column (or perhaps a set of columns) in Table 4. In setting forth the data in Table 4 it is possible to do more than to conform with the restrictions of Tables 1 and 2, namely, to see that the selections in any group of states are scattered more or less uniformly throughout the strata. For example, in the first column, in which there are to be two selections from group A in strata 5 to 8

TABLE 3. Work Sheet Showing Residual Probabilities for Each Designated Number of Selections

Cell designation		Selections in each cell		Probabilities of combinations to be successively deducted												
Strata	Group	Num.	Prob.	.1391	.0033	.0835	.0334	.0434	.0850	.2047	.0022	.0383	.0838	.1404	.0940	.0489
1–4	A	0	.3027	.1636	.1603	.0768	.0434	.0000								
		1	.6973						.6123	.4076	.4054	.3671	.2833	.1429	.0489	.0000
	B	1	.0850						.0000							
		2	.9150	.7759	.7726	.6891	.6557	.6123		.4076	.4054	.3671	.2833	.1429	.0489	.0000
	C	0	.2452							.0405	.0383	.0000				
		1	.7548	.6157	.6124	.5289	.4955	.4521	.3671				.2833	.1429	.0489	.0000
	D	0	.3671										.2833	.1429	.0489	.0000
		1	.6329	.4938	.4905	.4070	.3736	.3302	.2452	.0405	.0383	.0000				
5–8	A	1	.8609		.8576	.7741	.7407	.6973	.6123	.4076	.4054	.3671	.2833	.1429	.0489	.0000
		2	.1391	.0000												
	B	1	.8282	.6891			.6557	.6123		.4076	.4054	.3671	.2833	.1429	.0489	.0000
		2	.1718		.1685	.0850			.0000							
	C	1	.7953	.6562	.6529	.5694	.5360	.4926	.4076		.4054	.3671	.2833	.1429	.0489	.0000
		2	.2047							.0000						
	D	0	.5156	.3765	.3732	.2897			.2047	.0000						
		1	.4844				.4510	.4076			.4054	.3671	.2833	.1429	.0489	.0000
1–8	A	1	.1636		.1603	.0768	.0434	.0000								
		2	.8364	.6973					.6123	.4076	.4054	.3671	.2833	.1429	.0489	.0000
	B	3	.9132	.7741			.7407	.6973	.6123	.4076	.4054	.3671	.2833	.1429	.0489	.0000
		4	.0868		.0835	.0000										
	C	1	.0405								.0383	.0000				
		2	.9595	.8204	.8171	.7336	.7002	.6568	.5718	.3671			.2833	.1429	.0489	.0000
	D	1	.8827	.7436	.7403	.6568			.5718	.3671			.2833	.1429	.0489	.0000
		2	.1173				.0839	.0405			.0383	.0000				
9–13	A	0	.1233						.0383			.0000				
		1	.8767	.7376	.7343	.6508	.6174	.5740		.3693	.3671		.2833	.1429	.0489	.0000
	B	1	.7844	.6453	.6420	.5585	.5251			.3204	.3182		.2344	.0940	.0000	
		2	.2156					.1722	.0872			.0489				.0000
	C	1	.4102			.3267		.2833						.1429	.0489	.0000
		2	.5898	.4507	.4474		.4140		.3290	.1243	.1221	.0838	.0000			
	D	1	.6821	.5430	.5397		.5063	.4629	.3779	.1732	.1710	.1327	.0489			.0000
		2	.3179			.2344								.0940	.0000	

TABLE 3—*Continued*

14–17	A	0	.4298	.2907						.0860	.0838		.0000			
		1	.5702		.5669	.4834	.4500	.4066	.3216			.2833		.1429	.0489	.0000
	B	0	.4428		.4395	.3560		.3126	.2276			.1893		.0489		.0000
		1	.5572	.4181			.3847			.1800	.1778		.0940		.0000	
	C	1	.7761	.6370	.6337		.6003	.5569	.4719	.2672	.2650	.2267	.1429		.0489	.0000
		2	.2239			.1404								.0000		
	D	1	.3513			.2678	.2344							.0940	.0000	
		2	.6487	.5096	.5063			.4629	.3779	.1732	.1710	.1327	.0489			.0000
9–17	A	1	.5531	.4140					.3290	.1243	.1221	.0838	.0000			
		2	.4469		.4436	.3601	.3267	.2833						.1429	.0489	.0000
	B	1	.2272		.2239	.1404								.0000		
		2	.7728	.6337			.6003	.5569	.4719	.2672	.2650	.2267	.1429		.0489	.0000
	C	2	.1863					.1429							.0489	.0000
		3	.8137	.6746	.6713	.5878	.5544		.4694	.2647	.2625	.2242	.1404	.0000		
	D	2	.0334				.0000									
		3	.9666	.8275	.8242	.7407		.6973	.6123	.4076	.4054	.3671	.2833	.1429	.0489	.0000
1–17	A	3	.7167	.5776	.5743	.4908	.4574	.4140	.3290	.1243	.1221	.0838	.0000			
		4	.2833											.1429	.0489	.0000
	B	4	.1404											.0000		
		5	.8596	.7205	.7172	.6337	.6003	.5569	.4719	.2672	.2650	.2267	.1429		.0489	.0000
	C	4	.2268					.1834			.1812	.1429			.0489	.0000
		5	.7732	.6341	.6308	.5473	.5139		.4289	.2242			.1404	.0000		
	D	4	.9161	.7770	.7737	.6902	.6568		.5718	.3671			.2833	.1429	.0489	.0000
		5	.0839					.0435			.0383	.0000				

the arrangement is such that these are not taken from numerically adjacent strata. The same holds for the two selections from group C in strata 9 to 13, and the two selections from group D in strata 14 to 17. The two selections from group B in strata 1 to 4, on the other hand, are adjacent, due to the fact that all of stratum 3 is in group B; hence, whenever the selection for stratum 2 is taken from group B, the two Bs must appear in successive strata.

TABLE 4. Selection Patterns Showing the State Groups From Which the Various Sampling Units Will Be Selected

P = probability of selection of the pattern; *P cum* = cumulated probabilities of patterns.

P	.0200	.0061	.0038	.0195	.0124	.0657	.0116	.0002	.0031	...	.0489
P cum.	.0200	.0261	.0299	.0494	.0618	.1275	.1391	.0002	.0033	...	.0489
P cum.							.1391		.1424	...	1.0000
Stratum No.											
1	C	C	C	C	C	C	D	D	D	...	C
2	B	B	B	B	B	B	B	B	B		A
3	B	B	B	B	B	B	B	B	B		B
4	D	D	D	D	D	D	C	C	C		B
5	A	A	A	A	A	C	A	B	B		D
6	C	C	C	C	C	A	C	C	C		B
7	B	B	B	B	B	B	B	B	B		C
8	A	A	A	A	A	A	A	A	A		A
9	B	C	C	C	C	C	C	C	D		B
10	C	B	B	B	B	B	B	D	B		D
11	D	C	C	C	C	C	C	A	A		A
12	C	D	D	D	D	D	D	C	C		C
13	A	A	A	A	A	A	A	B	B		B
14	D	D	C	C	D	D	D	D	C		D
15	C	C	D	D	B	B	B	C	A		C
16	D	D	D	B	C	C	C	D	D		D
17	B	B	B	D	D	D	D	A	C	...	A

The table means that the pattern in the first column has a probability of .0200 of being chosen, the second one a probability of .0061, and so on. As indicated by the dots, many of the patterns are omitted here for lack of space. If the first pattern were selected, the sampling unit for the first stratum would be selected from those units in state group C, the sampling units for the second and third strata in state group B, the sampling unit for the fourth stratum in stage group D, and so on.

The patterns have been written down in such a way as to conform with Tables 1 and 2. The sum of the probabilities of selection of the patterns in which a certain letter appears on a given line (stratum) is equal to the probability shown for that letter and stratum in Table 1. For example, for C in stratum 1: .0200 + .0061 + .0038 + ... + .0489 = .4946. The patterns are grouped in sets, each set corresponding to a combination in Table 2, as indicated by the cumulated probabilities on the second line of the above table. Within each set within each group of strata (1–4, 5–8, etc.) each letter will occur the number of times required for that combination in Table 2. For example, in each pattern of the first set (*P* = .1391) in strata 1 to 4, there are zero As, two Bs, one C and one D. Moreover in writing down the patterns an attempt was made to avoid writing the same letter in two adjacent strata in the same pattern. Thus in each pattern the different letters (state groups) are well distributed.

Each of these patterns of course would have a chance of selection if ordinary stratified sampling were used. However the sum of the probabilities of all of them *under ordinary stratified sampling would be only* .0000058 *in contrast to the* 1.0000 *here.*

In writing down the selection patterns, as in Table 4, the manner in which Table 1 is used may be illustrated in the following way: According to Table 2, one selection in the first 4 strata is to be taken from units in group C in all but three of the combinations. Table 1 shows that the letter C must be entered opposite stratum 1 in selection patterns having a combined probability of .4946 and opposite stratum 4 in selection patterns having a combined probability of .2602. Entries for each letter can therefore be made one at a time for the first four strata until all of the patterns are completed as far as these strata are concerned. A similar procedure is then used for the other sets of strata.

In view of the small number of units in each of the first six strata (there are only three units each in the first two) a still more exacting system of controls seemed feasible for these strata. Here the process was extended to the setting down of combinations of individual sampling units with a probability of selection determined for each such combination. In this way, it was possible to emphasize desirable combinations of sampling units, that is, those yielding a good geographical scattering of the larger cities.

At this point a random selection was made between .0001 and 1.0000. Number .8184 was drawn, and if Table 4 were given in full the selected pattern could be found there, namely, the first one in which the cumulated probability equals or exceeds .8184.

One may rest at this point if he wishes, assuming that he has obtained the substantial part of whatever gains controlled selection may yield. He would then make his selection of one unit for each stratum at random among the sampling units of each selected group. That is what was done for purposes of the calculation of variances.

11. SECOND STAGE OF CONTROLLED SELECTION

On the other hand one may introduce further restrictions within the restrictions of the groups selected from Table 4. In this instance, it was decided to secure an approximate balance in respect to distribution among individual states and also in respect to per cent urbanization. At this stage this can only be done, however, by applying a procedure of controlled selection to the groups selected in step one. All the units in other cells of Table 1 are henceforth out of the picture altogether. Thus, the balancing of the ultimate sample in respect to these additional factors cannot be better on the average than that of the groups already selected.

A table somewhat comparable with Table 4 was prepared for the second stage of sample selection. It set forth the possible selection patterns for selection of sub-groups and again was not a unique system. Prior to the preparation of the table, sub-groups of units within the selected groups were formed on the basis of state, and within state, on the basis of whether the units were plus, average, or minus, in respect to per cent urbanization. For these purposes the probabilities of selection for each sub-group were scaled upward in such a way that the total probability for each group was 1.0000. Frequently, the units in a given sub-group belonged to only one of the states in the group; hence, the selection from the group could be drawn only from this one state. It would have been possible, of course, to have obtained a greater control of the sample distribution by states by having done additional work prior to the first selection. Moreover, at this, the second stage, it would be possible to introduce a measure of geographic control within the states if this were considered worth-while.

After a random draw was taken, based on this last table, there were eleven sets of sub-groups, some of which contained but two or three sampling units. (The individual sampling units for the first six strata had been automatically determined by the procedure of step one.) Additional random draws were taken within sub-groups until, finally, the selection of seventeen sampling units was completed.

12. COMPARISON OF VARIANCES

To obtain estimates of the variances for the controlled selection, 100 samples of 17 units each were drawn using Table 4. Within the selected groups individual units were chosen with probabilities proportionate to their 1940 populations, as required by the original sample design. This means that the samples were selected at the conclusion of step one of the control procedure; whatever further gains step two may have yielded were not measured since the selection of 100 samples by this method would have entailed considerably more work. The mean of the 17 units for each sample was obtained and then the variance among the 100 means. The variance of the stratified random selection was computed by means of the standard formula which utilizes the information for the entire population.

For the purposes of this comparison items were selected mainly from 1940 Census data. Items were sought for which data were published for each township (as well as for counties) because some of the sampling units have boundaries which cut across county areas along township lines. Another item included was the 1944 Presidential vote because of interest in this variable; in the states for which the vote was not published by townships, the vote for split counties was estimated by splitting it in proportion to the population.

Upon completion of the computations it was apparent that the items studied fell into two categories with respect to the magnitudes of the variances. Since in practice the selected first-stage sampling units are in turn subsampled, a "within" component of variance must be added to the between variance in order to secure a total variance comparable to those of estimates derived from an actual survey. For this purpose the within variance was computed using the formula appropriate for a simple random sample of n cases within each of the 17 first-stage units and assuming that the variance within each of these units was the same. When the between variances were compared with "total" variances it was found that the items studied fell into the two following classes: (1) those for which the between variance was an important part of the total variance and (2) those for which the between variance was a minor part of the total variance.

The results of the computations are presented in Table 5. In order to show the relative importance of the between component for the different items the between variance for stratified random sampling is also given as a percent of the total variance for cases of $n = 25$ and $n = 100$.

From this table it is evident that for items 1, 2, 3, and 4 the between component is an important part of the total variance for n's as small as 25 within the primary sampling units; while for items 5, 6, 7, and 8 the between component is relatively unimportant even for n's as large as 100. Now the practical value of a procedure designed to reduce the between component of the variance must be judged by its effect on those items for which the between component is important. For items for which the between component is small even a drastic reduction of the latter will have no great effect on the total variance. In Table 5 this means that the control procedure must be judged by its effects upon items 1, 2, 3 and 4.

Results of the computations showed that the use of controlled selection resulted in reductions of 23, 22, 32, and 11 per cent respectively in the between variances for items 1, 2, 3, and 4. In terms of the total variances the reductions are 14, 10, 13, and 5 per cent respectively for n's of 25, and 20, 17, 24, and 8 for n's of 100.

Gains of this kind clearly justify the additional work involved in carrying through this procedure. From one viewpoint the reduction of 20 per cent in the between variances is roughly equivalent to having spread a given number of inter-

TABLE 5. "Between" Component of Variances Derived From Controlled Selection and From Stratified Random Sampling[a]
(17 First-Stage Units in North Central States)

			Stratified random			Controlled selection	
				Between variance as per cent of total variance			Between variance as per cent of stratified random between variance
Item No.	Item	Population mean	Between variance	$n =$ 25	$n =$ 100	Between variance	
1	Average monthly rent ($)	21.62	1.015	.63	.87	0.785	77
2	Per cent dwellings vacant	5.70	1.129	.47	.78	0.879	78
3	Per cent foreign born	5.71	0.934	.42	.75	0.638	68
4	Per cent voting Democratic	45.98	3.896	.40	.73	3.452	89
5	Dwelling units per 100 persons	29.10	0.363	.07	.23	0.356	98
6	Per cent 65 years and over	8.20	0.125	.07	.22	0.080	64
7	Per cent adult females	31.83	0.165	.03	.11	0.146	88
8	Per cent 5 years and under	8.00	0.039	.02	.08	0.037	94

[a]In items 3, 6, 7, and 8 the categories are expressed as percentage of the total population (1940 Census).

The average monthly rent is from the 1940 Census, except that the within p.s.u. variance is unknown. A rough estimate of $16 for the standard deviation was used. The vacant dwelling units are expressed as per cent of total dwelling units (1940 Census).

Item 4 represents the votes cast for the Democratic Presidential candidate in 1944 as a per cent of the votes cast for all candidates, as given in official state manuals.

Item 5 represents the number of dwelling units per 100 persons in the population (both from the 1940 Census).

views over a number of primary sampling units one-quarter greater than originally contemplated. Of primary importance of course is the fact that for a given degree of accuracy a smaller sample is possible, insofar as the item is one for which the between component is important.

It is of interest to note that for items 5, 6, 7, and 8, for which the between component was unimportant, the reduction in variance was small, but for each of these items the procedure did result in a slight gain. Theoretically as will be shown, this procedure can result in an increased variance; and it is to be expected that for occasional items in which the between variance is small it will result in increases. Even so, of course, such increases could be expected to be of slight importance in comparison with the total variances for such items.

13. SOME PREVIOUS INVESTIGATIONS OF SELECTION TECHNIQUES

Earlier research workers seem to have dealt with more limited applications of controlled selection. There is, of course, a wide range of published material on stratification, including a fundamental contribution by Neyman (1934). Neyman also analyzed the usual type of purposive selection which he found to be generally inferior to stratified random sampling.

Previously Bowley (1926) and Jensen (1926) had reported their analyses of stratified random sampling and purposive selection but their findings had been far from conclusive. Jensen (1928) described the purposive selection of a sample from records of the 1923 Danish Agricultural Census and showed that it represented the population well in respect to distribution of several farm variables. Strand and Jessen (1943) compared the use of purposive and stratified random selection of townships in Iowa counties and concluded that "purposive selection does not provide samples of greater accuracy than stratified random selection" for situations of the type investigated. However, none of these investigators attempted to combine purposive selection with probability sampling.

Frankel and Stock (1942) suggested the use of a Latin square design which, within the framework of the modes of stratification used, extended the controls of stratification to a second dimension. The following year Tepping, Hurwitz, and Deming (1943) reported extensive analyses on techniques of this kind which they designated as "deep stratification." There is an important limitation of a Latin square design, namely that the probability of selection of each of the $L \times L$ cells into which the population units are grouped is the same. Tepping, Hurwitz, and Deming (1943) accordingly considered estimates, derived as though the probabilities for the various cells were equal, whether or not the combined probabilities for the units in the different cells were equal in all cases. That is, they considered the use of biased, as well as unbiased, estimates and in some cases found the bias to be an important source of error. Yates (1946) reported a selection procedure called balancing in which additional random selections are substituted for units originally drawn until finally "the mean value of the balanced factor in the sample is equal to the mean of the factor in the whole population." He stated that estimates derived from such a sample are "subject to some elements of bias" but concluded that "such bias is not likely to be of importance in practice."

14. CONTROLLED SELECTION AS AN EXTENSION OF PURPOSIVE SELECTION

Conceptually, the use of controls in selecting a sample may be viewed as an extension of the technique known as purposive selection (although perhaps involving more judgment). If, however, the estimates to be derived from the sample are to be unbiased, an additional step not ordinarily considered to be a part of purposive selection is required. In order that the sampling may be probability sampling, the sampler must select not just one but many purposive samples, until every unit in the universe is included in one or more samples. The number of samples in which each unit appears must be exactly proportionate to its assigned probability of selection.

After the complete set of purposive samples has been established, the random selection of one of them constitutes a probability sample. As has been seen, such a procedure is not in conflict with the use of stratification but, on the contrary, can be more readily accomplished after strata have been established.

The preceding describes what appears to be the ultimate in controlled selection. In the process of purposive selection one could, conceivably, use considerable judgment and also make numerous checks in regard to various known characteristics of the sampling units, finally establishing samples each of which was as nearly as possible in accord with the population as a whole with respect to each characteristic. Whether a procedure of this intricacy would ever be operationally feasible though is very questionable.

15. CONTROLLED SELECTION VIEWED AS SYSTEMATIC SAMPLING

Despite certain similarities of controlled selection to deep stratification and balancing there appears to be a still greater resemblance between this procedure and a method which has been increasingly discussed in the literature in the last few years, namely systematic sampling. Madow and Madow (1944) and Cochran (1946) have shown the theoretical conditions under which systematic sampling can be expected to be useful and have discussed the application of the method to certain real situations. In the research on systematic sampling to date, the emphasis has been upon the selection of a sample systematically from a more or less "naturally ordered" sequence of sampling units. Very recently, however, Madow (1949) has discussed the notion of selecting from the different strata in such a manner as to induce a negative intra-class correlation but his paper attacked the problem solely from a theoretical point of view.

The similarity between the highly controlled selection discussed above and systematic selection is apparent when it is considered that if the N units in the population are numbered in the proper sequence the selection of a sample systematically will automatically yield one of the purposive samples. In other words, under this condition the two methods become the same.

Systematic sampling, in fact, appears to be a good common denominator for various methods of probability selection. Unrestricted random selection may be viewed as systematic selection in which the sampling units in the universe have been ordered at random. Assuming a uniform sampling rate within all strata, stratified random selection is systematic selection in which the strata are placed in some sequence and the sampling units within each strata are randomly ordered. The selection with deep stratification, assuming equal probabilities for all cells, becomes a systematic one in which the cells within each stratum are ordered in a randomly chosen sequence to accord with the restriction of a Latin square and the sampling units within each cell are also randomly ordered.[3] Thus controlled selection may be viewed as systematic selection in which the ordering of the sampling units within strata has in part at least been done purposively.

Viewed in this way the vast amount of flexibility inherent in methods of controlled selection is evident. Moreover, the theory of systematic selection can be expected to provide valuable clues regarding the conditions under which various procedures of controlled selection may be useful.

16. THE UNDERLYING THEORY

As Madow and Madow (1944) have stated, systematic sampling may be viewed as cluster sampling, in which each possible sample is one of the clusters. If the expected intra-class correlation for a cluster of n units is negative,[4] less than $-1/(N-1)$ to be precise, the estimated mean has a smaller variance than that of an unrestricted random sample of n units. Moreover the variance decreases with decreasing values (increasing negatively) of the intra-class correlation.

In (proportionate) stratified random sampling the intra-class correlation is never positive, and usually it is less than $-1/(N-1)$. But Madow and Madow (1944) and Cochran (1946) have pointed out that under certain conditions a systematic sample will have a smaller variance than a stratified random sample (which is a systematic sample with the units ordered at random within strata); that is, that the intra-class correlation for the former will be less than that for the latter.

---

[3] In the selection of units with varying probabilities the analogy to systematic sampling holds only when exactly one unit is to be selected from each group. This is, however, the usual manner in which varying probabilities are utilized in practice. In such instances, one may proceed as follows: Express the probabilities assigned the various units in fractions having a common denominator, d. The sum of the numerators of these fractions is, of course, d. Then arrange the d chances of selection in a random order. It is well to recognize that with this procedure the various chances of selection relating to a single unit may be scattered throughout the sequence.

[4] As indicated in Footnote 3, units with varying probabilities may be viewed as having varying numbers of chances of selection, the sum of which was given as d. The problem of selection is then restated as that of selecting n units from a population in which the sum of the chances of selection for all units is N, in which case the number of units as such in the population becomes irrelevant. Thus, the conclusions to be reached apply alike to selection with equal and varying probabilities subject to the restriction that when varying probabilities are used one and only one unit is to be selected from each stratum or sub-group.

Now, unless the intra-class correlations corresponding to all of the possible arrangements of units within the strata are identical, the values corresponding to some of the arrangements must be less than the average of all of them, which is that of stratified random sampling. It follows that in practice there always exist orderings of the units which have a smaller intra-class correlation (and therefore smaller variance) than that of stratified random sampling. Equally inescapable is the fact that the ordering may increase the intra-class correlation above that of stratified random sampling.

Many questions are still to be answered. There is the problem of how to obtain optimum ordering in respect to a variable for which information is sought. Then there is the need to resolve the probable conflict between optimum orderings for two or more variables to be investigated in the same survey. For example, the optimum ordering for estimating one variable may result in less accuracy than would stratified random sampling as far as another variable is concerned. The question becomes one of finding an ordering of the units by means of which the intra-class correlations will be reduced on the average.

It is also pertinent to consider the possible reductions of the variances for different sizes of sample. Madow and Madow (1944) found that with a given ordering of the units, the effect of change in sample size was somewhat complex and that further investigation of this problem was necessary.

In summary, the theory shows that the use of additional controls may reduce the variance of estimates for some items in a survey but also that the result may be an increased variance for these or other items. It remains then for empirical studies and additional theoretical developments to determine what the results are likely to be in practice.

17. CONCLUDING REMARKS

The emphasis in this paper has been upon the possibilities for using controlled procedures of selection without violating principles of probability sampling. Although, conceivably it may appear to be desirable to reduce the probabilities for all "non-preferred" combinations to zero even at the risk of violating the assigned probabilities there is here contained no evidence on this question. The procedures outlined do eliminate the possibility of selection of vast numbers of combinations but among the millions remaining it may indeed be difficult to determine that any particular combinations will necessarily comprise inferior samples. At this writing there seems to be little justification in accepting less than complete objectivity in the final selection of first-stage units.

The procedures described illustrate a selection with uniform rates within strata of approximately equal size. The methods are sufficiently flexible, though, to permit their ready extension to disproportionate sampling with strata of different sizes.

Multi-stage sampling (the entire process) may itself be viewed as controlled selection in which the selection of first-stage units is simply the first phase in the process of selecting the ultimate physical units. In such a controlled selection the preferred combinations as a rule have larger variances than for stratified random sampling but the increased variance may be more than offset by reduced cost. Thus the preferred combinations of controlled selection are not necessarily the ones yielding the smallest variances because there are frequently cost differentials to be considered.

There are many applications of controlled selection in the later stages of multi-stage sampling and elsewhere. For example in selecting a sample of towns within selected first-stage units it may be worth-while to secure an approximate balance over the entire sample with respect to such characteristics as size of place, and average rental value and perhaps within individual first-stage units deliberately to secure an "unbalance" with respect to geographical location (this last in order to reduce travel from town to town within a county).

A problem that will require attention is the estimation of variances from sample data when controlled selection has been used. This is the problem of estimating variances for systematic sampling and while it presents difficulties it appears possible to develop approximations which may be so formulated as to be on the "safe side."

The results of the tests reported in this paper indicate that it is possible to obtain appreciable gains of precision through the use of the procedures of controlled selection proposed here. If the gains obtained in these tests are found to be typical for large scale surveys generally, the efficiency of sample surveys based on the relatively small number of sample areas can be increased through the use of these procedures. Additional tests are needed in relation to different variables, and also in relation to the use of ratio and other more complicated estimates.

5

SAMPLING ORGANIZATIONS AND GROUPS OF UNEQUAL SIZES

LESLIE KISH

A methodological issue arises whenever groups of elements of greatly differing sizes serve as observational units. The value of each unit is observed and assigned a single group value $\bar{Y}_\alpha$; but the values also have meaning for the parent population of elements. In studies of groups, organizations and ecological units, the unweighted group mean $\bar{Y}_g$ of the units is often computed automatically. But a mean weighted for significance should always be considered and usually preferred, particularly the element mean $\bar{Y}_e$, weighted by the numbers of elements in the units. For estimating $\bar{Y}_e$, it is efficient to select units with probabilities proportional to those numbers. Implications for several research designs are advanced. The difference $\bar{Y}_e - \bar{Y}_g = \bar{Y}_g RC_y C_n$ is discussed.

1. INTRODUCTION

One of the frightening statements made about American education, around 1957, was that half of the high schools offered no physics, a quarter no chemistry, and a quarter no geometry. It was later noted that, although these backward schools were numerous indeed, they accounted for only 2 per cent of all high school students. There were many more small schools than large ones, but the small proportion of large schools accounted for a large proportion of students. Moreover, the curricula and facilities of large and small schools can and do differ drastically. Hence, presenting average school characteristics gives a misleading picture of conditions facing the average student.

In this illustration, the population elements are students, since we are interested in their opportunities. The school serves as both sampling unit and obser-

---

Reprinted with permission from *American Sociological Review*, 30, 4, 564–572.
Presented in September 1964 to the annual meeting of the American Sociological Association in Montreal. Examples are drawn from research at the Survey Research Center, Institute for Social Research. Support was derived from the National Science Foundation Grant G–7571.

vational unit, since a single observation made on a school reveals the survey variables: the opportunities offered by the school's curriculum to all its students.

This issue arises whenever the group characteristic of each unit is observed and assigned a single value, which is associated with all elements comprising the group. The researcher is usually interested in the effect of the unit characteristic on individual elements. Nevertheless, many researchers have assigned equal weights to the observational units automatically and mistakenly, as if these were simple population elements. In my experience, once a researcher suffering from this confusion has been shown the difference, he will prefer to say, for instance, "High schools without physics courses account for 2 per cent of students," rather than "Half of the high schools offer no physics."

Several aspects of this problem are illustrated by the differences between the *group mean* $\bar{Y}_g$ and the *element mean* $\bar{Y}_e$ in the following situations.

(a) We want to estimate the prevalence of swimming pools in the high schools of a state. $\bar{Y}_g$ percent of the schools have swimming pools, but $\bar{Y}_e$ percent of the students go to schools with pools. $\bar{Y}_e$ is considerably larger than $\bar{Y}_g$, because large schools have pools more often than small schools.

(b) In a national voluntary organization $\bar{Y}_g$ per cent of the branches are in large metropolitan areas, but $\bar{Y}_e$ per cent of the members come from those areas. $\bar{Y}_e$ is much larger than $\bar{Y}_g$ because the branches have many more members in such areas. Most of us would prefer $\bar{Y}_e$ to $\bar{Y}_g$ as a measure of the extent to which this organization is metropolitan.

(c) To forecast industrial employment and mobility in a state, the heads of a sample of manufacturing plants are interviewed regarding their plans to expand, to stay in the state, or move out of it. The results can be presented in terms of $\bar{Y}_g$ per cent of plants; but normally data for plants that account for $\bar{Y}_e$ per cent of employees will be more useful. The two means can diverge widely, because large and small plants differ.

(d) In a certain industry $\bar{Y}_g$ per cent of firms operate with a specified type of organization (or leadership, or safety measures), and $\bar{Y}_e$ per cent of the employees are subject to it. I would generally prefer $\bar{Y}_e$ for measuring the prevalence of a given type of organization.

(e) To estimate the prevalence of museums in the cities of a country, one can choose between $\bar{Y}_g$, the proportion of cities with museums, and $\bar{Y}_e$, the proportion of people living in cities with museums.

These are actual issues I have encountered repeatedly in situations concerned with significant social research. In each case the researcher chose the element mean after the difference was explained to him. On the other hand, analyses based on the group mean often occur in the social science literature. I suspect that in most such instances the issue was not faced squarely, so that the group mean was chosen automatically although the element mean would have been more relevant.

Of course, in some situations the researcher may reasonably prefer and deliberately choose $\bar{Y}_g$ over $\bar{Y}_e$.[1]

2. WEIGHTED VERSUS UNWEIGHTED MEANS

Suppose that a population comprises A units, and that $\bar{Y}_\alpha$ is some value of the αth unit. From the values $\bar{Y}_\alpha$ of all A units in the population one can compute either the unweighted mean of the units, the *group mean*,

$$\bar{Y}_g = \frac{1}{A}\sum_{\alpha}^{A} \bar{Y}_\alpha, \tag{2.1}$$

or the *element mean*,

$$\bar{Y}_e = \frac{1}{A}\Sigma\frac{N_\alpha}{\bar{N}}\bar{Y}_\alpha = \Sigma\frac{N_\alpha}{\Sigma N_\alpha}\bar{Y}_\alpha = \frac{\Sigma N_\alpha \bar{Y}_\alpha}{\Sigma N_\alpha}, \tag{2.2}$$

using the numbers of elements in the units as weights. $\bar{N} = \Sigma N_\alpha / A$ is the mean unit size. Both means may be regarded as special cases of weighted means $\bar{Y}_w = \Sigma W_\alpha \bar{Y}_\alpha / A$, where $\Sigma W_\alpha = A$. For the group mean, $W_\alpha = 1$ is assigned arbitrarily to all units, while the element mean utilizes the weights $W_\alpha = N_\alpha / \bar{N}$.

I have centered the discussion around the numbers of elements used as weights, but the gist of the argument is equally relevant when the N_α represent any measure of importance or size for the units comprising a population. *Choice of a measure of importance or size is a substantive problem for the researcher.*

Note that $N_\alpha \bar{Y}_\alpha = \sum_{\beta}^{N_\alpha} Y_{\alpha\beta} = Y_\alpha$ can be considered the aggregate of a variable $Y_{\alpha\beta}$ which has the same value $Y_{\alpha\beta} = \bar{Y}_\alpha$ for all N_α values in the αth unit. Conversely, the unit aggregate Y_α may be considered divided into N_α equal portions $Y_\alpha / N_\alpha = \bar{Y}_\alpha$ among the N_α elements of the αth unit. Viewed as an element variable, $\bar{Y}_\alpha$ represents complete uniformity and homogeneity of the variable $Y_{\alpha\beta}$ within units. From this view the element mean is simply the mean of the $\Sigma N_\alpha = N$ elements in the population:

$$\bar{Y}_e = \frac{\Sigma N_\alpha \bar{Y}_\alpha}{\Sigma N_\alpha} = \frac{Y}{N}. \tag{2.2a}$$

Although (2.2) and (2.2a) are equivalent, I shall stress the form (2.2) of a weighted mean of units in discussing the choice between an unweighted group mean $\bar{Y}_g$ and a weighted element mean $\bar{Y}_e$ to represent a population of A units.

---

[1] See discussion in Appendix 1 of Townsend (1962).

The difference between the two means is

$$\bar{Y}_e - \bar{Y}_g = \frac{1}{A}\sum_{\alpha}^{A}\frac{N_\alpha}{\bar{N}}\bar{Y}_\alpha - \bar{Y}_g$$

$$= \frac{1}{\bar{N}}[\frac{1}{A}\sum_{\alpha}^{A} N_\alpha \bar{Y}_\alpha - \bar{N}\bar{Y}_g] = \frac{1}{\bar{N}}\text{Cov}(N_\alpha, \bar{Y}_\alpha)$$

$$= \frac{1}{\bar{N}} R\sigma_n\sigma_y = \bar{Y}_g RC_nC_y. \tag{2.3}$$

Here $\text{Cov}(N_\alpha, \bar{Y}_a)$ and $R = R_{ny}$ represent the covariance and the correlation coefficient between the variables N_α and $\bar{Y}_\alpha$. The symbols σ_n and σ_y denote standard deviations, and C_n and C_y denote coefficients of variation for the two variables. Thus

$$\sigma_y^2 = \sum_\alpha(\bar{Y}_\alpha - \bar{Y}_g)^2 / A \text{ and } C_y = \sigma_y / \bar{Y}_g.$$

The two means, $\bar{Y}_e$ and $\bar{Y}_g$, can differ emphatically when R and C_n are large. This occurs typically when the distribution of the sizes N_α is very skewed, so that a small proportion of units accounts for a large proportion of elements, and the variable $\bar{Y}_\alpha$ is strongly correlated with the sizes N_α. The element mean $\bar{Y}_e$ is greater or less than the group mean $\bar{Y}_g$, according to a positive or negative correlation R_{ny}.

All the examples of the first section dealt with group values $\bar{Y}_\alpha$ that are dichotomous: a school either has physics courses or not; it either has swimming pools or not; a city either has museums or not. Possession of the characteristic is denoted with $\bar{Y}_\alpha = 1$, and lack of it with $\bar{Y}_\alpha = 0$. Dichotomous variables are common in social research (probably too common) and they readily illustrate the large correlations R_{ny} that may exist between sizes N_α and the group values $\bar{Y}_\alpha$ of the units. For dichotomous variables the group mean is a proportion $\bar{Y}_g = P_g$, and its variance is $\sigma_y^2 = P_g(1 - P_g)$.

The entire argument is equally valid, however, for the frequent case when $\bar{Y}_\alpha$ is not dichotomous. For example, $\bar{Y}_\alpha$ may be the number of different physics courses offered by a school; or it may be the average number of physics courses taken by the school's students. For the branches of a voluntary organization, $\bar{Y}_\alpha$ may denote the size of the city, rather than an arbitrary "metropolitan" division; or it may denote the mean income of its members. For cities $\bar{Y}_\alpha$ may denote the number of museums, or the per capita attendance.

Proper representation of the group values $\bar{Y}_\alpha$ is a primary issue in many areas of social and economic research. The sources of group values vary. They may represent simply the means of individual values: for example, the mean income or the proportion of home owners, as characteristics of cities. They may be group values arising less directly from individuals: for example, the population sizes and densities of cities. Or, the values may belong specifically to the group, without direct effect from its elements: for example, the climate, altitude, or age of a city, its form of government, or the presence of museums. Whatever the origin of the group values, the choice between the element mean $\bar{Y}_e$ and the group mean $\bar{Y}_g$ should depend primarily on which conveys a more meaningful summary value.

Issues concerning the origin of group values are important, and more empirical studies should be made of measurements performed directly on social groups and organizations, rather than relying on the better known methods of measuring and averaging individual values. But methods of ascertaining and assigning group values are not my subject here. Group values may originate in measurements either on individuals (elements) or directly on groups; and the same measurements may also serve as individual values. Whatever its origin, a measurement can always serve in the double capacity of a group value and a value for each of the individuals comprising the group. A city may either be old or have citizens with a high average age. Its inhabitants can also be characterized either by the venerable age of their city or by the high mean age of its citizens. Four types of variables result from the two origins and the two uses of the data.

The use of variables for either group values or individual values must be distinguished from the origins of their measurements.[2] Origin refers to the *units of observation*; use refers to the *units of analysis*, which are often called the elements. Distinct from both is the issue of *sampling units*, which I shall discuss in the next section.

Analysis based on individual elements may utilize both element values and group values.[3] For example, one may correlate a personal health measure with city altitude, or a personal attitude measure with the city's latitude. Of course, both element and group variables may represent vectors of many variables.

When groups are the units of analysis, group values can be freely used, regardless of the origins of their measurement. For example, the mean of the element values within units becomes a single value for each unit. This may be denoted as $\bar{Y}_\alpha$ and used as a group value. But it may also be used as an individual value for all elements comprising the unit. The mean income of a city may be used to characterize either the city or each of its residents.

The terms group, organization, and unit are not distinguished here and are used interchangeably. I assume that each population element belongs to one and only one unit, that the population has been partitioned into a defined set of units without overlaps or gaps. The main argument can be extended to overlapping memberships of elements belonging to several units; but this extension would detract from the simplicity of the presentation.

---

[2] One area of confusion has been well noted: Robinson (1959); Goodman (1959). See also Davis, Spaeth and Hudson (1961).

[3] "In terms of the actual analysis the matter can be restated in the following terms: just as we can classify people by demographic variables or by their attitudes, we can also classify them by the kind of environment in which they live." Some of these issues are discussed by Kendall and Lazarsfeld (1950).

In this paper I deal with means, and comparison of means follows readily. More complex measures of association, such as regression, would require separate treatment, and the structural model would have to be specified in each case.

3. SAMPLING FOR GROUP VALUES

Choice of the proper mean should precede the sample design, since the design depends on which mean is more appropriate. We should first ask ourselves: If we had all A values of N_α and $\bar{Y}_\alpha$ for the entire population, which mean would we choose? Since the element mean will most often be chosen, I shall consider samples designed for it.

The sampling problem resembles that of selecting entire clusters of unequal sizes—but with some special features. First, evaluation of $\bar{Y}_\alpha$ requires only a single observation on each cluster. Second, the homogeneity is extreme, because $\bar{Y}_\alpha$ is the same for all elements within the units; the intraclass correlation is perfect. Third, variations in the unit sizes (N_α) must be accepted as given, because the units are fixed entities; they cannot be divided by subsampling.

Two further special assumptions are implicit in the following brief statistical treatment. First, that the variance of $\bar{Y}_\alpha$ is similar among units, regardless of unit size. Second, that the unit cost of obtaining $\bar{Y}_\alpha$ is similar for large and small units. Compared with the variations in the sizes N_α , differences in variance and cost factors often can be reasonably assumed small. When they are not, disproportionate allocation may be introduced.

If a units are selected with equal probability a/A, the simple mean of the sample $\bar{y}_{gs} = \sum_\alpha^a \bar{y}_\alpha / a$ is an unbiased estimate of $\bar{Y}_g$ (Kish, 1965, Section 2.8). To estimate the element mean $\bar{Y}_e$, one must introduce the weights N_α and use the ratio mean

$$\bar{y}_{es} = \sum_\alpha^a N_\alpha \bar{y}_\alpha / \sum_\alpha^a N_\alpha . \tag{3.1}$$

Its variance merely applies a formula for a ratio mean, appropriate to the method used for selecting the a units in the sample. If the N_α are grossly unequal, the weights will render this estimate inefficient: a few large selections will tend to dominate the estimate and its variance. In this extreme situation, selection with *probabilities proportional to size* (PPS) seems particularly appropriate.

If the units are selected with probabilities proportional to the N_α , an unbiased estimate of the population element mean is the simple mean

$$\bar{y}_{ep} = \frac{1}{a} \sum_\alpha^a \bar{y}_\alpha . \tag{3.2}$$

This demonstrates the chief advantage of selecting units with PPS to estimate the element mean: the ordinary self-weighting mean of sample observations is simple to compute, and is efficient under the special assumptions described above. If a selections have been drawn with replacement from the entire population, the variance of the sample mean may be estimated simply as

$$\text{var}(\bar{y}_{ep}) = \frac{1}{a(a-1)} \sum_{\alpha}^{a} (\bar{y}_{\alpha} - \bar{y}_{ep})^2. \tag{3.3}$$

This can readily be perceived by considering a selections drawn with replacement from a population of N elements, each with its own value $\bar{Y}_{\alpha}$. In this population N_1 elements have the value $\bar{Y}_1$, N_2 have $\bar{Y}_2$, N_{α} have $\bar{Y}_{\alpha}$, etc. To retain a simple presentation, I assume that the units are selected with replacement; if a unit is selected two or more times, it remains in the sample and in the estimate as often as it was selected.[4]

We typically use stratification in these situations, resorting to formulas appropriate for stratified samples. Suppose that the units have been sorted into H strata, and that a_h units have been selected with PPS from the hth stratum. The weight of the stratum is W_h, which ordinarily represents N_h/N, the proportion of elements it contains, and $\Sigma_h W_h = 1$. The element mean is estimated by

$$\bar{y}_{ep} = \sum_{h}^{H} W_h \bar{y}_h, \text{ where } \bar{y}_h = \frac{1}{a_h} \sum_{\alpha}^{a_h} \bar{y}_{h\alpha}. \tag{3.4}$$

Assuming that the a_h selections in the strata were made with replacement, the estimate of the variance is

$$\text{var}(\bar{y}_{ep}) = \sum_{h}^{H} \frac{W_h^2}{a_h(a_h - 1)} \sum_{\alpha}^{a_h} (\bar{y}_{h\alpha} - \bar{y}_h)^2. \tag{3.5}$$

The sample can be made self-weighting if the number of selections is made proportional to the stratum sizes ($a_h = kW_h$). Then for a total sample size of $a = \Sigma a_h$ we have

$$\bar{y}_{ep} = \frac{1}{a} \sum_{h}^{H} \sum_{\alpha}^{a} \bar{y}_{h\alpha}, \tag{3.6}$$

and

$$\text{var}(\bar{y}_{ep}) = \frac{1}{a^2} \sum_{h}^{H} \frac{a_h}{a_h - 1} \sum_{\alpha}^{a_h} (\bar{y}_{h\alpha} - \bar{y}_h)^2. \tag{3.7}$$

If $H = a/2$ strata of equal sizes are formed, and paired selections (1 and 2) are drawn from each stratum, the last formulas become

[4] If unconvinced by this argument, see another in Cochran (1963), Section 9.10.

$$\bar{y}_{ep} = \frac{1}{a}\sum_h^H(\bar{y}_{h1} + \bar{y}_{h2})$$

and

$$\text{var}(\bar{y}_{ep}) = \frac{1}{a^2}\sum_h^H(\bar{y}_{h1} - \bar{y}_{h2})^2. \tag{3.8}$$

Some large units may be larger than N/a, the designed population size per selection, and some almost as large. These should be taken into the sample with certainty; they will not contribute to the variance. At the other extreme, the strata may be filled with many units, each much smaller than the strata; for these it will matter little whether selection is with or without replacement, because the finite population correction $(1-f)$ may be disregarded in computing the variance. In between, the strata may contain units somewhat, but not much, smaller than the strata, where the factor $(1-f)$ should be considered.

Suppose that after selecting with probabilities proportional to the size measures N_α, one finds different desired "true" sizes n'_α for the selected units. The difference may be due to changes in size, discrepancies in the units of measurement, etc. Although the n'_α and N_α should be highly correlated, the differences may not be negligible; for example, one could underrepresent fast-growing units. Then weight each $\bar{y}_\alpha$ in the sample with $x'_\alpha = n'_\alpha / N_\alpha$ and, using $\bar{y}'_\alpha = \bar{x}'_\alpha \bar{y}_\alpha$, compute the ratio mean

$$\bar{y}_{ep} = \sum_\alpha \bar{y}'_\alpha / \sum_\alpha \bar{x}'_\alpha \tag{3.9}$$

For stratified selection, and with the selections proportional to stratum sizes, the ratio mean becomes

$$\bar{y}_{ep} = \sum_h\sum_\alpha \bar{y}'_{h\alpha} / \sum_h\sum_\alpha \bar{x}'_{h\alpha}. \tag{3.10}$$

Here $\bar{y}'_{h\alpha} = \bar{y}_{h\alpha}\bar{x}'_{h\alpha} = \bar{y}_{h\alpha} n'_{h\alpha} / N_{h\alpha}$, the variable corrected for the change in desired size from the measure of size. The variances of this ratio mean may be found in several textbooks.

With proper weighting, one can also estimate the group mean $\bar{Y}_g$, when required. If the units were selected with probabilities proportional to N_α, each selected mean should be weighted with $1/N_\alpha$. The ratio mean for a proportionate stratified selection would be

$$\bar{y}_{gp} = \sum_h\sum_\alpha(\bar{y}_{h\alpha} / N_{h\alpha}) / \sum_h\sum_\alpha(1/N_{h\alpha}), \tag{3.11}$$

and its variance is that of a stratified ratio mean.[5]

---

[5] The preceding arguments and formulas can be found in sampling texts. See, for example, Kish (1965, chaps. 6 and 7).

TABLE 1. Estimates of the Group Mean $\bar{Y}_g$ and the Element Mean $\bar{Y}_e$ From Samples Selected With Equal Probability and With Probabilities Proportional to Size (PPS)

Selection probabilities	Simple mean		Weighted mean	
Equal: a/A	$\bar{y}_{gs} = \sum_{\alpha}^{a} \bar{y}_{\alpha} / a$	$E(\bar{y}_{gs}) = \bar{Y}_g$	$\bar{y}_{es} = \sum_{\alpha}^{a} N_{\alpha} \bar{y}_{\alpha} / \sum_{\alpha}^{a} N_{\alpha}$	$E(\bar{y}_{es}) = \bar{Y}_e$
PP: $aN_{\alpha}/\Sigma N_{\alpha}$	$\bar{y}_{ep} = \sum_{\alpha}^{a} \bar{y}_{\alpha} / a$	$E(\bar{y}_{ep}) = \bar{Y}_e$	$\bar{y}_{gp} = \sum_{\alpha}^{a} \frac{\bar{y}_{\alpha}}{N_{\alpha}} / \sum_{\alpha}^{a} \frac{1}{N_{\alpha}}$	$E(\bar{y}_{gp}) = \bar{Y}_g$

Let us recapitulate the estimation of the group mean $\bar{Y}_g$ and the element mean $\bar{Y}_e$ from observations made on a sample of group values $\bar{y}_a$. If the groups are selected with equal probabilities, the simple mean estimates $\bar{Y}_g$; but to estimate $\bar{Y}_e$, the sample values $\bar{y}_a$ should be weighted by their sizes N_a. On the contrary, if groups are selected with PPS, the simple mean estimates $\bar{Y}_e$; but to estimate $\bar{Y}_g$, the values $\bar{y}_a$ should be weighted with $1/N_a$.

4. FURTHER SAMPLING CONSIDERATIONS

Selecting groups with PPS is particularly well suited to studies in which both individuals and the groups to which they belong are used as units in separate analyses. For example, in a study of a large organization some of the results concern individual members, others deal with the groups (units, branches) of the organization. By selecting groups with probability proportional to their size measures, and then subsampling elements within groups with probabilities inversely proportional to the same measure, one obtains an equal-probability selection of elements, also an equal number of elements per unit. The simple mean of individual values estimates their population mean $\bar{Y}$; and the simple mean of the sample group values $\bar{y}_a$ estimates the element mean $\bar{Y}_e$ of the group values (Kish 1965, Section 7.3).

Note, however, that selecting all N_{α} elements (or any constant fraction) from groups selected with PPS would not yield an equal probability of selection for individuals, nor a self-weighting mean for estimating $\bar{Y}$.

Subsampling with PPS is also generally efficient when the group values are computed as means of the n_{α} individuals selected from the group, $\bar{y}_{\alpha} = \sum_{\beta}^{n_{\alpha}} y_{\alpha\beta} / n_{\alpha}$. Because the group values $\bar{y}_{\alpha}$ are based on (approximately) equal numbers n_{α} of elements, they are subject to approximately equal variances. This typically enhances the efficiency of the statistical analysis of group values. These may be related to, and analyzed together with, other group values which can be measured directly on the groups.

Joint analysis of group variables $\bar{Y}_\alpha$ and individual variables $X_{\alpha\beta}$ is possible. Each population element possesses both kinds of variables. A sample of elements permits their joint analysis.

Results are often tabulated for domains defined by size classes of units. Researchers tend to define domains roughly equal in numbers of elements, because size is a measure of the domain's importance. The domains of large units typically contain fewer units than domains comprised of numerous small units. Under these conditions PPS selection has another advantage over selection with equal probabilities for all units. If the latter is used, the domains of large units, though important in numbers of elements, contain few units and receive very few selections. With PPS selection the several domains receive equal numbers of selections to the degree that they contain equal numbers of elements. These domains can also serve as strata. If the number of selections in domains based on size are proportional to their element sizes, and if variations in size are small within the strata, the efficiency of selecting with equal probabilities within strata will be roughly equal to that of selecting with PPS.

Suppose that one has selected with equal probability $(f = n/N)$ a sample of n persons directly from a population of N individuals, without using the groups in the selection. Now, suppose that this sample of individuals is also used to obtain the group variables $\bar{y}_\alpha$ concerning the groups to which they belong. For example, from a selection of n individuals one can obtain characteristics of the family, or county, or university to which they belong. For a group of size N_α, the expected representation in the sample is fN_α; the actual size n_α from the group is a random variable. The simple mean of the values of $\bar{y}_\alpha$ taken for all the n sample elements will be an unbiased estimate of the element mean $\bar{Y}_e$. To estimate the group mean $\bar{Y}_g$ one should use the weights $1/N_\alpha$ in the estimate $\bar{y}_g = \sum_\alpha(\bar{y}_\alpha / N_\alpha) / \sum_\alpha(1/N_\alpha)$, summed over the n sample cases.

For example, a sample of n voters can be selected with equal probabilities from an area selection of counties and segments. Suppose one ascertains for each sample voter a value $\bar{y}_\alpha$ about the Congressional District to which he belongs; for example, the vote of his Congressman on a bill. The simple mean of the n voters will estimate the element mean $\bar{Y}_e$ for Congressional Districts; in this mean each District is weighted by $N_\alpha / \bar{N}$, with N_α representing the number of voters. If a somewhat different weight n'_α is preferred, formulas (3.9) and (3.10) should be used. To estimate the simple mean $\bar{Y}_g$ of Congressional Districts one can weight the values of sample voters with $1/N_\alpha$. But it would be a mistake to accept the simple mean of all group values that happen to appear in the sample, whether once or several times. In a national area sample, Districts from large cities tend to come into the sample with high probabilities but low n_α; the reverse is the case for rural Districts with sparse populations.

5. A SPECIAL CASE

A curious special case of the unit variable $\bar{Y}_\alpha$ occurs when it represents the unit size N_α. Suppose, for example, we were to ascertain for each element in the population the size of his household (or his city, or his organization). For the variable N_α the element mean is

$$\bar{Y}_e = \frac{\Sigma N_\alpha N_\alpha}{\Sigma N_\alpha} = \frac{\Sigma N_\alpha^2 / A}{\bar{N}}$$

$$= \frac{\sigma_n^2}{\bar{N}} + \bar{N} = \bar{N}(C_n^2 + 1). \qquad (5.1)$$

This also follows from (2.3) for the special case when $N_\alpha = \bar{Y}_\alpha$; hence $\bar{Y}_g = \bar{N}$, $R = 1$, and $C_n = C_y$. (The variance can be computed, as for any ratio mean for the variables N_α^2 and N_α.) The variable N_α can be obtained from an equal probability selection from the $\Sigma N_\alpha = N$ population elements; the sample mean will estimate $\bar{Y}_e$. Clearly when the relative variance C_n^2 of unit sizes is great, this mean, sometimes called the "*contraharmonic mean*" of the unit sizes N_α, is much larger than the simple arithmetic mean $\bar{N}$. In other words, the variable representing the size of the unit to which elements belong has a larger mean than the mean size of the units. Although the mean number of adults per household is only 2.02, the mean number of household members is 2.24 for the average adult. The greater size ranges of large organizations produce more striking effects. In 1960, 50 million people lived in 130 U.S. cities that had 100,000 or more population; in this population, the average city size was 0.39 million, but the size of the city in which the average person lived was 2.0 millions. Using medians does not help: the median city size was 0.19 million, but the median person lived in a city of 0.62 million.

If we ask people, "How many siblings do you have?", the answer will be $N_\alpha - 1$; the mean of these is one less than the mean of N_α, or $\bar{N}C_n^2 + \bar{N} - 1$.

6. BIAS VERSUS VARIANCE IN GROUP ANALYSIS

Frequently our main research interest is in units rather than elements. The units of analysis are not persons, but cities, schools, or some other group, organization, or ecological unit. It is a common mistake in these situations to use $\bar{Y}_g$ automatically. Use of $\bar{Y}_g$ can only be justified as the result of an appropriate and deliberate choice of the specific set of uniform weights, 1/*A*. In most cases, however, I prefer $\bar{Y}_e$ to $\bar{Y}_g$, and the weights $N_\alpha / \bar{N}$ appear as most appropriate, or as reasonable approximations.

These ideas lead to selection with probabilities proportional to size (PPS) when a sample of units is needed. However, this method is subject to a severe test when the research is based on a relatively small number A of units, say a few dozen or hundred, that comprise an entire population. If data are readily available for all A units, sampling then would waste information.

Whenever the A unit values $\bar{Y}_\alpha$ represent all the necessary information without sampling error, the issue is simple. If $\bar{Y}_e$ is appropriate and $\bar{Y}_g$ is used in its place, the bias is known from (2.3):

$$\bar{Y}_g - \bar{Y}_e = \bar{Y}_g R C_n C_y. \tag{6.1}$$

This can be large, and it should be avoided if no reasonable countervailing arguments arise. If the A values represent a sample from a larger universe, however, the bias should be balanced against sampling errors. It may well be good research strategy to regard the A units as a simple random sample from a hypothetical infinite universe of such units. Then the variance of the element mean may be written[6] as approximately

$$\mathrm{Var}\left(\frac{1}{A}\sum_\alpha^A \frac{N_\alpha}{\bar{N}} \bar{Y}_\alpha\right) = \frac{\bar{Y}_g^2}{A}[C_y^2 + C_n^2 + 2RC_nC_y]; \tag{6.2}$$

or

$$\mathrm{Var}(\bar{Y}_e) = \mathrm{Var}(\bar{Y}_g) + \frac{\bar{Y}_g^2}{A}[C_n^2 + 2RC_nC_y]. \tag{6.3}$$

The second term expresses the excess of variance of $\bar{Y}_e$ over that for $\bar{Y}_g$; it will be positive when $C_n > -2RC_y$. We should balance this against the bias of $\bar{Y}_g$. This can be done most readily by comparing the two mean-squared-errors; the mean-squared-error equals Variance+Bias^2.

$$\begin{aligned}
&\mathrm{MSE}(\bar{Y}_g) - \mathrm{MSE}(\bar{Y}_e) \\
&= [\mathrm{Var}(\bar{Y}_g) + \mathrm{Bias}^2(Y_g)] - [\mathrm{Var}(\bar{Y}_e) + 0] \\
&= [\mathrm{Var}(\bar{Y}_g) + \bar{Y}_g^2 (RC_nC_y)^2] - [\mathrm{Var}(\bar{Y}_g) + \frac{\bar{Y}_g^2}{A}(C_n^2 + 2RC_nC_y)] \\
&= \bar{Y}_g^2[(RC_nC_y)^2 - \frac{1}{A}(C_n^2 + 2RC_nC_y)].
\end{aligned} \tag{6.4}$$

---

[6] The variance for a single random unit can be obtained from the formula for the product of two random variables:

$$\sigma^2(N_\alpha \bar{Y}_\alpha) = \bar{N}^2\sigma_y^2 + \bar{Y}_g^2\sigma_n^2 + 2\bar{N}\bar{Y}_g R\sigma_n\sigma_y.$$

For the mean of A random selections then, and after dividing by the mean values $\bar{Y}_g$ and $\bar{N}$ of the variables $\bar{Y}_\alpha$ and N_α, we get (6.2). See Kish (1965, Section 6.6.D).

The increase in the variance of $\bar{Y}_e$ decreases with A, the number of units, hence the bias of $\bar{Y}_g$ tends to predominate unless the correlation R is very small. If the correlation is small, the bias may become negligible, and the term C_n^2 / A can predominate in (6.4); but in this case the two means and their variances will be similar. If A is very small and the C_n^2 large, $\bar{Y}_g$ may appear more reliable than $\bar{Y}_e$; but for a very small "sample," other issues of inference also arise.

6

RETAINING UNITS AFTER CHANGING STRATA AND PROBABILITIES

LESLIE KISH AND ALASTAIR SCOTT

Survey samples are often based on primary sampling units selected from initial strata with probabilities, p_j, proportional to initial measures. However, later samples can be better served with new strata and new probabilities, P_j, based on new information. The differences between the initial and new strata and measures may be due to changes either in population distributions or in survey objectives. It is efficient to retain in the new sample the maximum permissible number of initial selections. Procedures are presented first for changing measures within fixed strata, then for strata with changed units. Modifications, improvements and simplifications are also introduced.

1. AIMS

Unequal selection probabilities are often assigned to sampling units. Our methods, though more generally applicable, are especially needed for the selection of primary sampling units for surveys. Often these are selected separately from many strata, with one selection from each stratum.

After the initial selection the units may be used for many surveys over several years. But as time passes, the needs of new surveys may be better served by new strata and new selection probabilities, based on new data, than by those used for the initial selection. The difference between initial and new data may be due to differential changes among the sampling units as revealed by the latest Census. (The censuses of 1970 make this a topical subject.) Or the difference may be due to

Reprinted with permission from the *Journal of the American Statistical Association* (1971), 66, 461–470.
This research was supported by GS–777 from the National Science Foundation. An earlier version, excluding Section 6, was given in Kish (1963). The authors are grateful for suggestions from Irene Hess, Vinod K. Sethi, Roe Goodman, Walter M. Perkins, Ivan Fellegi, Benjamin Tepping, the editors and referees.

changes in survey objectives and populations; for example, a sample initially designed for households and persons may later be required to serve a survey of farmers, or college students. *Obviously our methods are also applicable to designing simultaneously a related group of samples with differing objectives.* But the chief need for the methods is for differences arising in time, and we shall find it convenient to refer to initial and new selections and strata.

We may prefer new data for measures and for strata when these are available and differ considerably from the initial data. A brand new selection would mean changing most of the units. But continued use of the initial units has several advantages. First, each primary unit in the sample may represent considerable investment in office work, and especially in a trained force of local interviewers. Second, using the same units can decrease considerably the standard errors of comparisons between periodic surveys. Third, changing the entire field force can introduce discontinuity into a time series. Fourth, panel studies depend on stable units.

Fortunately, one can change to new probabilities and strata, yet retain most of the initially selected units. The desired new probabilities are attained as products of two probabilities, combining operations in two phases: the initial probabilities of selection, and conditional probabilities of selection in the new phase. In the latter our methods greatly increase the probabilities of reselection of initially selected units, with corresponding decreases for the others. We propose several methods of reselection for diverse situations, balancing two criteria: maximal retention of selected units, and practical simplicity of reselection procedures.

Practical operations must be designed to guard against selection biases for or against any of the units. Specifically, *both the assignment of the new probabilities and the shifting to new strata for any of the units must be entirely independent of their having been initially selected.* This independence must be guaranteed either by a "blindfold" ignorance of the identities of initially selected units, or by objective criteria which strictly ignore those identities. However, we can have all the freedom needed for meaningful stratification. The set of units shifted from an initial stratum to a new stratum may be chosen by any desirable criteria, objective or subjective, which ignore initial selection. Our proposed methods do not assume that the sets of shifted units are random samples either from the initial strata or within the new strata. We have similar freedom in the assignment of new measures of size.

When selection probabilities and strata are changed, the situation may also require changes in the boundaries of sampling units. Some may be split, and others combined. The initial selection probability, p_j, of a unit may be divided among several portions according to specified rules; similarly, several p_j may be combined into one new unit. The creation of entirely new units is noted with $p_j = 0$, and the elimination of initial units with $P_j = 0$, as if the population vanished. Retaining of initial selections in a new sample depends on their one-to-one identification with new units. Hence such unique identification for most of the units and for most of the measures p_j is needed for retaining most of the initial selections. Thus unit K (with initial probability p_k) can be joined to another unit J (with initial probability p_j); such combining of units has the same effect as if all people moved out of K and into J, thus creating $P_k = 0$. The new (combined) unit will be identified with the unit J and retain p_j as its initial probability; the new probability P_j of the combined unit will be based on the sum of new measures of both units. If the initial probabilities, p_k, of the dropped units (with $P_k = 0$) are small, correspondingly little is lost in the average probability of retaining.

Our methods are designed for single selections from strata. In places we generalize to two or more selections per stratum. However, for two or more selections without replacement and with unequal probabilities these methods become very complicated, even for changing probabilities within fixed strata, as may be seen in a method devised by Fellegi (1966). Furthermore, there exist too many methods with diverse joint probabilities of selection, and these might require diverse complex treatments. (This is one example of additional penalties incurred by complex selection methods.) However, the method can be applied to single selections from random half-strata (Kish 1965a); this may serve as a reasonable approximation for others resembling it.

Because of their complexity, methods for controlled selection (multiple stratification) are not tackled here. We suspect that the sacrifice of gains attainable from controls is small when changes in selected units are held to a small portion of the total.

Section 2 describes the basic procedure for changing probabilities within fixed strata. This is modified in several ways in Section 3 and extended to include modest changing of strata. In Section 4 we present a simple adjustment for growth. A simple, practical and near-optimal method for changed strata is given in Section 5; this method is justified in the final Section 8. Section 6 describes an alternative method for changed strata. Section 7 discusses briefly some closely related issues.

The procedures of Sections 5–8 are more complex than those of Sections 2–4. But they are necessary, because the changes (of objectives or of census data) which motivate changes in measures for units also tend to motivate shifting of units between strata for two reasons. First, strata grow at different rates, creating undesirable inequalities between sizes of the strata. Second, the homogeneity of strata based on the initial data used for stratification is generally diminished with the new data. Desired sizes and homogeneity, based on new data, may be restored with new stratification.

2. BASIC PROCEDURE (A) WITHIN FIXED STRATA

We first present a procedure for changing from initial probabilities, p_j, to new probabilities, P_j, when all units in each stratum come from the same initial stratum. The probabilities are based generally on some measures of size, so that $p_j = m_j / \Sigma m_j$ and $P_j = M_j / \Sigma M_j$. This is a simple version essentially of a method first presented by Keyfitz (1951). There are $D+I$ units in the new stratum, of which D receive decreases in probability, and I receive increases or remain unchanged. Subscripts distinguish the two sets of units, so that

$$P_d < p_d \text{ and } P_i \geq p_i, \tag{2.1}$$

where

$$(d = 1,2,\ldots,D)$$

and

$$(i = D+1, D+2,\ldots,D+I).$$

The new probabilities, to which we want to change, sum to unity in the stratum:

$$\Sigma_{I+D} P_j = \Sigma_D P_d + \Sigma_I P_i = 1. \tag{2.2}$$

The rules for changing probabilities follow:

(a) If an initially selected unit shows an increase in probability or no change ($P_i \geq p_i$), retain it in the sample as if selected with P_i.
(b) If an initially selected unit shows a decrease in probability ($P_d < p_d$), retain it in the sample with a probability P_d / p_d. Thus the compound probability of original selection and retention is made

$$p_d \times P_d / p_d = P_d. \tag{2.3}$$

The unit is dropped with probability $1 - P_d/p_d$. If the unit has been omitted from the new stratum, it may be considered as having $P_d = 0$, and dropped with certainty.
(c) If a decreasing unit is dropped from the sample, select a replacement from the increasing units with probabilities $(P_i - p_i) / \Sigma(P_i - p_i)$ proportional to their increases. This insures that the probability of selecting an increasing unit is brought up to P_i from its initial value p_i. The probability of *not* having selected a unit in Steps (a) and (b) is:

$$1 - \Sigma_I p_i - \Sigma_D P_d = \Sigma_I P_i - \Sigma_I p_i = \Sigma_I (P_i - p_i). \tag{2.4}$$

The total probability of selecting the ith increasing unit is the sum of the disjoint probabilities for initial or new selections:

$$p_i + (\Sigma P_i - \Sigma p_i) \bullet \frac{(P_i - p_i)}{\Sigma(P_i - p_i)} = P_i. \tag{2.5}$$

In (2.3) and (2.5) we see that the desired probabilities are obtained for both decreasing and increasing units. The solution is obviously optimal in the sense that the only selected units not retained in the sample are those that must be dropped with Step (b), to convert decreasing units from p_d to P_d. These rules for single selections from strata can be readily extended also to two or more selections *with* replacement.

3. EXTENSIONS OF THE BASIC PROCEDURE

Into the basic Procedure A, one can introduce further considerations of efficiency. Two of these modifications (a) and (b) take advantage of flexibilities derived from knowing that generally it is neither necessary nor possible to have precise measures of size.

(a) Important changes in probabilities may be confined to a small proportion of units for which P_j / p_j departs greatly from 1. For the great majority of units these changes may be small enough, so that neglecting them adds little to the variance. To these numerous units we can reassign arbitrarily the initial probabilities, $P_j = p_j$. This flexible procedure reduces the number of units switched and also the work needed for revising office records for selected units with changed probabilities.

This procedure was used in changing from 1940 to 1950 Census measures for the 54 strata of the national sample of counties of the Survey Research Center of the University of Michigan (Kish and Hess 1959a). Increases under $P_j / p_j = 1.1$ were declared negligible, and $P_j = p_j$ reassigned. The procedure consisted of four steps:

1. Compute the provisional probabilities P_j' as usual.
2. For units for which $P_j' / p_j \geq 1.1$, declare $P_i = P_j'$ accepted as new probabilities with increases. Compute the sum $\Sigma(P_i - p_i)$ of these increases in the stratum.
3. From the smallest values of P_j' / p_j, take enough decreasing units to balance exactly the sum of the increases, $\Sigma(P_d - p_d) + \Sigma(P_i - p_i) = 0$; for these decreasing units assign $P_d \doteq P_d'$, with marginal adjustments to balance the decreases exactly against the increases.
4. For all other units, reassign the old probabilities, $P_j = p_j$.

(b) We need not merely accept changes from the one decennial Census to another; we can also project them forward into the middle of the period of use of the Census frame. This may be especially worth doing for the fastest and steadiest growing units to improve a sample of units serving throughout the intercensal period.

(c) Faced with small probabilities of change $(1 - P_d / p_d)$ in several strata, we need not draw independently within each stratum. Instead we can control the variation in the number of needed changes by cumulating the probabilities of change from one stratum to another, then applying an interval of one after a random start. Thus the actual number of changes can be controlled within a fraction of the expected number of changes. This deliberate sacrifice of independence has no practical disadvantages we can conceive.

3.1 A Simple Procedure (B) for Changed Strata

The basic procedure of Section 2 can also be adopted to accommodate modest changes in strata. If all units in the initial stratum are also present in the new stratum, we have $\Sigma_D p_d + \Sigma_I p_i = 1$ for the initial probabilities, similar to (2.2) for the new P_j's. But we neither needed nor used that new condition. Hence we may remove from a new stratum some of the initial units—perhaps those that grew too much and those that no longer fit the stratum's criteria; these units may be considered as having $P_d = 0$. Contrariwise, units moved into a new stratum are treated as newly created, having initial probabilities $p_i = 0$. In other words, moved units are assigned $p_i = 0$ in their new strata and $P_d = 0$ in their initial strata. For units moved into other strata, having been initially selected is disregarded both in their old and their new strata.

Each of the new strata needs to be identified uniquely with one of the initial strata, for which proper rules are needed. However, initial strata may be abolished. Also new strata may be created in which all units are treated as having $p_i = 0$.

This simple extension of the basic procedure may be satisfactory as long as the proportion of initial units to which $p_i = 0$ is assigned is not great. However, when considerable proportions of units are shifted from initial to new strata, these proportions represent a loss in previously selected units, a loss which should be drastically reduced. This is what the methods I and II described in Sections 5 and 6 can do.

Example 3.1. The table below is a simple example for the steps and probabilities if any one of the four units was initially selected.

A is a decreasing unit. Give it a chance of .2 /.6 of dropping, for a combined probability of .6 × .4/.6 = .4.

B was moved to another stratum. It is a "decreasing" unit, whose initial selection must be eliminated and need not appear on the list of this stratum.

C is an increasing unit. Its original selection probability is increased by .3 /.5 if a decreasing unit is dropped: .1 + .5 (.3 /.5) = .4.

D was moved from another stratum, and its initial selection there must be disregarded. Thus its probability is .0 + .5 (.2 /.5) = .2.

Unit	p_j	P_j	$P_i - p_i$	P_d/p_d
A	.6	.4	—	4/6
B	(.3)	(.0)	—	0
C	.1	.4	.3	—
D	.0	.2	.2	—
Total	—	1.0	.5	—

4. A SIMPLE PROCEDURE (C) FOR OVERGROWN UNITS

A simpler procedure than A can handle a problem confined chiefly to large growth in a small proportion of primary units. For example, suppose we have a sample of blocks selected with the initial probabilities, p_j, proportional to the initial sizes, n_j. Suppose also that for a new sample one is willing to retain the original probability, p_j, for all blocks, except for those that have more than doubled in size. That is, if the new size $N_j \leq 2n_j$, we accept the initial p_j. These ordinary units, which do not qualify as growth units, remain subject to the initial selection rates in two stages, $n_j/(b/f) \times b/n_j = f$, where f is the overall uniform sampling fraction and b is the planned size of the subsample; but we are ready to accept subsample sizes up to $2b$ for units up to $N_j = 2n_j$ in size.

For the *growth units* with $N_j > 2n_j$, we want new probabilities P_j proportional to N_j. Of course, instead of 2, some other factor may be chosen. Generally, place into new *growth strata* the portion $(N_j - n_j)$, denoting the size increase of primary units designated as *growth units*. Then select from these growth units a sample with probabilities proportional to the values $(N_j - n_j)$. In most situations the constant of proportionality will be the same (b/f) as for the initial selection, to preserve the uniform overall sampling fraction f. Thus an element can be selected in either of two selection processes. First there was the initial selection which gave the unit a probability of $n_j f/b$ of selection. Then the unit had another chance of $(N_j - n_j)f/b$ of selection in the growth stratum. When either selection occurs, a subsampling fraction of b/N_j is applied, so that the overall selection probability of elements in the two distinct processes is f, as desired:

$$\frac{n_j f}{b} \bullet \frac{b}{N_j} + \frac{(N_j - n_j) f}{b} \bullet \frac{b}{N_j} = f. \tag{4.1}$$

The planned size of the subsample is b if the unit gets selected in either one of the two processes; the actual size depends on how well the new measure N_j resembles the actual size. If the unit gets selected twice, which seldom occurs in practical situations, the planned subsample of size b must be doubled.

The procedure amounts to splitting the growth units into two parts, consisting of an initial fraction n_j/N_j and a growth fraction $(N_j - n_j)/N_j$, then subjecting these to separate selections. But this preliminary splitting need not be carried out actually. With an equivalent procedure we may simply apply the subsampling fraction b/N_j to the entire unit if selected once, and the fraction $2b/N_j$ if selected twice.

This procedure may be applied to two or more selections within strata, with or without replacement. Further, the method can readily utilize several sampling fractions. For example, the sampling fraction f may be halved to $f/2$ in the growth units, then compensated with weights of 2 in the estimation process. This may be desirable if element costs in growth units are four times greater.

5. METHOD I FOR CHANGED STRATA

In the general case, a new stratum is composed of units from several strata. The word *stratum* will denote a new stratum, and any collection of units in a stratum coming from a common initial stratum will be called a *set*. A stratum may contain one, two, or more sets; let these be denoted by $A_1, A_2, \ldots, A_k, \ldots, A_n$. Each set contains either zero or one initial selection.

The new probabilities P_j sum to one, $\Sigma_I P_i + \Sigma_D P_d = 1$, with i and d denoting again increasing and decreasing units. The initial probabilities, p_j, of units within the set A_k sum to $p_k = \Sigma_{A_k} p_j$; the sum of the measures over the stratum $\Sigma_n p_k = p. \neq 1$ generally.

The procedures for applying this method are rather simple, especially for strata with either zero or only one initial selection. In practice this will tend to be true of the great majority of strata for either of two reasons: if single sets comprise most of the strata, because only moderate portions of units have been shifted among the strata, or if the new sample represents an expansion of the old sample. Then Rules 1–3 yield easy solutions for most strata, and Rule 4 for the remaining requires a little work. To be more specific, in a sample of 100 strata, Rule 4 may ordinarily need to be applied in one or two, or none at all. In the other 98 or so strata, the initial selections are obtained according to Rules 1–3 with negligible effort. The expected work per stratum is very small. The rules are justified in Section 8.

Use Rules 1–4 to select a *set* containing an initial selection. Then Rules 2, 3 and 4 require conversions from initial to new probabilities, with Procedure A for *units*. This is applied to preliminary probabilities $p_j^* = p_j / p. = p_j / \Sigma_n \Sigma_{A_k} p_j$ over all units in the entire new stratum. Then:

(a) Increasing units $(p_i^* \leq P_i)$ are accepted;
(b) Decreasing units $(p_d^* > P_d)$ are retained only with P_d / p_d^*, and dropped with $1 - P_d / p_d^*$;
(c) If dropped, a new selection is made from all increasing units in the entire new stratum with probabilities $(P_i - p_i^*) / \Sigma(P_i - p_i^*)$.

For example, let a stratum contain a single set of three units with initial probabilities {.2, .2, .1} and new probabilities {.2, .5, .3}. The preliminary probabilities will become {.4, .4, .2}. If the first unit was an initial selection, it will be dropped with probability 1 − .2 /.4 = 0.5. If dropped, a new selection is made with probability (.5 − .4) / {(.5 − .4) + (.3 − .2)} for the second unit, and a similar probability for the third unit. The second and third units are both increasing; if either was the initial selection, it is retained automatically.

Rules 1–4 for designating the set with an initial selection follow:

Rule 1. If none of the sets contains an initial selection, select one unit directly with P_j from the entire stratum.

Rule 2. If only one of the sets contains an initial selection, accept it as the preliminary selection, then convert from preliminary p_j^* to new P_j.

Rule 3. If there are only two sets in the stratum *and* both A_1 and A_2 contain initial selections, select set A_1 with probability

$$y_1 = p_2 /(p_1 + p_2). \tag{5.1}$$

From the selected set accept the initial selection, then convert from preliminary p_j^* to new P_j. Note that the sets are selected with odds *inversely* proportional to their initial probabilities.

Rule 4. When the stratum contains *more* than two sets, *and* two or more sets contain initial selections, order all sets *objectively* into a nested series of paired groups, one or two sets for a group, then one or two groups forming a higher group, etc. Then begin with the highest pair and work down the dichotomies with these rules. If only one group contains an initial selection, take it. If both groups contain initial selections, choose G_1 with a probability symbolized with:

$$y_1 = [p_2 + (r_1 - r_2)]/(p_1 + p_2), \tag{5.2}$$

where

$$r_1 = (p_1 / p_1') = (p_{11} + p_{12})/(p_{11}' + p_{12}' - p_{11}'p_{12}'),$$

and

$$r_2 = (p_2 / p_2') = (p_{21} + p_{22})/(p_{21}' + p_{22}' - p_{21}'p_{22}').$$

Note that p_i is the *measure* of group G_i: the sum of initial probabilities of the groups (and sets and units) it contains; whereas p_i' is the probability that G_i contains one or more (at least one) initial selection(s). Then $r_i = p_i / p_i'$; and $(r_1 - r_2)$ is an adjustment of the *inverse* odds used for selecting G_i from a pair. After a group G_i is selected, use the same rule to select one of the two subgroups it contains. Continue until a set is selected. In the selected set accept the initial selection, then convert from preliminary p_j^* to new P_j.

Suppose first that the stratum contains four sets, which have been paired into two groups, G_1 and G_2. Then $p_1 = p_{11} + p_{12}$ denotes the initial probabilities of the pair of sets in the first group and $p_2 = p_{21} + p_{22}$ in the second group. Since each G_i contains a single set,

$$p_{ij}' = p_{ij} \text{ and } p_i' = p_{i1} + p_{i2} - p_{i1}p_{i2}.$$

If G_1 contains a single set, then $p_1 = p_{11}$ and $p_{12} = 0$; also $p_1' = p_1$ and $r_1 = 1$. Similarly $r_2 = 1$ if G_2 contains a single set. If both G_1 and G_2 are single sets $r_1 = r_2 = 1$, and we have Rule 3.

Now in general, G_i contains groups G_{i1} and G_{i2} with $p_i = p_{i1} + p_{i2}$, and $p_{ij} = p_{ij1} + p_{ij2}$, the measures (initial probabilities) in the two groups contained in G_{ij}. Then $p_i' = p_{i1}' + p_{i2}' - p_{i1}'p_{i2}'$ and $p_{ij}' = p_{ij1}' + p_{ij2}' - p_{ij1}'p_{ij2}'$. If G_{ijk} contains a single set, then $p_{ijk}' = p_{ijk}$; but if it contains two sets, then $p_{ijk}' = p_{ijk1} + p_{ijk2} - p_{ijk1}p_{ijk2}$; and if it contains two composite groups, then $p_{ijk}' = p_{ijk1}' + p_{ijk2}' - p_{ijk1}'p_{ijk2}'$, etc.

Example 5.1 for Rule 4. Stratum of five sets, of which four sets, marked with (*), contain initial selections.

p_{ij}	Indexes			p_i			p_i'	
.20*	1	11		.35	.20			.32
.15		12			.15			
.10*	2	21	211	.24	.18	.10	.172	.2217
.08*		21	212			.08		
.06*		22	221		.06	.06	.06	

$$p_1' = p_{11} + p_{12} - p_{11}p_{12} = .20 + .15 - .20 \times .15 = .32$$
$$p_2' = p_{21}' + p_{22} - p_{21}'p_{22} = .172 + .06 - .172 \times .06 = .2217$$
$$p_{21}' = p_{211} + p_{212} - p_{211}p_{212} = .10 + .08 - .10 \times .08 = .172$$
$$r_1 = p_1 / p_1' = (.35/.32) = 1.0938$$
$$r_2 = p_2 / p_2' = (.24/.2217) = 1.0825$$
$$r_{21} = p_{21} / p_{21}' = (.18/.172) = 1.0465$$
$$r_{22} = p_{22} / p_{22}' = (.06/.06) = 1.0000$$

First choose p_1 with

$$y_1 = [p_2 + (r_1 - r_2)]/(p_1 + p_2)$$
$$= [.24 + (.0113)]/(.35 + .24) = .4259.$$

If this succeeds, accept set p_{11} = .20 automatically, since set p_{12} has no initial selection. If it fails, choose p_{21} with

$$y_{21} = [p_{22} + (r_{21} - r_{22})]/(p_{21} + p_{22}) = [.06 + (.0465)]/(.18 + .06) = .4438.$$

If this fails, accept set p_{221} = .06. If it succeeds, choose set p_{211} = .10 (over set p_{212} = .08) with

$$y_{211} = p_{212} /(p_{211} + p_{212}) = .08/(.10 + .08) = .4444.$$

Example 5.2 for Rule 4. Stratum of ten sets of which three sets, marked with (*), contain initial selections.

p_k	Indexes					p'_k			
.50*	0								
.09	1	11	111	1111			.1628	.2681	.3718
.08				1112					
.07			112	1121			.1258		
.06				1122					
.05*		12	121	1211			.0880	.1417	
.04				1212					
.03			122	1221	12211	.0494	.0589		
.02					12212				
.01*				1222		.0100			

p' for

$$1221 = .02 + .03 - .0006 = .0494$$

$$122 = .0494 + .0100 - .0005 = .0589$$

$$121 = .05 + .04 - .0020 = .0880$$

$$112 = .07 + .06 - .0042 = .1258$$

$$111 = .08 + .09 - .0072 = .1628$$

$$12 = .0880 + .0589 - .0052 = .1417$$

$$11 = .1628 + .1258 - .0205 = .2681$$

$$1 = .2681 + .1417 - .0380 = .3718$$

$$r_0 = 1.0000$$

$$r_1 = .45 / .3718 = 1.2103$$

$$r_{121} = .09 / .0880 = 1.0227$$

$$r_{122} = .06 / .0589 = 1.0187$$

First choose set $p_0 = .50$ with

$$y_0 = (.45 - .2103)/(.50 + .45) = .2523.$$

If this fails, go to $p_{12} = .15$, because p_{11} contains no initial selections. Choose $p_{121} = .09$ with

$$y_{121} = (.06 + .0040)/(.09 + .06) = .4267.$$

If this succeeds, take $p_{1211} = .05$; if it fails go directly to $p_{1222} = .01$.

These rules would become cumbersome if carried far. But this will seldom be necessary—only when the stratum has many sets, and two or more of them contain selections.

An objective rule is needed for pairing the groups, independent of the presence of selections in the sets, to prevent bias in accord with that presence; or any ordering by a person completely ignorant of that presence. A practical rule may speed up the search:

1. A set containing more than half of the p_j values of the stratum establishes the first split.
2. Order all others by size; then divide them successively into two equal numbers of sets, putting odd numbers into the second half. For example, a stratum composed of a majority set 0, plus 9 other sets (numbered in decreasing size) would be divided as follows: {0} {[(1, 2) (3, 4)] [(5, 6) (7, 8/9)]}. If there were 0 plus 7 sets, the last subgroup would be (7).

Though not quite optimal, the method is nearly so in many actual situations. It does find an initial selection in any stratum which has one or more. Its weakness comes in strata with two or more initial selections: Rules 3 and 4 can select a decreasing unit and then drop it, instead of retaining an increasing unit present in another set. This marginal inefficiency will often be small; in changing a national sample of 54 units (Kish 1963), only three of them had to be dropped altogether; these because of decreased measures, which may be necessary under any method.

6. METHOD II FOR CHANGED STRATA

This method requires more work than Method I in most situations, but its advantages may prevail in some situations.

1. It can retain some initial units which may be dropped by Method I; if these are substantial, it may be preferred. Specifically, Method II will always retain an increasing initial selection, whereas Method I may fail to do so. The advantages of Method II over I may be greatest when changes in measures are drastic, and when the needed computations are inexpensive compared to changing even a few extra units.
2. It is easier to apply when several initial selections are present in the new strata than when most strata have zero or one selection. The reverse is true for Method I.
3. Its justification is more straightforward.

As before, a new *stratum* is composed of sets A_k (k = 1, 2, ..., n) which are portions of initial strata. The sets must be ordered according to a rule that is independent of our knowledge of which sets contain initial selections. Any such ordering will do. We may look for an objective rule that is also useful in reducing

Example 6.1. Method II applied to five sets in two distinct orderings.

(A) The five sets are arranged in order $P_{k\text{-}1} > P_k$. This order tends to reduce computations, with low values of Q_k coming early.

Final P_j	.25	.15	.17	.13	.07	.10	.04	.04	.05
Initial p_j	.35	.20	.24	.18	.10	.15	.06	.06	.08
Q_k	$Q_1 = .60$		$Q_2 = .288$			$Q_3 = .2448$	$Q_4 = .2154$		$Q_5 = .1982$
$r_{kj} = p_jQ_{k\text{-}1}$	.35	.20	.144	.108	.060	.0432	.0147	.0147	.0172

(B) The five sets are arranged in order $P_{k\text{-}1} < P_k$. This order tends to improve slightly the chances of retention.

Final P_j	.05	.04	.04	.10	.17	.13	.07	.25	.15
Initial p_j	.08	.06	.06	.15	.24	.18	.10	.35	.20
Q_k	$Q_1 = .95$	$Q_2 = .87$		$Q_3 = .77$	$Q_4 = .40$			$Q_5 = .18$	
$r_{kj} = p_jQ_{k\text{-}1}$	.08	.057	.057	.1305	.1848	.1386	.077	.14	.08

Note that the probability (6.6) of failing to find a selection in any of the five sets is $Q_5 = .18$ with ordering (B), lower than $Q_5 = .1982$ with ordering (A). With ordering (A) the first set has two decreasing units, $p_j > P_j$. If the set has no initial selection, we proceed to the second set after computing $Q_1 = 1 - .25 - .15 = .60$, then $r_{21} = .6 \times .24 = .144$, $r_{22} = .6 \times .18 = .108$, and $r_{23} = .6 \times .10 = .060$. For all three units $r_{2j} < P_j$ and become increasing. Thus if the second set has no initial selection, we proceed to the third set after computing $Q_2 = .60 - .144 - .108 - .060 = .288$; and so on through the other sets if necessary, noting that all the r_{kj} values gave increasing values early.

With ordering (B) the first set has a decreasing unit; if it is not an initial selection we go to the second set with $Q_1 = 1 - .05 = .95$. There $r_{21} = r_{22} = .95 \times .06 = .057$, both decreasing units. Hence, if the second set contains no initial selection, we go to the third set with $Q_2 = .95 - .04 - .04 = .87$. For its only unit $r_{31} = .87 \times .15 = .1305$, and this is also decreasing; if it is not an initial selection we go to the fourth set with $Q_3 = .87 - .10 = .77$, etc.

---

the work of finding an initial selection; for example, ordering sets from larger to smaller sums of probabilities: $P_1 > P_2 > \ldots > P_k \ldots > P_n$, where $P_k = \Sigma_{A_k} P_j$ is the sum of new probabilities for units within the set A_k. Again $\Sigma_n P_k = 1$.

We use subscripts 1, 2, k, n to denote probabilities for sets and subscripts j, i, d for units. We continue to use capitals for new probabilities and lower case for initial probabilities.

Step 1. Start with A_1: if it contains an initial selection apply to it Rules (a) and (b) of Procedure A. If the initial selection is (a) increasing

$(p_i \le P_i)$, or (b) decreasing $(p_d > P_d)$ but retained with a probability operation of P_d/p_d, that initial selection becomes the new selection from the stratum.

Step 2. If A_1 yields no new selection, go to A_2. If it contains an initial selection, apply Rules (a) and (b); but in the place of the initial probabilities p_j we must use the modified probabilities

$$r_{2j} = p_j Q_1 = p_j (1 - \Sigma_{I_1} p_i - \Sigma_{D_1} P_d), \tag{6.1}$$

where Q_1 is the probability that A_1 failed to yield a new selection. Note that (a) is applied to "increasing" units when $r_{2i} = p_i Q_1 \le P_i$; (b) is applied to "decreasing" units, when $r_{2d} > P_d$, with the probability operation $P_d / r_{2d} = P_d / p_d Q_1$ of retaining.

Step 3. If A_2 yields no new selection, go to A_3. If it contains an initial selection apply Rules (a) and (b) with the modified probabilities

$$r_{3j} = p_j Q_2 = p_j (Q_1 - \Sigma_{I_2} r_{2i} - \Sigma_{D_2} P_d), \tag{6.2}$$

where Q_2 is the probability that A_1 and A_2 failed to yield a new selection. Note again that r_{3j} is used to distinguish "increasing" from "decreasing" units, and in the operation P_d / r_{3d}.

Step 4. If $A_1, A_2, \ldots, A_{k-1}$ yield no new selection, keep repeating the procedure to set A_k. We apply Rules (a) and (b) of Procedure A to A_k with the modified probabilities

$$r_{kj} = p_j Q_{k-1} = p_j (Q_{k-2} - \Sigma_{I_{k-1}} r_{(k-1)i} - \Sigma_{D_{k-1}} P_d), \tag{6.3}$$

where Q_{k-1} is the probability that $A_1, A_2, \ldots, A_{k-1}$ have failed to yield a new selection, and $r_{(k-1)i} = p_i Q_{k-2}$. The set A_k has D_k units with $r_{kd} > P_d$ and I_k units with $r_{ki} \le P_i$. For the first set, A_1, $r_{1j} = p_j$ and $Q_o = 1$.

Step 5. If no unit has been selected after all A_n sets in the stratum have been examined, choose a unit from the "increasing" units, defined by $r_{ki} \le P_i$, with probabilities proportional to the increases $P_i - r_{ki}$.

6.1. Some Strategic Features

1. Our chief aim is to retain whenever possible one initial selection in each stratum. If an increasing initial selection is present in any of the sets of the new stratum, an existing unit will be retained.
2. If all initial selections are decreasing ($p_d > P_d$), the probability of dropping them all is less than $\Pi\,(1 - P_d/p_d)$; it is equal to $(1 - P_d/p_d)$ when the first set contains the single decreasing selection. An optimal method would require ordering the sets to minimize Q_n, the probability of no retention after all n sets. But this would be messy and hardly worthwhile since the

nonoptimality of Method II is less than that of Method I, which also may often be marginal in practice. A simple rule for getting high values of Q_n may be to arrange the sets in increasing order of decreases: the value of $(p_d - P_d)$ summed within the sets. It is questionable whether this advantage is worth giving up the order $P_{k-1} > P_k$ suggested earlier to reduce computations.

3. Reducing computations is our second aim. If an initial selection is present in any set, computing the Q_k and r_{kj} values will seldom be necessary. It is not needed in A_1; if A_1 is large, Q_1 is easily less than one; hence, $r_{2j} = p_j Q_1 < P_j$ for almost all units, and this holds increasingly in later sets because $Q_k < Q_{k-1}$. This process is speeded up if we arrange the sets from larger to smaller sums of probabilities, so that $P_k < P_{k-1}$.
4. However, if no initial selection is present and retained in the n sets and we must apply Step 5 to obtain a new selection, then we may have a fair job in computing the n pairs of Q_k and r_{kj}. Hence we may prefer Method I of Section 5 when many strata lack initial selections.

6.2. Justification of the Procedure

There are $D = \Sigma D_k$ "decreasing" units and $I = \Sigma I_k$ "increasing" units in a new stratum, summed over the (k=1, 2, ..., n) sets. Each set comes from an initial stratum from which one initial selection was made with probability p_j. One new selection is to be made from the stratum with probabilities P_j, hence $\Sigma_I P_i + \Sigma_D P_d = 1$.

For any "decreasing" unit in A_k the overall selection probability is the product of three probabilities:

$$P_r[\text{initial selection}] \bullet P_r[\text{no selection before } A_k] \bullet P_r[\text{retaining}]$$

$$= p_d \bullet Q_{k-1} \bullet P_d / r_{kd} = p_d \bullet Q_{k-1} \bullet P_d / p_d Q_{k-1} = P_d. \tag{6.4}$$

For any "increasing" unit in A_k the overall selection probability is the sum of two disjoint probabilities:

$$P_r[\text{initial selection}] \bullet P_r[\text{no selection before } A_k]$$

$$+ P_r[\text{no selection through } A_n] \bullet (P_i - r_{ki}) / \Sigma_I (P_i - r_{ki})$$

$$= p_i Q_{k-1} + (1 - \Sigma_I Q_{k-1} p_i - \Sigma_D P_d) \bullet (P_i - r_{ki}) / \Sigma_I (P_i - r_{ki})$$

$$= r_{ki} + (\Sigma_I P_i - \Sigma_I r_{ki}) \bullet (P_i - r_{ki}) / \Sigma_I (P_i - r_{ki}) = P_i. \tag{6.5}$$

The probability of failing to get a selection in or before set A_n is

$$Q_n = 1 - \Sigma_I Q_{k-1} p_i - \Sigma_D P_d = \Sigma_I P_i - \Sigma_I r_{ki}. \tag{6.6}$$

The single summation over I denotes summation for the $(k-1)$th set, as in (6.2), then over all k sets.

6.3 Thoughts on Optimization

Even with the best ordering of the sets, Method II does not always maximize the probability of retaining an initial unit. To achieve this, we must use a unit-by-unit, rather than a set-by-set, sequential scheme in which all the increasing units are considered first, with corresponding modifications to the initial probabilities of the remaining units, then all the modified increasing units, etc. This is extremely complicated in most practical situations, and the gains are so small as to almost never be worthwhile.

7. METHODS FOR EQUAL INITIAL PROBABILITIES

Our methods have been designed chiefly for selecting units with unequal probabilities. They seem to become too complex for practical work with more than one selection from a stratum. But when units have equal probabilities within strata, the symmetries of selection probabilities permit multiple selections. Some practical methods may be noted.

The simplest example is initial selection with simple random sampling of n units from the entire population of N units; that is, all $\binom{N}{n}$ combinations are equally probable. These can be sorted into arbitrary strata with procedures that guarantee no effect from the initial selections on the sorting. The n_h selections can be accepted as equivalent to random selections within the hth stratum. The stratum samples then can be increased or decreased at random to obtain the numbers of random selections needed from each stratum. The symmetries of simple random selections are sufficient to justify this method, sometimes called "random quotas" (Kish 1965a).

A proportionate stratified random sample is selected with constant $f = n_h / N_h$ for all strata, and all $\binom{N_h}{n_h}$ combinations within strata are equally likely. If units are shifted after initial selection into new strata, they carry along their equal initial probabilities f; hence, all initial selections found within the new strata may be retained up to the desired new sampling fractions $f'_h = n'_h / N'_h$, which may vary in the new design.

Let us now expand the preceding to a stratified random sample when the fractions $f_h = n_h / N_h$ vary between strata, while the $\binom{N_h}{n_h}$ remain equally likely within strata. If we shift from initial selections to new strata, the symmetries of equal f_h remain within the sets (portions of initial strata) in the new strata. If two or more sets had equal initial probabilities f_h, they may be considered a single set, within which substitutions of initial selections are allowed. The new sampling fraction f'_h may be applied to each set within the stratum to determine how many may be retained from it. Further units may be selected from sets not already complete.

We obtain the desired probabilities of selection for all units despite their changes of strata. However, the methods of the preceding two paragraphs will result in joint probabilities for pairs of selections which differ from those that would be given by new selections within the new strata. The proposed procedure tends to retain a stratification within the sets, which would be disregarded in the variance computations within the new strata. (The resulting overestimation of the variance would likely be negligible.) That computation assumes randomness within the new strata. If this must be had, first select the n'_h at random from all N'_h in the new stratum; then substitute initial for new selections *only within sets*. But this modification will sacrifice some initial selections.

A special case of the preceding is a single selection with $f_h = 1/N_h$ from each of many initial strata, with some N_h equal and others not. When shifted to new strata, the units with similar initial f_h may be considered as belonging to a single set, within which an initial selection may be used instead of a new selection with $1/N'_h$. This is equivalent to a procedure recently proposed (Cohen 1969); it is also a special application of Method III (see Section 8) to equal probabilities within sets. However, this procedure sacrifices initial selections present in sets not chosen by the new selection, which can be retained with Methods I or II of Section 5 or 6.

Suppose an initial selection made with equal probabilities f needs conversion to a single selection from each stratum, with variable probabilities P_j. Since the entire stratum is a single set with symmetry of initial probabilities, we need only apply a modified Procedure A.

(a) Selections with $P_j \geq f$ should be retained; they also receive additional selection probabilities $(P_j - f)$;
(b) Selections with $P_j < f$ should be retained only with applied probability P_j/f;
(c) If two or more selections remain after (a) and (b), choose one with equal probability;
(d) If no selections remain, choose one with P_j.

Step (d) will seldom be necessary if the initial selection with f is large enough. (This could also be the basis for a method for two or more selections from strata.)

8. JUSTIFICATION OF METHOD I

Consider first a simple two-step selection procedure. The sets $A_1, A_2, \ldots, A_k, \ldots, A_n$ comprise a stratum; the new probabilities of the units in set A_k sum to $P_k = \Sigma_{A_k} P_j$ and $\Sigma P_k = 1$. These probabilities could be obtained in two operations of selection; first a set from the stratum, then a unit within the set: $P_j = P_k \times P_j/P_k$. The second operation actually consists of accepting a previously selected unit with its initial probability p_j, then converting to P_j with Procedure A of Section 2. The Census Bureau (1967) switched from a 333 area sample based on 1950 data to a 357 area sample based on 1960 data and described its procedure as follows:

> For generality, consider a revised stratum in the 357 area design made up of parts from more than one stratum in the 333 area design. In principle, one of these parts can be selected with probability proportionate to 1960 size. If, then, the selected part does not include a PSU previously selected in the 333 area design, a selection can be made from PSUs in that part with probability proportionate to 1960 size. On the other hand, if the selected part contains a sample PSU from the 333 area design, selection from the part can be made by the Keyfitz Method (Keyfitz 1951) maximizing the chance of retaining the previously selected PSU.

This procedure, Method III, is simple and straightforward, but it can be far from optimal; it fails to utilize initial selections present in sets not chosen in the first operation. This disadvantage is moderate only when single sets predominate within the strata; i.e., when shifts between strata are modest.

A modification of the preceding procedure would use the initial probabilities $p_k = \Sigma_{A_k} p_j$ of the units in the sets. We consider these as measures of size; not as probabilities, because their sum $p. = \Sigma p_k$ over the new stratum is not generally unity. A selection in two operations can be symbolized with $p_k/p. \times p_j/p_k = p_j/p.$; first a set, then a unit within the set would select a unit with probabilities proportional to the measures p_j. The second operation p_j/p_k is made if the set does not contain an initial selection; but if an initial selection is present, it is accepted as an equivalent operation. Finally, the preliminary probability $p_j^* = p_j/p.$ is converted to the new probability P_j with Procedure A applied to the entire stratum. But if none of the sets contained an initial selection, an equivalent operation would be to select directly with P_j from the entire stratum. This indirect method is a modification of the preceding direct Method III; it serves only to introduce Method I. Method I retains more of the initial selections because it obtains the probability $p_k/p.$ for set A_k with an equivalent operation which manages to select a set containing an initial selection if any are present in the stratum. Method I must be now shown to be equivalent to the indirect method.

Let us begin with a stratum containing two sets only, with measures $p_1 + p_2 = p.$. (The presence of only one set merely means that $p_2 = 0$.) We shall obtain the odds p_1/p_2 for choosing A_1, as desired. Now note that p_1 and p_2 also represent for each set the probability that the set contains the initial selection from the initial stratum. Hence, if we assume independence between the two probabilities p_1 and p_2, we have the following probabilities for the presence of an initial selection:

$$
\begin{array}{ll}
(1-p_1)(1-p_2) & \text{in neither set,} \\
p_1(1-p_2) & \text{in set } A_1 \text{ only,} \\
(1-p_1)p_2 & \text{in set } A_2 \text{ only,} \\
p_1 p_2 & \text{in both sets.}
\end{array}
\qquad (8.1)
$$

Our strategy is to obtain the odds p_1/p_2 for A_1 over all four situations by achieving it separately for no selection, and jointly over the situations of one or two selections. In other words:

(a) We need to maintain the odds p_1/p_2 for choosing A_1 over the combination of four situations.
(b) It is convenient to have the odds p_1/p_2 when neither set contains an initial selection.
(c) When only one set contains a selection, we want that set selected with certainty.
(d) When both A_1 and A_2 contain initial selections, we choose odds that will overcome the imbalance introduced by (c) and reestablish (a).

We solve for y_1 in the expression that equates the desired odds p_1/p_2 to the compound probabilities of the occurrence of the situation and the selection of the set:

$$\frac{p_1(1-p_2)\cdot 1+(1-p_1)p_2\cdot 0+p_1p_2\cdot y_1}{p_1(1-p_2)\cdot 0+(1-p_1)p_2\cdot 1+p_1p_2\cdot(1-y_1)}=\frac{p_1}{p_2} \tag{8.2}$$

Hence

$$(1-p_2)+p_2y_1=(1-p_1)+p_1(1-y_1),$$

and

$$y_1=p_2/(p_1+p_2). \tag{8.3}$$

Thus A_1 should be chosen with the inverse odds p_2/p_1, establishing Rule 3 of Section 5 for initial selections in both sets comprising the stratum. Rule 2 for retaining the only set containing a selection is part of (8.2). For the disjoint situation of neither set containing an initial selection, the same probability $p_k/p.$ is needed, followed by the preliminary selection p_j/p_k and then conversion from $p_j/p.$ to P_j. But instead of these three steps we select with P_j directly, giving Rule 1.

Direct extension of the method seems too complex even for three or four sets. But one can arrange a larger number of sets into a series of dichotomies, then apply in sequence a procedure similar to that for two sets. A stratum containing four sets can be divided into group G_1 containing sets A_{11} and A_{12}, and group G_2 containing A_{21} and A_{22}. First choose G_1 with odds p_1/p_2, then the set A_{i1} with p_{i1}/p_{i2}. The overall probability of selecting unit j in set A_{ik} must be made $p_j/p. = p_i/(p_1+p_2)\cdot p_{ik}/(p_{i1}+p_{i2})\cdot p_j/p_{ik}$.

We design an equivalent procedure for preserving the odds p_1/p_2 and p_{i1}/p_{i2}. The situation of no initial selections is treated as a disjoint case with the simple Rule 1. A new problem is posed by the fact that a group G_i can contain either one or two selections. The probabilities for the two groups ($i = 1, 2$) are respectively: $p_i' = p_{i1}+p_{i2}-p_{i1}p_{i2}$. But we can deal more conveniently with the odds p_1/p_2 than with p_1'/p_2'. Hence we adjust accordingly the odds $y_1/(1-y_1)$ for choosing between the two groups when they both contain selections; we set y_1 to attain the desired odds p_1/p_2 jointly for double and single selections:

$$\frac{p_1'(1-p_2')+p_1'p_2'y_1}{(1-p_1')p_2'+p_1'p_2'(1-y_1)}=\frac{p_1}{p_2}=\frac{r_1p_1'}{r_2p_2'}, \tag{8.4}$$

where $r_1 = p_1 / p_1'$ and $r_2 = p_2 / p_2'$. Then

$$r_2(1-p_2')+r_2p_2'y_1=r_1(1-p_1')+r_1p_1'(1-y_1),$$

and

$$y_1=[p_2+(r_1-r_2)]/(p_1+p_2). \tag{8.5}$$

Thus $(r_1 - r_2)$ represents an adjustment of the probability $p_2 / (p_1 + p_2)$ for choosing Group G_1 over G_2 when each is composed of two sets, and each group contains a selection or two. If only one group contains a selection or two, we take it with certainty. Having established the joint odds p_1 / p_2 for the groups, we can now choose within it with certainty a single selection, or with odds p_{i2}/p_{i1} for A_{i1} if both sets have selections.

Thus Rule 4 is established for four sets. The presence of only three sets means that G_1 contains only set A_{11}, $p_1 = p_{11}$, $p_1' = p_1$, and $r_1 = 1$. The presence of only two sets means that $r_1 = r_2 = 1$, and Rule 4 becomes Rule 3.

When there are more than four sets in the stratum, and two or more selections are present, the formation of dichotomies is continued as needed, with an objective rule. Now any of the four probabilities p_{ij} ($i = 1, 2$ and $j = 1, 2$) can be considered as having been composed of two sets. In general we must write

$$p_i' = p_{i1}' + p_{i2}' - p_{i1}'p_{i2}', \tag{8.6}$$

where p_{ij}' must denote the probability that the group A_{ij} contains one or more selections. If A_{ij} contains a single set, $p_{ij}' = p_{ij}$; if it contains two sets, $p_{ij}' = p_{ij1} + p_{ij2} - p_{ij1}\, p_{ij2}$. But if A_{ijh} in turn contains two sets, $p_{ij}' = p_{ij1}' + p_{ij2}' - p_{ij1}'\, p_{ij2}'$. This rule can be continued with the understanding that $p_{i...s}' = p_{i...s1}' + p_{i...s2}' - p_{i...s1}'\, p_{i...s2}'$, and $p_{i...st}' = p_{i...st}$ when it contains a single set. When only one of the subgroups contains an initial selection, we take it. When both contain some, we choose $A_{i...1}$ with probability

$$y_{i...1}=[p_{i...2}+(r_{i...1}-r_{i...2})]/(p_{i...1}+p_{i...2}),$$

where

$$r_{i...s} = p_{i...s} / p_{i...s}'. \tag{8.7}$$

Note finally that the adjustment factor $(r_1 - r_2) > p_1$ would result in the impossible choice of $y_1 > 1$; similarly $(r_2 - r_1) > p_2$ would mean $y_1 < 0$. But it can be shown that these can occur, at the worst, only if p_1 or p_2 would exceed 3, and seldom then. Now the p_i values arise from probabilities, but here they are measures only and may exceed 1 if the new stratum contains sets that sum to more than one initial stratum. But it would be rare for p_i to exceed 3 and for other conditions to coincide as well. The p_i' values are probabilities, but the ratios $r_i = p_i / p_i'$ could hardly become sufficiently abnormal.

Benjamin Tepping has pointed out a problem of some theoretical annoyance. After units are moved between strata, the *joint* selection probabilities with the two-step procedure differ from the joint probabilities $P_{hi}P_{hj}$ with independence between strata h and h'. Hence variance estimates based on such independence will not be theoretically unbiased. However, any bias is bound to be very small because it depends on a third-degree difference: first the portions involved in changes; second, the small changes of joint probabilities; third, the correlation between changes in probabilities and sizes of differences between pairs of units. The latter is bound in practice to be small, and of the haphazard and compensating kind. It is bound to be small compared with the overestimate of variances due to "collapsing" strata, itself often considered negligible. Only a fanatic devotee of unbiasedness would introduce appropriate correcting factors into the estimates of the variance.

7

MULTIVARIATE AND MULTIPURPOSE STRATIFICATION

LESLIE KISH AND DALLAS W. ANDERSON

The number of strata $L = \Pi_i L_i$ (i = 1, 2, ..., k) depends on the number of stratifying variables k and the number of classes L_i from each; but strata are limited by cost restrictions on the number of selections, especially when these are primary selections in cluster sampling. Several stratifying variables often seem desirable both to reduce variances and to create domains. We quantify the gains of multivariate stratification in tables for 12 multivariate normal structures, with L= 12 and 36 and for three empirical studies. Even greater gains are shown for multipurpose designs, which are most common in surveys.

Key words: Stratification; multivariate stratification; multipurpose; sample design; survey sampling.

1. INTRODUCTION

We aim here to bring closer to the needs of survey practice the theoretical results of stratified sampling. These have been concerned, alas, mostly with the effects of a single stratifying variable acting on a single study variable. However, the multivariate situation is more common, with several variables both available and desirable for stratification. Further, most surveys are also multipurpose: several or many variables, and many more statistics, compete for attention as the principal objectives of survey efforts. It is poor practice to try to tailor sample designs to the simple bivariate theory of one stratifier for one study variable.

It is usually better to utilize several variables rather than just one; this often would be true even if the best one were known. This more or less common wisdom appears in print in several places, but our aim is to explore this problem systematically and quantitatively. We shall also relate it as a matter of course to the needs of

---

Reprinted with permission from the *Journal of the American Statistical Association* (1978), 73, 24–34.

Research was supported by Joint Statistical Agreement No. 75–7 with the U.S. Bureau of the Census. The authors are grateful for suggestions from Professor Graham G. Kalton, the Editor and a referee.

multipurpose designs, and we present tables that should be directly useful for designing samples.

There have been several papers on multipurpose allocation for fixed strata, but that is not our concern here. Rather, we are concerned mainly with the related choices of the numbers of stratifying variables and of the numbers of class intervals from each variable. The two are related through common practical restrictions on the total number of strata. If L_i is the number of strata from the ith variable, the total number of strata is around $L = \Pi_i L_i$. For a fixed total, L, using several variables reduces the number L_i of classes available from each, and using one or two large L_i also reduces the L_i from other variables.

On the one hand, the number of strata actually used may be less than $\Pi_i L_i$ because some of the cells can be small or empty—especially in the corners. On the other hand, some large cells may be split further. Introducing asymmetries into the cells is used in practice to increase further the utility of multivariate stratification. But these techniques must be rather specific, and outside the scope of this brief yet general treatment. In any case, it must be clear that the numbers of strata, hence of stratifying variables and of their classes, are limited by cost restrictions on the numbers of sample selections. These restrictions become particularly relevant in selections of primary clusters, when stratification is especially important, as discussed in Section 2. We also need to note two other aspects of stratification: the choice of class boundaries and the allocation of sample units to strata; this is done briefly in Section 3.

The multivariate and multipurpose model and restrictions are described in Section 4. Section 5 presents our main results: the relative efficacies of multivariate stratification. These results and tables also have similar relevance in two other contexts: in matching for control in data analysis and in grouping data for regression analysis (Anderson, Kish, and Cornell 1980).

Section 6 compares the efficacies of multivariate stratification and of stratification with principal components. In Section 7 the advantages of multivariate stratification are presented with sets of empirical data.

Our discussions are confined to variances for estimates of means, but samples are also used for other estimates: differences between subclass means, regression coefficients, and other analytical statistics. They deserve separate treatments, but we cannot burden ourselves or the reader with them here.

Proportionate stratification of elements reduces the variance of the mean from σ^2/n to $\Sigma W_h \sigma_h^2/n$, where W_h and σ_h^2 are the weights and element variances of the strata (disregarding finite population corrections). From the components of the total element variance, $\sigma^2 = \Sigma W_h \sigma_h^2 + \Sigma W_h (\bar{Y}_h - \bar{Y})^2$, the relative gain is $\Sigma W_h (\bar{Y}_h - \bar{Y})^2 / \sigma^2$. With a linear regression model, for large L the relative gain equals R_1^2, where R_1 is the correlation coefficient between the survey and stratifier variables (Cochran 1963; Sethi 1963). This has been generalized to multivariate regression with R_p^2, the square of the multiple regression coefficient (Anderson, Kish, and Cornell 1980).

With optimal allocation, the relative gains are further increased with $\Sigma W_h (\sigma_h - \bar{\sigma})^2$, the variance between standard deviations (Cochran 1963; Kish 1965a). This term could complicate the presentation. But our computations use approximately optimal stratum boundaries, and with those boundaries, the proportionate gains are close to optimal gains, as discussed in Section 3.

2. STRATIFIED SELECTION OF CLUSTERS

Efficient stratification has much more effect on the selection of clusters than of elements, for several reasons. These are noted briefly here, because the distinct problems of stratification of clusters are often neglected in basic literature and in naïve practice.

First, the proportion of the variance gained by the partition into strata is often considerably greater for larger clusters than for elements. In the ratio expressing relative gains the numerator expresses the variance between stratum means as $\Sigma W_h(\bar{Y}_h - \bar{Y})^2$, where W_h and $\bar{Y}_h$ are the weights and means of the strata. The denominator expresses the variance between sampling units; this is σ^2 for element sampling, but the unit variance is reduced to σ^2/b for clusters of b random elements. For correlated elements, this may be expressed as $(\sigma^2/b)[1 + \delta(b-1)]$, where δ is the intraclass correlation; in practical designs the denominator is much less than σ^2 when b is large (Hansen, Hurwitz, and Madow 1953, Vol. I, Sec. 7.4). We may also view this effect within the framework of correlation between stratifying and study variables and note that correlations are higher for means of grouped observations ("ecological correlation" of group means) than for elements.

Second, considerably more data are usually available for larger sampling units than for elements. Available here must refer to identified units on the lists (frames) from which selections can be readily made. For administrative units like cities and counties and for institutions, many stratifying variables are available. For elements, such as persons or families, few variables are available in forms useful for stratified selection. Thus multivariate stratification is more feasible for selecting clusters.

Third, multipurpose design becomes more necessary for clusters, especially for the primary units of multistage samples. These units are often used for many surveys covering diverse subjects over several years. For these units the conflict between multiple purposes and cost restrictions on numbers of units becomes especially severe. Desire for several dimensions of stratification imposed on severely limited numbers of units often leads to only two selections, or to single selections, per stratum, and even to more strata than units with "controlled selection" methods (Goodman and Kish 1950; Kish 1965a; Hess, Riedel, and Fitzpatrick 1975).

Three other characteristics of stratification for clusters merit attention here. First, the means of large clusters tend toward normality even for common variables, such as dichotomies, which are not normal for populations of elements. This is among the justifications for our use of multivariate normal distributions for much of the numerical results. Second, cluster samples also tend to use proportionate samples from roughly equal-sized strata; also to select equal-sized subsample clusters from unequal-sized primary units. This consideration influenced some of our empirical work and imparts wide utility to its simple design. Third, partition of the sample into domains of analysis has special implications for cluster samples (Kish and Frankel 1974). When strata are also domains, each contains fewer units, and further stratification with other variables may seem especially important.

3. CHOICES OF VARIABLES, CLASSES, AND BOUNDARIES

The basic principles of stratification are well known for univariate situations. We concentrate on those aspects and modifications that have special relevance to multivariate and multipurpose situations.

First, we note that stratification is often used for creating distinct domains for analysis. This basic aspect may be overlooked because theoretical treatments tend naturally to emphasize the mathematically tractable use of stratification for reducing variances for the entire sample. The two aspects are not entirely unrelated, because interest in differences between statistics for domains should be related to gains from stratification; however, domains are often determined chiefly by exogenous considerations and fail to coincide with the best choices for the strata. In particular, several or many variables may be used for creating domains, well beyond the needs of reducing variances for the entire sample. Furthermore, the number of domains from any variable needed for analysis may also be larger than the numbers of classes needed for reducing overall variances. This conflict may be especially great for clustered selections, where both the availability and the effects of stratifying variables are greater. In this framework, the needs for multivariate stratification appear in sharper light.

The choices of boundaries for strata are often limited by needs for creating domains which have substantive meanings to the analysts of the data. They are also limited by the nature of the available data, especially when these are categorical data. Nevertheless, one can begin with the knowledge that reducing the within-stratum variances of units, $\Sigma W_h \sigma_h^2$, reduces the variances of means.

For boundaries of continuous variables, Ekman's rule for creating strata with equal values of $W_h \sigma_h$ is known to perform well (Cochran 1961; Cochran 1963). A practical procedure for obtaining these boundaries is the cumulative $\sqrt{f}$ rule, and this has been shown to perform practically as well as the theoretically optimal boundaries, both for small and large numbers of strata. Note in Table 1 that the proportions of the gains obtainable with optimal boundaries are above 0.80 for three strata, about 0.90 for four, about 0.95 for six, and about 0.98 for ten. In Table 2 we see that the cumulative $\sqrt{f}$ rule is close to both proportionate and Neyman optimals in boundaries, and even closer in areas. Its gains have been shown to be within 1/1000 of the gains from both optimals (Anderson, Kish, and Cornell 1976).

Here we are chiefly concerned with ratios for comparing gains from one versus two stratifying variables. Those ratios are bound to be even less sensitive to changes of boundaries than the proportions of Table 1. The ratios may be slightly biased in favor of single stratifiers, because the boundaries are (nearly) optimal for them. The gains from multivariate stratification may be somewhat increased with

TABLE 1. Proportions of the Maximal Reduction of the Variance Using Stratification Obtainable with 2 to 10 Strata and (a) Optimal Boundaries or (b) Strata of Equal Areas[a]

		Number of strata								
Distribution	Boundaries	2	3	4	5	6	7	8	9	10
Uniform	a & b	0.750	0.889	0.938	0.960	0.972	0.980	0.984	0.988	0.990
Normal	a	0.637	0.810	0.883	0.920	0.942	0.956	0.965	0.972	0.977
	b	0.637	0.793	0.860	0.897	0.920	0.935	0.945	0.952	0.960
Exponential	a	0.648	0.820	0.891	0.927	0.948	0.961	0.969	0.975	0.980
	b	0.481	0.650	0.735	0.787	0.823	0.847	0.875	0.881	0.893

[a]Data from Aigner, Goldberger, and Kalton (1975).

TABLE 2. Areas and Deviates by Type of Stratum Boundaries for 10 Strata for the Standard Normal Distribution[a]

Type	Areas				Deviates			
Equal width	0.60	0.70	0.80	0.90	0.25	0.52	0.84	1.28
Cumulative $\sqrt{f}$	0.64	0.77	0.88	0.96	0.36	0.74	1.19	1.81
Optimal, Neyman ($\rho = 0.95$)	0.65	0.79	0.90	0.97	0.39	0.81	1.29	1.91
Optimal, proportionate	0.67	0.81	0.92	0.98	0.43	0.88	1.38	2.03

[a]Area = 0.50; deviation = 0.0.

boundaries specifically optimal for this case, especially for correlated stratifiers; but we did not explore that refinement.

The advantages of multivariate stratification can be appreciated even for a single study variable. For any stratifier, the gains in reducing the variance within strata show rapidly diminishing returns with few strata, because little variance is left within the strata, as seen in the asymptotic approaches in Table 1. Those proportions refer to the variances of the stratifying variable, and they leave unaffected the residual variance of the study variables. When R is the correlation between stratifying and survey variables, the gains approach not 1 but only R^2; the extra gains can become a minute portion of the residual variance $(1 - R^2)$. It seems reasonable to expect that beyond some limited number of strata on the first stratifier, other stratifiers can yield larger gains. Our empirical results chiefly compare one-way to two-way stratification because it would be cumbersome to tabulate more. However, advantages do not cease with two stratifiers. After a few strata (three to five) from each of the two stratifiers, a third should be investigated, then a fourth, etc.

These advantages should exist even where the first stratifier turns out to be the best; however, often the relative powers of several potential stratifiers may not be clear at all, except for some periodic studies with the same variables. The relationships of stratifiers to new study variables can only be guessed; furthermore, they can change over time. Safety may be sought in the use of several stratifiers.

Finally and chiefly, the advantages of several stratifiers are much greater for multipurpose surveys. For several study variables, each may have a different best (first) stratifying variable. Then compromises between the purposes are much more likely to be satisfied with several stratifiers, as we shall see.

The gains from stratifiers depend first on their relations to the survey variables and second on their own intercorrelations. These are general statements that will be quantified with several examples, but exact mathematical formulations depend on further restrictions. For linear correlations, the variance is reduced in the proportion $[K_\alpha R^2 + (1 - R^2)]$ (Sethi 1963). But K_α depends both on the numbers of strata and on the distributions involved; its values are shown for three distributions in Table 1. Cochran (1963, Sec. 5A.7) uses a rough lower limit of $K_\alpha = L^{-2}$, where L is the number of strata for univariate linear correlations. Results in his Table 5A.13 coincide with those for the uniform distribution in our Table 1; but his 0.212 for $R = 0.90$ and $L = 6$, for example, contrasts with 0.237 or 0.255 for (a) or (b) for the normal distribution in Table 1. For $L = 3$ the contrasts are our (a) 0.344 and (b) 0.358 against his limit of 0.280. The contrasts are greater for the exponential distribution. For lower values of R^2, the values $(1 - R^2)$ predominate.

4. RESTRICTIONS ON THE MULTIVARIATE NORMAL MODEL

The effects of two stratifiers x_1 and x_2 on a single survey variable y can be represented, in the simplifying context of trivariate correlations, with three coefficients: ρ_{1y}, ρ_{2y}, and ρ_{12}. For general utility, each should have several values, even within the context of one assumed distributional structure. With two survey variables, there are five correlations to vary. Not only do the computational tasks become great, but the presentation of results also takes several tables, even though we condensed their essence from a greater volume.

It is difficult to present results that will have general utility for many situations, in the presence of many parameters, and each with a wide range of possible values. Special choices of parameters were necessary to conserve our resources. We hope that this condensed presentation is useful and convincing.

(a) Multivariate normal distributions may be the most natural and useful standard, especially for considering stratification of clusters.

(b) For stratum boundaries, we chose optimal stratification for proportionate sampling, and we approximated with the cumulative $\sqrt{f}$ rule.

(c) For total number of strata, we chose $L = 36$ and $L = 12$ to allow for reasonable comparisons that will be useful for realistic problems of design.

(d) The multipurpose situation was reduced to two purposes because more would become too unwieldy. However, the scale of relative weights can be used to roughly approximate the results from more purposes. Furthermore the extreme weights of one and zero in the tables give results for single purposes.

(e) To measure the relative efficacy of two-way stratification on a survey variable y_g, we use V_{2g} / V_{0g}, the ratio of the variance, V_{2g}, with two stratifiers utilized with a specified combination $L_1 \times L_2 = L$ of numbers of strata, over the unstratified variance, V_{0g}. The ratio V_{1g} / V_{0g} expresses the similarly standardized variance using a specified one-way stratification $L \times 1$ with the same total number of strata. For the gth survey variable, the relative efficacy of two stratifiers with $L_1 \times L_2 = L$ strata over the single stratifier with $L \times 1$ would be $(V_{2g} / V_{0g}) / (V_{1g} / V_{0g}) = V_{2g} / V_{1g}$. This is a special case, with $I_g = 1$, of the more general multipurpose assessment of efficacy, in which we use I_g to denote the relative importance of the gth survey variable, with $\Sigma I_g = 1$. The weighted joint-relative efficacy is then measured for two stratifiers over two survey variables, a and b in our computations, and for p stratifying variables over all survey variables ($g = 1, 2, \ldots$), with

$$E_2 = \frac{I_a V_{2a} / V_{0a} + I_b V_{2b} / V_{0b}}{I_a V_{1a} / V_{0a} + I_b V_{1b} / V_{0b}}$$

and

$$E_p = (\sum_g I_g V_{pg} / V_{0g}) / (\sum_g I_g V_{1g} / V_{0g}). \qquad (4.1)$$

These efficacies measure the importance-weighted standardized variances of multiple stratification over the weighted variances that the best stratifier would give for each survey variable.

(f) Even for two purposes and two stratifiers we would face a complex problem of presentation with five correlation parameters. For the sake of comprehension, we reduced these to three parameters by assuming some symmetry: $\rho_{1a} = \rho_{2b}$ is denoted by ρ_{1y} in the tables, and $\rho_{1b} = \rho_{2a}$ is denoted by ρ_{2y}. We assume that $\rho_{1y} \geq \rho_{2y}$, i.e., x_1 is better than x_2 for y_a, while x_2 is better than x_1 for y_b. This structure explores the effects of conflicts between stratifiers with differing effects on the two survey variables. The case when x_1 is better for both y_a and y_b can be viewed as the unipurpose case with $I_a = 1$ on the bottom rows of the tables.

(g) For values of ρ_{1y}, we chose values of 0.95, 0.75, 0.50, and 0.25; lower values would show weak gains and weaker comparisons. For ρ_{2y}, we considered the same values, though most of our comparisons assume the relationship that $\rho_{1y} > \rho_{2y}$. Note that we have six pairs of values satisfying that relationship. Values of $\rho_{1y} = \rho_{2y}$ have optimal numbers of splits near $\sqrt{L}$ for each stratifier (Anderson 1976).

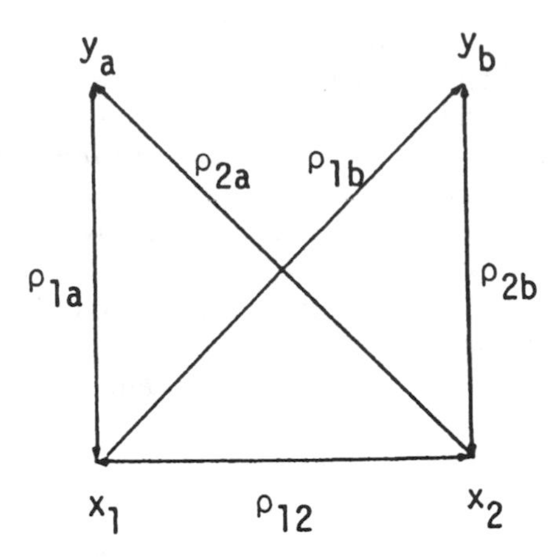

Figure 1. Correlation structure for multipurpose investigation.

(h) For correlations ρ_{12} between stratifiers, we chose the values 0.90, 0.60, and 0.30. The requirement that $\rho_{1y}^2 + \rho_{2y}^2 + \rho_{12}^2 < 1 + 2\rho_{12}\rho_{1y}\rho_{2y}$ is necessary, however, for the correlation matrix to be positive definite, because all three $|\rho| < 1$, all three $\sigma^2 > 0$, and $(\rho_{12} - \rho_{1y}\rho_{2y})^2 / (1 - \rho_{1y}^2)(1 - \rho_{2y}^2) = \rho_{12.y}^2 < 1$ (Yule and Kendall 1937, Sec. 12.25). Thus some high (low) values of ρ_{12} are not possible for low (high) values of ρ_{1y} and ρ_{2y}. For the case when $\rho_{1y} > \rho_{2y}$, there are $6 \times 3 = 18$ combinations. Of these, five were ruled out by the requirement of positive definiteness. From the remaining 13 possible combinations, we omit the least interesting, $\rho_{1y} = 0.50$, $\rho_{2y} = 0.25$, $\rho_{12} = 0.60$, to permit a more concise presentation of the tabled results.

Thomsen (1977) shows, for the single-purpose case, similar advantages of bivariate stratification for the normal, rectangular, and exponential distributions. His values are based on mathematical models with asymptotic approximations, which our computations avoid. He used only $\rho_{1y} = 0.85$ and $\rho_{2y} = 0.50$, with $\rho_{12} = 0.0$ in three tables and $\rho_{12} = 0.20$ in one. But correlations between stratifiers are probably not low in practice, and they do affect the results.

5. MULTIVARIATE AND MULTIPURPOSE RESULTS FOR NORMAL DISTRIBUTIONS

For a simple introduction, we deal in Table 3 with the special cases when $\rho_{1y} = \rho_{2y}$; this means, with our earlier restrictions, that all four correlations of the two stratifiers with the two survey variables are equal. We note gains of two-way over

TABLE 3. Ratios of Variances for Bivariate/Univariate Stratifications[a]

$\rho_{1y} = \rho_{2y}$	ρ_{12}	6×2	4×3
0.25	0.30	0.98	0.98
0.50	0.30	0.90	0.88
0.75	0.30	0.62	0.54
0.25	0.60	0.99	0.99
0.50	0.60	0.96	0.95
0.75	0.60	0.86	0.83
0.25	0.90	1.00	1.00
0.50	0.90	1.01	1.01
0.75	0.90	1.02	1.04
0.95	0.90	1.13	1.27

[a]Based on trivariate normal distributions. Denominator is for 12 strata: 12×1. Numerator is for bivariate $L_1 \times L_2 = L_2 \times L_1 = 12$ strata.

one-way stratification in much of the table, especially when $\rho_{1y} = \rho_{2y}$ are high and the correlations ρ_{12} between stratifiers are not. Because of the symmetry, the gains are similar for both survey variables. These values for $L_1 \times L_2 = 12$ may be compared to those of Table 5 where $\rho_{1y} > \rho_{2y}$. Values for 36 strata to compare with Table 4 were not computed; those would favor more strongly bivariate stratification.

Comparisons become more interesting and complicated when we look at Table 4 where $\rho_{1y} > \rho_{2y}$.

(a) First we may inspect the bottom rows, which represent, with $I_a = 1$, situations where there is one survey variable (or where both survey variables have similar correlations), and we know which stratifier is better. Even in these situations, bivariate stratifications with 18×2, 12×3, or 9×4 are usually better than one-way stratifications 36×1, which are the denominators of the ratios. The ratios vary around 0.90 but become about 0.75 in Table 4B and Table 4D.

(b) We may note much greater gains in the same columns (18×2, 12×3, and 9×4) in the top rows that represent cases where the second variable y_b gets all the weight, $I_b = 1$ and $I_a = 0$. Further, these top lines can also represent cases when $I_a = 1$, but $\rho_{2y} > \rho_{1y}$. Thus they show how well those compromise stratifications can reduce bad losses due to either one of two mistakes: choosing either the wrong stratifier or the wrong weights for survey variables.

(c) Large gains for two-way stratification show up as soon as one leaves the bottom line that represents extreme confidence about both ρ's and I_g's. With decreasing confidence, the optima shift rapidly close to the center, to the allocations 9×4 and 6×6. At $I_a = I_b = 4/8$ the optimal maximum is

TABLE 4. Ratios of Variances for Bivariate/Univariate Stratifications[a]

I_a	2×18	3×12	4×9	6×6	9×4	12×3	18×2	I_a	2×18	3×12	4×9	6×6	9×4	12×3	18×2
A.	$\rho_{1y} = 0.95, \rho_{2y} = 0.75, \rho_{12} = 0.90$							B.	$\rho_{1y} = 0.95, \rho_{2y} = 0.75, \rho_{12} = 0.60$						
0/8	0.20	0.20	0.22	0.30	0.44	0.56	0.73	0/8	0.18	0.17	0.17	0.20	0.27	0.35	0.53
1/8	0.30	0.27	0.28	0.33	0.46	0.57	0.73	1/8	0.24	0.21	0.20	0.22	0.28	0.36	0.54
2/8	0.41	0.36	0.34	0.37	0.48	0.58	0.74	2/8	0.33	0.26	0.24	0.25	0.30	0.38	0.55
3/8	0.56	0.47	0.43	0.43	0.50	0.60	0.75	3/8	0.43	0.33	0.29	0.28	0.33	0.40	0.56
4/8	0.76	0.62	0.54	0.49	0.54	0.62	0.76	4/8	0.57	0.42	0.36	0.33	0.36	0.42	0.57
5/8	1.02	0.82	0.69	0.59	0.59	0.65	0.77	5/8	0.77	0.54	0.45	0.39	0.40	0.45	0.60
6/8	1.42	1.11	0.92	0.72	0.66	0.69	0.79	6/8	1.05	0.72	0.58	0.48	0.46	0.50	0.63
7/8	2.04	1.58	1.27	0.93	0.78	0.76	0.82	7/8	1.50	1.01	0.79	0.62	0.56	0.58	0.68
8/8	3.20	2.45	1.93	1.33	0.99	0.89	0.89	8/8	2.33	1.54	1.18	0.88	0.75	0.73	0.77
C.	$\rho_{1y} = 0.95, \rho_{2y} = 0.50, \rho_{12} = 0.60$							D.	$\rho_{1y} = 0.95, \rho_{2y} = 0.50, \rho_{12} = 0.30$						
0/8	0.13	0.14	0.16	0.20	0.27	0.35	0.53	0/8	0.10	0.09	0.10	0.13	0.18	0.26	0.42
1/8	0.21	0.19	0.19	0.22	0.29	0.37	0.54	1/8	0.16	0.13	0.12	0.14	0.20	0.26	0.43
2/8	0.30	0.25	0.24	0.25	0.31	0.38	0.55	2/8	0.23	0.17	0.16	0.16	0.21	0.28	0.44
3/8	0.42	0.33	0.30	0.29	0.34	0.41	0.57	3/8	0.33	0.23	0.20	0.19	0.23	0.29	0.45
4/8	0.59	0.44	0.38	0.35	0.38	0.44	0.59	4/8	0.46	0.31	0.25	0.22	0.25	0.31	0.46
5/8	0.83	0.60	0.50	0.43	0.44	0.49	0.62	5/8	0.66	0.43	0.33	0.28	0.29	0.34	0.48
6/8	1.23	0.86	0.69	0.56	0.53	0.56	0.67	6/8	0.98	0.62	0.47	0.36	0.35	0.39	0.51
7/8	2.00	1.35	1.06	0.81	0.72	0.71	0.76	7/8	1.59	0.98	0.72	0.53	0.46	0.47	0.57
8/8	4.00	2.66	2.03	1.48	1.20	1.09	1.02	8/8	3.20	1.93	1.39	0.96	0.76	0.71	0.73
E.	$\rho_{1y} = 0.95, \rho_{2y} = 0.25, \rho_{12} = 0.30$							F.	$\rho_{1y} = 0.75, \rho_{2y} = 0.50, \rho_{12} = 0.90$						
0/8	0.11	0.12	0.13	0.16	0.22	0.29	0.45	0/8	0.53	0.50	0.49	0.52	0.61	0.69	0.81
1/8	0.17	0.16	0.16	0.18	0.23	0.30	0.46	1/8	0.59	0.55	0.54	0.55	0.63	0.70	0.81
2/8	0.25	0.21	0.20	0.21	0.25	0.31	0.47	2/8	0.66	0.61	0.58	0.59	0.65	0.72	0.82
3/8	0.36	0.27	0.25	0.24	0.28	0.34	0.48	3/8	0.75	0.68	0.64	0.62	0.67	0.73	0.83
4/8	0.51	0.37	0.31	0.29	0.31	0.37	0.51	4/8	0.84	0.75	0.70	0.66	0.70	0.75	0.84
5/8	0.73	0.51	0.42	0.36	0.37	0.41	0.54	5/8	0.95	0.83	0.76	0.71	0.73	0.77	0.85
6/8	1.10	0.74	0.59	0.48	0.46	0.49	0.59	6/8	1.07	0.93	0.84	0.76	0.76	0.79	0.87
7/8	1.86	1.22	0.95	0.73	0.65	0.64	0.70	7/8	1.21	1.05	0.93	0.82	0.80	0.82	0.88
8/8	4.22	2.70	2.04	1.50	1.23	1.13	1.05	8/8	1.38	1.18	1.04	0.90	0.84	0.85	0.90

TABLE 4, *continued*

G. $\rho_{1y} = 0.75, \rho_{2y} = 0.50, \rho_{12} = 0.60$							
0/8	0.59	0.59	0.60	0.62	0.65	0.69	0.78
1/8	0.64	0.64	0.64	0.65	0.68	0.72	0.79
2/8	0.71	0.69	0.68	0.69	0.71	0.74	0.81
3/8	0.78	0.74	0.73	0.73	0.75	0.77	0.83
4/8	0.86	0.81	0.79	0.78	0.79	0.81	0.86
5/8	0.95	0.88	0.85	0.83	0.83	0.85	0.89
6/8	1.06	0.97	0.93	0.89	0.89	0.89	0.92
7/8	1.18	1.07	1.01	0.97	0.95	0.95	0.96
8/8	1.33	1.18	1.11	1.05	1.02	1.01	1.00

H. $\rho_{1y} = 0.75, \rho_{2y} = 0.50, \rho_{12} = 0.30$							
0/8	0.52	0.51	0.50	0.51	0.54	0.58	0.68
1/8	0.57	0.54	0.54	0.54	0.57	0.60	0.69
2/8	0.62	0.59	0.57	0.57	0.60	0.63	0.71
3/8	0.69	0.63	0.61	0.61	0.63	0.66	0.73
4/8	0.76	0.69	0.66	0.65	0.66	0.69	0.76
5/8	0.83	0.75	0.71	0.69	0.70	0.72	0.78
6/8	0.93	0.82	0.77	0.75	0.75	0.76	0.81
7/8	1.03	0.90	0.85	0.81	0.80	0.81	0.85
8/8	1.16	1.00	0.93	0.88	0.86	0.86	0.89

I. $\rho_{1y} = 0.75, \rho_{2y} = 0.25, \rho_{12} = 0.60$							
0/8	0.44	0.43	0.43	0.46	0.50	0.56	0.68
1/8	0.50	0.48	0.47	0.49	0.53	0.58	0.70
2/8	0.57	0.53	0.52	0.53	0.56	0.61	0.71
3/8	0.66	0.60	0.57	0.57	0.60	0.64	0.74
4/8	0.76	0.67	0.64	0.62	0.64	0.67	0.76
5/8	0.88	0.77	0.71	0.68	0.69	0.72	0.79
6/8	1.03	0.88	0.81	0.76	0.75	0.77	0.83
7/8	1.21	1.02	0.92	0.85	0.83	0.84	0.87
8/8	1.45	1.19	1.07	0.97	0.93	0.92	0.93

J. $\rho_{1y} = 0.75, \rho_{2y} = 0.25, \rho_{12} = 0.30$							
0/8	0.47	0.48	0.48	0.50	0.53	0.58	0.67
1/8	0.53	0.52	0.52	0.54	0.57	0.60	0.69
2/8	0.60	0.58	0.57	0.58	0.60	0.63	0.72
3/8	0.68	0.64	0.63	0.62	0.64	0.67	0.75
4/8	0.78	0.72	0.69	0.68	0.69	0.72	0.78
5/8	0.89	0.81	0.77	0.75	0.75	0.77	0.82
6/8	1.03	0.92	0.87	0.83	0.82	0.83	0.87
7/8	1.21	1.05	0.99	0.94	0.92	0.91	0.93
8/8	1.44	1.23	1.14	1.07	1.03	1.02	1.01

K. $\rho_{1y} = 0.50, \rho_{2y} = 0.25, \rho_{12} = 0.90$							
0/8	0.73	0.70	0.68	0.69	0.74	0.79	0.87
1/8	0.77	0.73	0.71	0.71	0.75	0.80	0.88
2/8	0.81	0.76	0.73	0.73	0.76	0.81	0.88
3/8	0.85	0.79	0.76	0.75	0.78	0.82	0.89
4/8	0.89	0.83	0.79	0.77	0.79	0.83	0.89
5/8	0.94	0.87	0.82	0.79	0.81	0.84	0.90
6/8	0.98	0.90	0.85	0.81	0.82	0.85	0.90
7/8	1.03	0.95	0.89	0.84	0.84	0.86	0.91
8/8	1.09	0.99	0.92	0.86	0.85	0.87	0.92

L. $\rho_{1y} = 0.50, \rho_{2y} = 0.25, \rho_{12} = 0.30$							
0/8	0.79	0.79	0.80	0.80	0.82	0.83	0.87
1/8	0.82	0.82	0.82	0.82	0.83	0.85	0.88
2/8	0.86	0.85	0.84	0.84	0.85	0.87	0.90
3/8	0.89	0.87	0.87	0.87	0.87	0.88	0.91
4/8	0.92	0.90	0.90	0.89	0.90	0.90	0.92
5/8	0.96	0.93	0.92	0.92	0.92	0.92	0.94
6/8	1.00	0.97	0.95	0.94	0.94	0.95	0.96
7/8	1.04	1.00	0.99	0.97	0.97	0.97	0.97
8/8	1.09	1.04	1.02	1.00	0.99	0.99	0.99

[a]Based on trivariate normal distributions. Denominator is for 36 strata for first stratifier: 36×1. Numerator is for bivariate $L_1 \times L_2 = 36$ strata. $I_a = 1 - I_b$, the relative weight for first survey variable.

TABLE 5. Ratios of Variances for Bivariate/Univariate Stratifications[a]

I_a	2×6	3×4	4×3	6×2	I_a	2×6	3×4	4×3	6×2	I_a	2×6	3×4	4×3	6×2
A.	0.95, 0.75, 0.90				*B.*	0.95, 0.75, 0.60				*C.*	0.95, 0.50, 0.60			
0/8	0.32	0.45	0.59	0.75	0/8	0.26	0.31	0.39	0.55	0/8	0.20	0.27	0.35	0.53
1/8	0.41	0.51	0.63	0.77	1/8	0.33	0.35	0.42	0.57	1/8	0.27	0.31	0.38	0.55
2/8	0.52	0.59	0.68	0.79	2/8	0.41	0.41	0.45	0.59	2/8	0.36	0.37	0.42	0.57
3/8	0.67	0.70	0.74	0.82	3/8	0.51	0.47	0.50	0.61	3/8	0.47	0.44	0.47	0.60
4/8	0.85	0.83	0.83	0.85	4/8	0.65	0.56	0.56	0.65	4/8	0.63	0.54	0.54	0.63
5/8	1.11	1.01	0.94	0.90	5/8	0.83	0.67	0.64	0.69	5/8	0.87	0.69	0.64	0.69
6/8	1.47	1.26	1.10	0.97	6/8	1.09	0.84	0.75	0.75	6/8	1.24	0.92	0.80	0.78
7/8	2.02	1.65	1.35	1.07	7/8	1.49	1.09	0.93	0.85	7/8	1.92	1.34	1.09	0.93
8/8	2.98	2.33	1.78	1.25	8/8	2.19	1.53	1.24	1.02	8/8	3.56	2.36	1.79	1.32
D.	0.95, 0.50, 0.30				*E.*	0.95, 0.25, 0.30				*F.*	0.75, 0.50, 0.90			
0/8	0.15	0.20	0.27	0.43	0/8	0.16	0.22	0.29	0.45	0/8	0.58	0.62	0.70	0.81
1/8	0.21	0.23	0.29	0.44	1/8	0.22	0.25	0.31	0.46	1/8	0.64	0.67	0.73	0.83
2/8	0.28	0.27	0.32	0.46	2/8	0.30	0.30	0.34	0.48	2/8	0.71	0.72	0.76	0.84
3/8	0.38	0.33	0.35	0.48	3/8	0.40	0.36	0.39	0.51	3/8	0.79	0.77	0.79	0.86
4/8	0.51	0.40	0.40	0.51	4/8	0.54	0.45	0.45	0.54	4/8	0.87	0.83	0.83	0.87
5/8	0.70	0.51	0.48	0.55	5/8	0.76	0.58	0.54	0.60	5/8	0.97	0.90	0.88	0.89
6/8	1.00	0.69	0.60	0.61	6/8	1.11	0.79	0.69	0.68	6/8	1.09	0.98	0.93	0.92
7/8	1.55	1.01	0.81	0.73	7/8	1.79	1.20	0.98	0.85	7/8	1.22	1.07	0.99	0.94
8/8	2.88	1.78	1.33	1.01	8/8	3.73	2.38	1.80	1.33	8/8	1.37	1.18	1.05	0.97
G.	0.75, 0.50, 0.60				*H.*	0.75, 0.50, 0.30				*I.*	0.75, 0.25, 0.60			
0/8	0.62	0.65	0.69	0.78	0/8	0.55	0.56	0.60	0.68	0/8	0.47	0.51	0.56	0.68
1/8	0.67	0.69	0.73	0.80	1/8	0.60	0.60	0.62	0.70	1/8	0.53	0.55	0.59	0.70
2/8	0.73	0.74	0.76	0.82	2/8	0.65	0.64	0.65	0.72	2/8	0.60	0.60	0.63	0.72
3/8	0.80	0.79	0.80	0.85	3/8	0.71	0.68	0.69	0.75	3/8	0.68	0.66	0.67	0.75
4/8	0.88	0.85	0.85	0.88	4/8	0.77	0.73	0.73	0.77	4/8	0.78	0.72	0.72	0.78
5/8	0.96	0.91	0.90	0.91	5/8	0.85	0.78	0.77	0.80	5/8	0.89	0.80	0.78	0.82
6/8	1.06	0.98	0.96	0.95	6/8	0.94	0.85	0.82	0.84	6/8	1.03	0.90	0.86	0.86
7/8	1.18	1.07	1.02	0.99	7/8	1.04	0.92	0.88	0.88	7/8	1.21	1.02	0.95	0.92
8/8	1.31	1.17	1.10	1.04	8/8	1.15	1.01	0.95	0.93	8/8	1.43	1.18	1.07	0.99
J.	0.75, 0.25, 0.30				*K.*	0.50, 0.25, 0.90				*L.*	0.50, 0.25, 0.30			
0/8	0.50	0.53	0.58	0.67	0/8	0.76	0.77	0.80	0.87	0/8	0.81	0.82	0.83	0.87
1/8	0.56	0.58	0.61	0.70	1/8	0.79	0.79	0.82	0.88	1/8	0.83	0.84	0.85	0.88
2/8	0.62	0.63	0.65	0.72	2/8	0.83	0.82	0.84	0.89	2/8	0.86	0.86	0.87	0.90
3/8	0.70	0.68	0.70	0.76	3/8	0.87	0.84	0.85	0.90	3/8	0.90	0.89	0.89	0.91
4/8	0.79	0.75	0.75	0.79	4/8	0.91	0.87	0.87	0.91	4/8	0.93	0.92	0.92	0.93
5/8	0.90	0.83	0.82	0.84	5/8	0.95	0.90	0.89	0.91	5/8	0.96	0.94	0.94	0.95
6/8	1.04	0.93	0.90	0.89	6/8	0.99	0.93	0.91	0.92	6/8	1.00	0.97	0.96	0.96
7/8	1.20	1.05	1.00	0.96	7/8	1.04	0.97	0.93	0.93	7/8	1.04	1.00	0.99	0.98
8/8	1.41	1.21	1.12	1.05	8/8	1.09	1.00	0.95	0.94	8/8	1.08	1.04	1.02	1.00

[a]Based on trivariate normal distributions. Denominator is for 12 strata for first stratifier: 12 × 1. Numerator is for bivariate $L_1 \times L_2 = 12$ strata. $I_a = 1 - I_b$, the relative weight for first survey variable. For correlation structure refer to Table 4.

always at 6 × 6, due to the symmetrical correlational structure of the model; otherwise it would be somewhat noncentral, but other conclusions would not be much affected. The optima are not sharp and 9 × 4 and 4 × 9 are almost as good as 6 × 6. Thus 9 × 4 will serve well when one has some preference for I_a but lacks strong confidence about the stratifiers.

(d) The southwest corners of $I_a = 1$ and 2 × 18 show how badly one would lose if he used mostly the wrong stratifier. Even worse would be choosing entirely the single wrong stratifier; this can be symbolized by $(1\mathscr{L} \times 36\mathscr{s})$ / $(36\mathscr{L} \times 1\mathscr{s})$ and would equal $(1-\rho_{2y}^2)/(1-\rho_{1y}^2)$. The relative efficacies would be 0.44/0.10 = 4.4 for Table 4A and 0.75/0.10 = 7.5 for Table 4C.

(e) We may summarize how the choices of weights and of numbers of strata interact by representing symbolically the four corners of the tables, with $\mathscr{L}$ and $\mathscr{s}$ for large and small correlations with the two survey variables:

$\dfrac{18\mathscr{L} \times 2\mathscr{s}}{1\mathscr{L} \times 36\mathscr{s}}$	$\dfrac{2\mathscr{L} \times 18\mathscr{s}}{1\mathscr{L} \times 36\mathscr{s}}$
$\dfrac{2\mathscr{L} \times 18\mathscr{s}}{36\mathscr{L} \times 1\mathscr{s}}$	$\dfrac{18\mathscr{L} \times 2\mathscr{s}}{36\mathscr{L} \times 1\mathscr{s}}$

In the southeast corner we find moderate gains or only small losses by going from 36 × 1 to 18 × 2. This compromise buys good gains going to the other variable in the northeast corner with 2 × 18 instead of 1 × 36. Even larger gains for the other variable are obtained in the northwest corner with 18 × 2 instead of 1 × 36. The southwest corner shows large losses for the first variable with 2 × 18 instead of 36 × 1. The entire center area shows moderate to good gains when both survey variables are considered jointly.

(f) The absolute magnitudes of the gains and losses are diverse for the different sets of correlational structures. The denominators for 36 × 1 strata are 0.10 for $\rho_{1y} = 0.95$, 0.44 for $\rho_{1y} = 0.75$, and 0.75 for $\rho_{1y} = 0.50$. The ratios shown over those bases show great differences between the tables. The advantages of bivariate stratification are generally the greatest for large correlations ρ_{1y} and ρ_{2y} and for smaller correlations ρ_{12} between stratifiers.

(g) Essentially similar results are observed in Table 5 for the totals of $L_1 \times L_2 = L = 12$ strata. The chief differences are in the last rows for $I_a = 1$, denoting full confidence in both purpose and stratifiers. In that situation even 6 × 2 cannot do as well as 12 × 1 on the single stratifiers with very high values of correlation.

6. COMPARISON OF BIVARIATE STRATIFICATION WITH PRINCIPAL COMPONENTS

The techniques of principal components analysis, factor analysis, and cluster analysis have been suggested as methods for combining several variables available for stratification (Hagood and Bernert 1945; Berry 1961; Pocock and Wishart 1969; Golder and Yeomans 1973; Heeler and Day 1975; and Yeomans and Golder 1975). These techniques seem particularly attractive with the advent of modern computing technology. Little work has been done, however, to investigate the efficacy of these techniques relative to the standard approaches used with stratification variables. Here we focus on comparisons of bivariate stratification to principal components. See also Hansen, Hurwitz, and Madow (1953, Vol. I, Sec. 9.10).

First, principal components do not seem to us useful for categorical stratifiers. Second, for quantitative stratifiers that are far from linear, the method will not yield their potential gains. Third, the resulting strata are not readily interpretable as analytical domains. Fourth, the method relates to the internal structure of variates rather than their effects on the survey variables. Fifth, multivariate stratification can usually yield higher gains, as in our results.

However, the method may be useful to reduce a large number of variables to a smaller subset which is amenable to multivariate stratification. The principal component can give higher gains than a single variable, as in our results, especially if the single variable is not much stronger than the others. These aspects may deserve deeper investigation than we were able to undertake.

The first principal component performs best in our results for stratifying variables of equal strength; it performs as well as the best linear combination with which it is equivalent (Anderson 1976). The latter cannot be used generally because it requires knowledge of correlations between stratifying and survey variables, whereas the principal component is computed from correlations only between the stratifiers. Perhaps with knowledge gained in periodic studies, the gains of the best linear combination may be approached.

We summarize the comparisons in Table 6 of bivariate stratification with stratification based on principal components; these may be buttressed with others (Anderson 1976). We believe that with large departures from linear relationships the case for bivariate stratification and multivariate stratification should be even stronger than here.

(a) Bivariate stratification in general yields greater gains in reducing the variance than the first principal component. It is better more often and with greater ratios. The advantages are more obvious for 36 strata (or more) than for 12 (or less).

(b) Using 6×6 bivariate strata is always better than the principal component. This seems almost always true for 9×4 and usually true even for 12×3. The situation is not clear for 18×2, unless we are confident about both the purpose and the stratifier by staying near $I_a = 1$. For 12 strata using 4×3

TABLE 6. Ratios of Variances for Bivariate/Principal Component Stratifications

A. ($L_1 \times L_2 = 36$ Strata)[a]

Correlations			$I_a = 1.0$				$I_a = 0.5$				$I_a = 0.0$			
ρ_{1y}	ρ_{2y}	ρ_{12}	18×2	12×3	9×4	6×6	18×2	12×3	9×4	6×6	18×2	12×3	9×4	6×6
0.95	0.75	0.90	0.37	0.37	0.41	0.55	0.84	0.69	0.60	0.55	1.32	1.01	0.80	0.55
0.95	0.75	0.60	0.78	0.73	0.75	0.89	1.56	1.14	0.97	0.89	2.34	1.55	1.19	0.89
0.95	0.50	0.60	0.29	0.31	0.35	0.43	0.72	0.54	0.47	0.43	1.16	0.77	0.59	0.43
0.95	0.50	0.30	0.38	0.37	0.39	0.49	1.01	0.68	0.55	0.49	1.65	0.99	0.72	0.49
0.95	0.25	0.30	0.23	0.25	0.27	0.33	0.59	0.43	0.36	0.33	0.94	0.60	0.45	0.33
0.75	0.50	0.90	0.67	0.64	0.63	0.67	0.85	0.76	0.70	0.67	1.03	0.88	0.78	0.67
0.75	0.50	0.60	0.86	0.86	0.87	0.90	1.00	0.94	0.91	0.90	1.13	1.01	0.95	0.90
0.75	0.50	0.30	0.98	0.95	0.94	0.96	1.12	1.02	0.98	0.96	1.27	1.09	1.02	0.96
0.75	0.25	0.60	0.60	0.59	0.59	0.62	0.76	0.67	0.64	0.62	0.93	0.76	0.68	0.62
0.75	0.25	0.30	0.72	0.72	0.73	0.76	0.87	0.80	0.77	0.76	1.02	0.88	0.81	0.76
0.50	0.25	0.90	0.81	0.77	0.75	0.76	0.88	0.82	0.78	0.76	0.96	0,87	0.81	0.76
0.50	0.25	0.60	0.91	0.91	0.92	0.93	0.96	0.94	0.93	0.93	1.01	0.97	0.94	0.93
0.50	0.25	0.30	0.95	0.95	0.95	0.96	1.00	0.97	0.96	0.96	1.04	0.99	0.98	0.96

TABLE 6 *continued*

B. ($L_1 \times L_2 = 12$ Strata)[a]								
Correlations			$I_a = 1.0$		$I_a = 0.5$		$I_a = 0.0$	
ρ_{1y}	ρ_{2y}	ρ_{12}	6×2	4×3	6×2	4×3	6×2	4×3
0.95	0.75	0.90	0.56	0.79	0.95	0.92	1.33	1.04
0.95	0.75	0.60	1.03	1.24	1.61	1.39	2.20	1.54
0.95	0.50	0.60	0.42	0.57	0.78	0.66	1.13	0.75
0.95	0.50	0.30	0.56	0.73	1.07	0.86	1.58	0.98
0.95	0.25	0.30	0.33	0.45	0.63	0.52	0.92	0.59
0.75	0.50	0.90	0.73	0.79	0.88	0.84	1.03	0.89
0.75	0.50	0.60	0.90	0.95	1.01	0.98	1.13	1.01
0.75	0.50	0.30	1.01	1.04	1.14	1.07	1.26	1.10
0.75	0.25	0.60	0.64	0.69	0.78	0.72	0.92	0.76
0.75	0.25	0.30	0.75	0.81	0.88	0.84	1.01	0.87
0.50	0.25	0.90	0.83	0.84	0.90	0.86	0.96	0.88
0.50	0.25	0.60	0.92	0.94	0.97	0.95	1.01	0.96
0.50	0.25	0.30	0.96	0.97	1.00	0.98	1.04	0.99

[a]Based on trivariate normal distributions. Numerator is for bivariate stratification using $L_1 \times L_2$ strata. Denominator is for stratification based on principal components. $I_a = 1 - I_b$ denotes the relative importance of first purpose.

also seems more efficient in general, but the advantages seem smaller. For the 6 × 2 bivariate case the situation is mixed. See Table 6B.

(c) For I_a near 1, denoting confidence on both purpose and stratifier, bivariate stratification appears always better. Even for the compromise situation of I_a=0.5, the bivariate method appears better, except for the extremes of 18×2. For 12 strata the advantages are weaker, but lie in the same directions.

(d) The ratios exhibit great diversity between rows, which refer to the covariance structure. The advantages of the bivariate appear greater when the difference between ρ_{1y} and ρ_{2y} is large, and then when the correlation ρ_{12} between stratifiers is also large.

(e) The principal component appears best when ρ_{1y} and ρ_{2y} are similar, both are high, and ρ_{12} is low. It has been shown under some assumptions that for $\rho_{1y} = \rho_{2y}$ the principal component is the same as the best linear combination of stratifiers, which is optimal (Anderson 1976).

We should add that our condensed table can be readily extended. The ratios can be interpolated linearly for any value from $I_a = 1$ to $I_a = 0$, for any fixed $L_1 \times L_2$ and ρ_{1y}, ρ_{2y}, ρ_{12}. The ratios are symmetrical, so that values for 2 × 18, 3 × 12, and 4 × 9 for I_a are the same, respectively, as values for 18 × 2, 12 × 3, and 9 × 4 for $1 - I_a$. The ratios for 6 × 6 are constant for all values of I_a. The range of ratios from $I_a = 1$ to $I_a = 0$ gets larger with departures from 6 × 6 toward the extremes (18 × 2 and 2 × 18).

7. SOME EMPIRICAL RESULTS

Three sets of data were examined, which were presented in detail elsewhere (Anderson 1976). The most important of these concerned variables for 3,107 counties from the 48 contiguous states, as given in the *County and City Data Book, 1972* (U.S. Bureau of the Census 1973). Four variables were chosen first:

(a) Percent of families with $25,000 or more of income (1969),
(b) Percent of families below the low income level,
(1) Percent of persons 25 years old and over who completed four years of college or more,
(2) Percent of housing units with 1.01 or more persons per room.

Variables (a) and (b) were designated as survey variables, and (1) and (2) were designated as stratification variables. Investigation of multipurpose design was aided by the fact that stratifier (1) was more effective for (a), and (2) for (b). The weighted correlation coefficients were: $\rho_{a1} = 0.75$, $\rho_{b2} = 0.74$, $\rho_{a2} = -0.36$, $\rho_{b1} = -0.46$, and $\rho_{12} = -0.35$. For the first principal component, it was $\rho_{ap} = -0.73$ and $\rho_{bp} = 0.69$. The negative signs are merely artifacts of the directions of (b) and (2).

Each of the four percentages was skewed, with the medians falling near (a) 10, (b) 25, (1) 18, and (2) 16. The cumulative $\sqrt{f}$ rule was used to establish the stratum boundaries. However, the rule was not applied to numbers of counties but to relative frequencies of measures of size based on the 1970 Census count of the total population. Thus we simulated the common use for selecting counties in order to obtain ultimate clusters of about equal numbers of dwellings and persons. Our stratification effects apply directly to the common methods for selecting counties.

The boundaries of strata for real data involve random variations—unlike the smooth distributions of Section 4. This is a source of random variations that may be noted in the tables.

Table 7 for variable (b), similar to another for variable (a), is taken from Anderson (1976). For any combination of $L_1 \times L_2 = L$ strata, the four lines compare the variances of four stratification techniques as proportions of the unstratified variance; each is based on the same number L of strata. Note several features of the two-way stratification on the top lines. It is always much better than the weaker stratifier found on the second line. Compared to the stronger stratifier on the third line, it is worse for few strata but becomes better for more strata when $L_2 > L_1$ (above the diagonal). Compared to the principal component, it is worse for few strata but catches up for more strata and for $L_2 > L_1$. Moreover, the principal component was less effective in the other examples.

Table 8 presents reassuring similarities to the results in Tables 4 and 5. It shows the advantages of multipurpose designs of two-way ($L_1 \times L_2 = L$) over one-way stratification ($L \times 1$) with variable (1), which is better for variable (a). The extremes of 0/8 and 8/8 for I_a show results for the single aims of variables (b) and (a), respectively. We note several features of two-way stratification. First, it is sensibly and generally superior to one-way stratification. Second, its superiority increases with numbers of strata. Third, it is always superior in the middle range, and even more, of course, for weights adverse to the one-way stratification. Fourth, it is not badly inferior for moderate numbers of strata even for the extreme weight of 8/8, most favorable to one-way stratification. Fifth, it is inferior only for extreme weights for (a), combined with few strata (2×2) or adverse stratification (2×5).

A second empirical study concerned some data on burned patients collected in 1968–72 from 13 hospitals on 4,145 surviving burned patients, 1,148 females and 2,997 males, made available by the National Institute for Burn Medicine located at Ann Arbor, Michigan. The survey variable (a) was the number of weeks from admission to discharge. Three stratifiers were used, and their strengths in variance reduction with 25 strata were: (1) percent total burn (0.27), (2) burn-depth ratio (0.11), and (3) age (0.09). In this case the two-way stratification of variables (1) and (2) or (1) and (3) are just as good as but not better than (1) alone. For $5 \times 5 =$ 25 strata, both combinations yield variance reductions of 0.28. This is only negligibly better than 0.27 for (1) alone, but considerably better than 0.11 or 0.09 for (2) or (3), respectively. One of these could be chosen in the usual situation of ignorance about the relative strengths of stratifiers. Here the dependability of two-way stratification may be viewed as insurance against ignorance (Anderson 1976).

TABLE 7. Proportions of Gains for Four Techniques of Stratification for USA Counties[a]

	L_2			
L_1	2	3	4	5
2	0.61	0.54	0.45	0.44
	0.76	0.71	0.68	0.68
	0.48	0.47	0.44	0.44
	0.53	0.45	0.41	0.40
3	0.58	0.52	0.41	0.41
	0.71	0.68	0.68	0.67
	0.47	0.45	0.43	0.44
	0.45	0.41	0.40	0.38
4	0.56	0.50	0.40	0.40
	0.68	0.68	0.67	0.66
	0.44	0.43	0.42	0.43
	0.41	0.40	0.38	0.37
5	0.54	0.47	0.39	0.37
	0.68	0.67	0.66	0.66
	0.44	0.44	0.43	0.43
	0.40	0.38	0.37	0.37

[a]Each cell shows four ratios of variances over unstratified variance: (1) Bivariate $L_1 \times L_2 = L$ strata; (2) Univariate $L \times 1$ strata over weaker stratifier; (3) Univariate $1 \times L$ over stronger; (4) Principal component with L strata.
Data from U.S. Bureau of the Census (1973).

TABLE 8. Ratios of Bivariate/Univariate Stratifications for USA Counties[a]

I_a	2 × 2	3 × 2	2 × 3	3 × 3	5 × 2	2 × 5	4 × 4	5 × 5
0/8	0.80	0.82	0.76	0.76	0.79	0.65	0.60	0.56
1/8	0.84	0.84	0.80	0.80	0.81	0.70	0.63	0.59
2/8	0.88	0.87	0.84	0.84	0.83	0.75	0.67	0.63
3/8	0.93	0.90	0.89	0.89	0.85	0.81	0.71	0.68
4/8	0.98	0.93	0.94	0.94	0.87	0.88	0.75	0.73
5/8	1.04	0.97	1.00	0.99	0.90	0.96	0.80	0.78
6/8	1.11	1.01	1.06	1.06	0.93	1.05	0.86	0.85
7/8	1.19	1.05	1.13	1.13	0.96	1.15	0.92	0.93
8/8	1.29	1.11	1.21	1.21	1.00	1.27	1.00	1.03

[a]Numerator is variance for $L_1 \times L_2 = L$ strata; L_1 = strata for 1, the better stratifier for a; L_2 = strata for 2, the better stratifier for b. Denominator is variance for $L \times 1$ strata, univariate for 1. $I_a = 1 - I_b$ = the relative weight of importance of variable a. Data from U.S. Bureau of the Census (1973).

The advantages of multivariate stratification should not stop with two stratifiers. We introduced a third stratifier (4): a dichotomy of type of hospital. The combinations of $4 \times 3 \times 2 = 24$ for variables (1), (2), and (4) or (1), (3), and (4) produce reductions of 0.35, in place of the 0.28 for the $5 \times 5 = 25$ strata for two-way stratification.

We also tried principal components on several variables, including those stratifiers mentioned in the preceding paragraphs, and obtained 0.19 for the first principal component and 0.03 for the second for 25 strata. Combined into 5×5 strata they yield a reduction of 0.21, which compares unfavorably with the 0.28 for the two-way stratifications of either (1) and (2) or (1) and (3). We also found that the best linear combination of (1), (2), and (4) reduces the variance by 0.36, but the knowledge needed to use it is not ordinarily available in surveys.

A third body of data concerned observations on the right eye of 1,000 persons, 315 women and 685 men, between 20 and 35 years old (Anderson 1976). For our investigation, the study variable was (a) axial length. Three stratifiers were used, of which (1) ocular refraction was strong, but (2) age and (3) sex were very weak. In this extreme case, the single stratifier (1) was more powerful up to 25 strata than the two-way stratification of (1) and (2) with 5×5 strata; the latter reduced the variance by 0.50, and the former by 0.59. This was almost as powerful as the best linear combination with 0.61. The first principal component was a poor stratifier with 0.28.

APPENDIX

Let $\Phi(z)$ and $\Phi^{-1}(p_0)$ denote the cumulative distribution function (cdf) and inverse cdf, respectively, of a standard normal random variable. The approximations for $\Phi(z)$ and $\Phi^{-1}(p_0)$ used in this article are presented in Abramowitz and Stegun (1965).

Suppose the random vector $\mathbf{z}^T = (z_1, \ldots, z_p)$ has a p-dimensional multivariate normal distribution with standardized marginal variates. Let the joint frequency function be $\phi_p(\mathbf{z})$, and denote the cdf by

$$\Phi_p(h_1,\ldots,h_p) = \int_{-\infty}^{h_1} \cdots \int_{-\infty}^{h_p} \phi_p(\mathbf{z})d\mathbf{z}\,. \qquad \text{(A.1)}$$

Curnow and Dunnett (1962) note that in certain special cases the integral (A.1) can be reduced to a single integration, e.g., when the correlation coefficient ρ_{ij} can be expressed as $\rho_{ij} = \alpha_i\alpha_j$, $i \neq j$, where $-1 \leq \alpha_i \leq 1$. In this special case, the integral becomes

$$\Phi_p(h_1,\ldots,h_p) = \int_{-\infty}^{\infty} \prod_{i=1}^{p} \Phi[(h_i - \alpha_i y)/(1-\alpha_i^2)^{1/2}]\,\phi(y)dy. \qquad \text{(A.2)}$$

Let $g(y)$ denote the integrand from (A.2). Note that $g(y)$ is zero with negligible error outside of the closed interval [-10, 10]. So for all practical purposes

$$\int_{-\infty}^{\infty} g(y)dy = \int_{-10}^{10} g(y)dy. \tag{A.3}$$

Define $y_k = -10 + (k/10)$, $k = 0, \ldots, 200$. From Kaplan (1952) the trapezoidal rule is

$$\int_{-10}^{10} g(y)dy \sim (1/20) \cdot [g(-10) + 2g(y_1) + \ldots + 2g(y_{199}) + g(10)], \tag{A.4}$$

where the width of each trapezoid is 1/10. Expressions (A.3) and (A.4) can be used to evaluate (A.2) numerically.

Our interest is in the case of the bivariate normal. Observe that for $p = 2$ (A.2) holds without assuming any special correlation structure. For $\rho_{12} \geq 0$ set $\alpha_1 = \alpha_2 = (\rho_{12})^{1/2}$ and for $\rho_{12} < 0$ set $\alpha_1 = -(|\rho_{12}|)^{1/2} = -\alpha_2$. Application of the trapezoidal rule (A.4) gives an approximation for the bivariate normal cdf which performs quite well, as is indicated by Table 9. This table gives approximate and exact values of $\Phi_2(h_1, h_2)$ for selected h_1, h_2, and ρ_{12}.

Comparison of the approximate and exact values from Table 9 suggests that the trapezoidal approximation may be accurate to five or six decimal places—at least in the lower left quadrant of the plane. Additional numerical work not presented here indicates this degree of accuracy in the other quadrants as well.

TABLE 9. Trapezoidal Approximation for $\Phi_2(h_1, h_2)$

h_1	h_2	ρ_{12}	Approximate values	Exact values[a]
-2.0	-2.0	0.50	0.004053	0.004053
-1.0	-1.0	0.80	0.097637	0.097637
-0.8	-0.3	-0.25	0.054093	0.054093
-0.3	-0.9	0.15	0.085871	0.085871
-0.2	-1.7	0.60	0.040167	0.040167
-1.0	-1.2	0.95	0.104068	0.104068
-1.0	-1.2	-0.95	0.000000	0.000000
-0.3	-0.9	-0.15	0.055407	0.055407
-0.2	-1.7	-0.60	0.001684	0.001684
-3.0	-1.9	0.20	0.000138	0.000138

[a]Exact values from National Bureau of Standards (1959).

ISSUES IN INFERENCE FROM SURVEY DATA

KEITH RUST AND MARTIN FRANKEL[1]

Throughout his career Leslie Kish was both an advocate of the need for reliable and robust techniques for drawing inferences from survey data, and also an innovator in developing and implementing such techniques. He published many papers on such topics, and the three included in this volume cover a range of these activities both temporally and in content.

The paper "Inferences from Complex Surveys" co-authored with Martin Frankel, is arguably the most frequently and widely cited of Leslie's published manuscripts. There are several reasons for this. The manuscript covers a broad range of subtopics in the area of inference. The authors discuss the impact of sample design on sampling error, and the use of design effects to summarize this in a given case. They cover relatively simple estimators, and then break new ground by discussing the impact of complex sample design on inferences drawn from models fitted to the data. The associated discussion of the paper, originally published together with the manuscript, illustrates that this was a topic of considerable debate, and was something that had not been addressed substantially in the literature previously. Finally, Kish and Frankel reported on the results of empirical simulations that compared the results of three alternative methods of calculating sampling errors for complex statistics in practice.

This particular aspect of the paper was highly significant for several reasons. The three different methods then currently available (and still the three methods used in most survey applications) are described in one place, and their empirical properties compared. These methods were termed by Kish and Frankel as the Taylor Expansion Method, Balanced Repeated Replication, and Jackknife Repeated Replication. Until Wolter (1985) was published, this was the primary published material to present this information in sufficient detail that practitioners could see how the methods were to be implemented in practice. The simulations that Kish and Frankel conducted provided the reassuring result that the three

[1] Keith Rust, Westat, Rockville, MD 20850, and Joint Program in Survey Methodology, University of Maryland. keithrust@westat.com.
Martin Frankel, Baruch College, CUNY, New York, NY 10010, martin_frankel@baruch.cuny.edu

methods gave very similar results in practice. They also began the consideration as to which of the methods might be superior in various applications. In the period since then a number of researchers have undertaken theoretical work on asymptotic results, and carried out further empirical simulations, to expand our knowledge in this field. This continues to be an area of active research (see, e.g., Rao and Shao 1999).

A further reason that the paper was seminal is that it represented the culmination of over 15 years of research by Kish and colleagues at the ISR, and others, into developing practical methods for estimating and presenting sampling errors for surveys. It is probably fair to say that, at that time in 1973 when the manuscript was read before the Royal Statistical Society, the Institute for Social Research (ISR), as a result of Kish's influence, led the world in this activity. In his 1957 paper Kish (1957) discussed the use of confidence intervals for clustered samples, but there was little evidence that the methods proposed were generally appropriate, and theory was lacking. Thus Kish and his colleagues began work on the unsolved problem, among practicing survey sampling statisticians, of how to make appropriate inference from stratified clustered samples. Initially Kish and his students attempted to extend the concept of the Central Limit Theorem to provide a theoretical solution to the problem, but this proved intractable. They then turned to the ideas of McCarthy (1966) on balanced half-samples (another term for BRR) and Tukey (1958) regarding the jackknife. They then obtained survey data from the U.S. Current Population Survey, which they used to simulate a realistic population. From this they drew extensive simulations of clustered samples, to investigate the properties of the three variance estimation methods, with special emphasis on the properties of confidence intervals created using these approaches. This was likely the first time that a computer was used to draw repeated samples from a real population.

In discussing the paper, Professor Durbin noted that "in practice, very few survey practitioners do in fact ever calculate sampling errors as a routine part of their work." This situation has clearly changed over the period since 1973. Many would attribute that to advances in computing. However, frankly, advances in computing have contributed to the calculation of inappropriate sampling errors at least as much as to calculations using the methods described by Kish and Frankel. In fact it is the influence of Leslie Kish that has arguably been the greatest single factor in the promulgation of appropriate methods of calculating sampling errors for complex surveys, especially outside national statistical offices. Kish and colleagues were the first to develop specialized computer software available to general users, the Sampling Error Program Package (Kish, Frankel, and Van Eck 1972). This drew attention to the need for practitioners to use specialized methods, and often specialized software, to carry out their analyses appropriately.

Kish persuaded Frankel that they should submit their work for publication in the *Journal of the Royal Statistical Society*. Kish felt that the manuscript would receive a thoughtful review from statisticians who would appreciate the issues facing practitioners. The authors were delighted to be invited to read the paper before a meeting of the Society.

The second of the three papers on inference included here dates from 1992. In the years between 1974 and 1992 Kish was active in a number of other aspects of survey research, in particular, sample design issues such a rolling samples, and the issue of small area estimation. However, he remained heavily involved in research and pedagogy concerning variance estimation and other inference issues. The World Fertility Survey publication "Sampling Errors for Fertility Surveys" (Kish, Groves, and Krotki 1976) provided a valuable exemplar for those seeking practical approaches to variance estimation. Kish also headed work in the development of statistical software for calculating sampling errors (as part of the OSIRIS package).

The paper "Weighting for Unequal *Pi*" provided an important contribution in Kish's efforts to address the fact that weighting survey data is a routine activity carried out by survey practitioners but essentially unaddressed in textbooks on survey sampling. At the time the paper was published in 1992 the only notable exception was Kish's own *Survey Sampling*, published in 1965.

The paper provided a simple and succinct description of the reasons why weighting is an issue for inference from survey samples, and also described procedures and issues for implementing weighting in practice in large-scale surveys. This provides a very useful reference for the practitioner confronted with an analyst who does not understand why this issue arises in the analysis of survey data when it is not addressed, at least in this form, in other areas of statistical analysis.

The paper did more than provide a handy reference that describes and justifies standard survey practice. Kish addresses the issue of when the requirement to apply weights to ensure approximate unbiasedness of survey estimates comes at the expense of undue sampling variance. It seems likely that this was the first time that this issue had been so explicitly and directly addressed. In the section "Balancing Variance Increases against Biases," Kish develops a straightforward mathematical approach to compare the mean square errors of weighted unbiased estimates with unweighted ones. This treatment demonstrated that the key issue is whether the squared correlation between the weights and the variable of interest is greater than the reciprocal of the effective sample size. If such a relationship occurs then weighting is to be preferred. A more formal and detailed approach to this issue has been developed subsequently by Korn and Graubard (1999, Chapter 4).

Kish goes on to point out that the effects of weighting, and different weighting procedures, vary widely across different kinds of statistics, within the same survey, as well as across surveys. This parallels the point, discussed below, that design effects vary in this way also. Kish often warned students and colleagues to be alert to the misconceptions that a given survey has a single design effect, or that weighting had equal effects on all analyses of a given survey.

Kish also raises the issue of the use of weight trimming and shrinkage algorithms in developing survey weights. Since the concept of survey weighting itself was hardly addressed in the literature prior to 1992, it is hardly surprising that this somewhat obscure but widespread practice had not been significantly addressed either. In this paper Kish merely raised the issue and outlined what these procedures are designed to achieve, in the context of the rest of the paper. This remains an undeveloped research area in the field of survey methodology.

Kish's 1995 paper, "Methods for Design Effects," presented him with an opportunity to review and clarify the concepts, uses, and misconceptions surrounding the formulation and calculation of design effects. The design effect has become a key concept in the design of surveys. It is also referred to in the context of survey analysis, and sometimes the distinction is made of calling it a misspecification effect when used in this context (Skinner, Holt, and Smith 1989).

Kish introduced the term "design effect" in *Survey Sampling* (Kish 1965). In his 1995 paper, Kish briefly reviews the history of the concept prior to 1965. Previous texts on survey sampling had referred to the concept, sometimes labeling it as "ratios of variances" (Hansen, Hurwitz, and Madow 1953; Kish 1957). With encouragement from John Tukey, Kish introduced the term because he viewed the concept as key both in the evaluation of alternative sample designs, and in the consideration of the effects of the design on analysis. He also used to relate that he thought it was important to give key and innovative concepts meaningful labels. Otherwise, he observed, they often ended up being named after someone (not necessarily their original developer). This was not especially helpful to newcomers to the concept. The "Kish grid," used for selecting a respondent from a household at random, seems to provide a demonstration of a case where Kish failed to come up with a good label quickly enough!

Kish defined the design effect of a statistic as the ratio of the actual sampling variance of the statistic, under the particular sample design in question, to the variance of the same statistic that would have been obtained had a simple random sample of the same size been used. The misspecification effect replaces this hypothetical denominator with the expected value of the variance estimator appropriate for a simple random sample but applied to the design in question. Thus, formally, the design effect is of interest to those designing surveys, while the misspecification effect is of interest to those analyzing complex survey data. Kish never made this distinction in any formal way. For *epsem*, or self-weighting, designs, the distinction is negligible, and Kish could have believed that making the distinction would create more confusion than enlightenment among the broader community of analysts and statisticians. Most practitioners today use the term "design effect" to denote either concept, or both simultaneously.

In "Methods for Design Effects" Kish reviewed the uses of design effects in practice. He suggested that a more useful definition of the design effect is in fact the square root of the quantity defined above. He uses the label *deff* for the ratio of sampling variances, and *deft* for the ratio of standard errors. Kish did not explicitly explain this preference. It is reasonable to suppose, however, that it derives from the fact that this concept fits more closely with the widely accepted idea of using the standard error, rather than the sampling variance, as the reported measure of reliability of statistics derived from a sample. Thus *deft* tells the user what the effect of the complex design is on the standard error.

Kish argues that the principal reasons for calculating design effects from survey data are an essential part of any process of generalization of sampling errors. Thus they are important in reports that do not present a sampling error for every estimate presented. They are also important in cases where an estimate of sampling error is provided for every published estimate. The reader may take differences or averages or ratios of published statistics, and needs a means of gaining a general idea about the magnitude of sampling error for such derived quantities. Information about design effects, especially of the kind that Kish advocates, can be useful for this purpose. It is also the case that individual estimates of sampling error are themselves subject to sampling error, sometimes substantial. By analyzing and summarizing design effects, an alternative, more robust, estimate of sampling error is available. That is, a suitable average design effect, multiplied by the simple random sampling estimate of the standard error, may be substantially more accurate than the direct estimate of the standard error that reflects the design.

Kish cautions of the need to calculate a variety of estimates of design effects, for different population subgroups and for different kinds of statistics. As he points out, inevitably these will vary considerably within a survey (unless they are all very close to 1.0), and it is important to avoid the notion of "the" design effect for a given survey. The paper also addresses the issue of separating the effects of stratification and clustering from those of differential weighting due to differential sampling rates, or nonresponse, or other adjustments to the survey weights. Kish points out that generalizations of design effects are more often effective when considered net of these weighting effects.

As well as sharing common ground related to making estimates and inferences from complex survey data, these three papers, and the others that Kish wrote on these topics, clearly share a common philosophy. There are several aspects to the philosophy that can be identified. First, issues of practical concern are paramount. It is the consequences for the actual analysis of real surveys that must be of chief concern, rather than niceties of mathematical exposition. "The best is the enemy of the good," one of Kish's favorite aphorisms, and the verb "to satisfice" that he devised, both come to mind in this context. Second, one should generalize as much as possible when considering alternative approaches or presenting results of technical evaluations, but always recognize that such generalizations are bound to be limited, having important exceptions. Third, empirical calculations and simulations, in abundance if possible, are an important tool for advancing knowledge for the practice of analyzing survey data. Fourth, conceptual innovations must be accompanied by practical advances if they are to become part of established practice. Kish was among the first to recognize that the boundaries of what is undertaken on a routine basis by analysts of survey data are largely determined by the capabilities of the computer software available to them. Fifth, there are many contributors to advances that are of practical consequence. Kish's papers often point out relationships among the works of diverse authors that would not have generally been recognized until he pointed them out. And sixth, things that are useful to know are best summarized in numbered lists!

8

INFERENCE FROM COMPLEX SAMPLES

LESLIE KISH AND MARTIN RICHARD FRANKEL

Summary: The design of complex samples induces correlations between element values. In stratification negative correlation reduces the variance; but that gain is less for subclass means, and even less for their differences and for complex statistics. Clustering induces larger and positive correlations between element values. The resulting increase in variance is measured by the ratio *deff*, and is often severe. This is reduced but persists for subclass means, their differences, and for analytical statistics. Three methods for computing variances are compared in a large empirical study. The results are encouraging and useful.

Key words: Clusters; complex sample; sampling error; design effect; *BRR*; jackknife; inference; standard error; replication; intraclass correlation; sample design; sampling variance; subclass analysis; stratification; replication

1. INTRODUCTION

Standard statistical methods have been developed on the assumption of simple random sampling. The assumption of the independent selection of elements (hence independence of observations) greatly facilitates obtaining theoretical results of interest. It is essential for most measures of reliability used in probability statements, such as $\sigma/\sqrt{n}$, chi-squared contingency tests, analysis of variance, the nonparametric literature and standard errors for regression coefficients. Assumptions of independence yield the mathematical simplicity that becomes more desirable—and at present necessary—as we move from simple statistics such as means, to the complex statistics typified by regression analysis. Independence is

---

Reprinted with permission from the *Journal of the Royal Statistical Society* (1974), 36, Ser. B, 1–22. Discussion of the paper appears on pp. 22–37.

Read before the Royal Statistical Society at a meeting organized by the Research Section on Wednesday, October 17th, 1973, Professor J. Gani in the Chair.

This research was supported by Grant 3191X from the National Science Foundation.

often assumed automatically and needlessly, even when its relaxation would permit broader conclusions.

Although independence of sample elements is typically assumed, it is seldom realized in the procedures of practical survey work. Randomization of the sample would be unnecessary if the population itself were randomized, but "well-mixed urns" are seldom provided by nature or created by man. This uneasy situation exists widely in the social sciences. In the natural and physical sciences it is typically even more difficult to achieve complete randomization of the sample over the target population, but more often it may be somewhat reasonable to assume its existence in the population—although not entirely and not always.

Much research is actually and necessarily accomplished with complex sample designs, especially in social, health, economic and business studies. It is often economical to select existing clusters or natural groupings of elements. These are characterized by relative homogeneities within the clusters that negate the assumption of the independence of sample elements. The assumption may fail mildly or badly; hence standard statistical techniques result in mild or bad underestimates in reported probability intervals. Overestimates can seldom be severe.

Survey sampling was developed, mostly in the social sciences, censuses and agriculture during the past half century, to provide statistical techniques and theory for the complex selection methods needed for large-scale surveys. It was developed (1) for samples which were not simple random samples, and (2) for finite populations rather than hypothetical infinite populations. It was developed chiefly for descriptive statistics, that is for means, proportions and aggregates. Early landmarks in the literature of sampling were papers by Tchuprow (1923), Neyman (1934), Mahalanobis (1944) and Yates (1946). Five classics in five years—Yates (1949), Deming (1950), Cochran (1953), Hansen, Hurwitz, and Madow (1953) and Sukhatme (1954)—outlined the boundaries that have largely defined and confined developments of methods in this subdiscipline of statistics.

The literature of survey sampling concentrates on providing estimates $\bar{y}$ and standard errors, $\text{ste}(\bar{y})$. The estimate $\bar{y}$ may be for an aggregate like $\hat{Y} = Fy$, where F is a constant and $y = \Sigma y_j$ is the sample sum over elements of a variable; or it may be a simple mean y/n of the sample elements, or a weighted mean, $\Sigma w_j y_j / \Sigma w_j$; or it may be a ratio, regression or difference estimator of the mean or aggregate. Further, $\text{ste}(\bar{y}) = \sqrt{\text{var}(\bar{y})}$ is the estimated standard error, computed from the sample data in accord with the complexities of the sample design. The function of these statistics is to provide statistical intervals of the type $\bar{y} \pm t_p \text{ste}(\bar{y})$ for inference about population values of the estimated $\bar{Y}$.

We think it imperative and urgent to extend these statements to more complex statistics. More and more, researchers are able to obtain data from complex samples, and to write computer programs for complex analytical statistics. We need methods for dealing properly with complex statistics from complex samples. We need statistics for probability statements. Such statements are symbolized here with $b \pm t_p \text{ste}(b)$, where b is some complex statistic, and ste(b) is its computed standard error. For example, b can be the difference of two subclass means or a regression coefficient. Inferences based on standard errors are acceptable on the assumption that survey samples are large enough to yield the needed approximate normality in spite of the nonindependence of the observations. Standard errors should be computed in accord with the complexity of the sample designs; neglect of that complexity is a common source of serious mistakes (Kish 1957; Kish and Frankel 1970). On the other hand, trying to obtain more exact but more complicated statistics than standard errors would become too difficult for complex selection designs.

When discussing the independence of observations, we deliberately neglect here the problems of sampling without replacement from finite populations. A learned literature devoted to these problems has recently arisen, but is limited essentially to the relatively simple problems of means and aggregates. We believe that the important theoretical issues of representing defined finite populations concern all statistical applications, not only survey sampling. We welcome similar recent views from C. R. Rao (1971): "Unfortunately the same situation prevails in other areas and considerable literature in statistics is devoted to an examination of the foundations of statistical methodology." The theoretical implications are important, pervasive and subtle, but practical effects are usually small, and we may safely ignore them in the present discussion.

For our discussions of inferential procedures, we propose dividing into three levels of complexity both the selection methods and the statistical estimates, as shown in Figure 1. The divisions are arbitrary but useful. Among estimators beyond sample means, the analysis of subclasses is common, presents fewer problems than more complex measures of relations, yet provides analogies and conjectures about them. Among selection methods, the stratification of elements generally has effects that are simpler and weaker than those of clustered sampling.

STATISTICS

Selection Methods	1 Means and totals of entire samples	2 Subclass means and differences	3 Complex analytical statistics, e.g., coefficients in regression
A. Simple random selection of elements			
B. Stratified selection of elements		Available	Conjectured
C. Complex cluster sampling		Available	Difficult: *BRR, JRR, TAYLOR*

Figure 1. The present status of sampling errors. Row A is the domain of standard statistical theory, and Column 1 of survey sampling.

Standard statistical theory continues to supply new and improved inference procedures for Row A, always assuming independent observations. In contrast, the literature of survey sampling is mostly confined to Column 1, with theoretical discussions in cell A1 about its finite populations. In stratified element sampling, solutions are clear and simple for means and totals of entire samples (cell B1); they are also fairly simple for subclass means and their comparisons (cell B2, discussed in Section 2). For the more complicated analytical statistics used for relations between variables (cell B3), the solutions seem theoretically difficult and unclear, but rather simple conjectures appear reasonable, with some empirical justification. In clustered samples, for simple means and totals (cell C1), and for subclasses and their comparisons (cell C2), the answers are usually relatively simple and useful (discussed in Section 3). Our main concern (discussed in Sections 4 and 5) must be with complex analytical statistics from clustered samples (cell C3). We have some useful results, but we also have suggestions for further work. All four of the areas (cells B2, B3, C2 and C3), and cell C3 in particular, present challenging problems in need of both theoretical and empirical contributions. They are of utmost important to statistical applications, and of great difficulty and variety.

2. STRATIFIED SAMPLES OF ELEMENTS (B2 AND B3)

The problems of subclasses are common and not difficult here. There are some useful and surprising results, especially for the kind of domains we shall call *crossclasses*—subclasses that cut across the strata used in selection. In crossclasses, the M_{ch} subclass members of the cth subclass among the N_h elements of the population in the hth stratum are distributed *roughly proportionately*; so that $\bar{M}_{ch} = M_{ch} / N_h$ in the stratum roughly equals $\bar{M}_c = M_c / N = \Sigma M_{ch} / \Sigma N_h$ in the population. This is typical of most subclasses used in analyses of survey data. Conversely, the case when the subclass can be placed into separate strata before selection belongs to standard sampling theory B1. So does the situation when the data can be adjusted after selection with post-stratification weights M_{ch} /M from known values of M_{ch}. But for most subclasses the values of M_{ch} are unknown, and the sample of elements m_{ch} of subclass members among the n_h selected at random from the N_h in the hth stratum is a random variable. This common situation has drastic effects on the behaviour of the sample, as was first noted by Yates (1953).

The most drastic effect is on the variance of the simple crossclass aggregate $\hat{Y}_c = \Sigma y_{ch} N_h / n_h$, where $y_{ch} = \Sigma^{m_{ch}} y_{chj}$, which is the crossclass aggregate of the m_{ch} sample element values y_{chj} selected at random from the hth stratum. Here the effect of using $\hat{Y}_c$ for a subclass would be to increase the element variance approximately from σ^2_{ch}, the variance of element values of the cth subclass members around their mean $\bar{Y}_{ch} = Y_{ch} / M_{ch}$ in the hth stratum, to $[\sigma^2_{ch} + (1 - \bar{M}_{ch})\bar{Y}^2_{ch}]$; the element relvariance is increased by $(1 - \bar{M}_{ch})$. This drastic loss is well known and generally avoided in practice by using some other estimators such as $\Sigma M_{ch} \bar{y}_{ch}$.

Also well known is the variance for the mean $\bar{y}_c = \Sigma y_{ch} / \Sigma m_{ch}$ of a *proportionate* stratified sample of elements:

$$\text{var}(\bar{y}_c) = \frac{1-f}{fM_c} \sum_h W_{ch} \{\sigma_{ch}^2 + (1 - \bar{M}_{ch})(\bar{Y}_{ch} - \bar{Y}_c)^2\}, \tag{2.1}$$

where

$f = n/N = n_h/N_h = f_h$, $W_{ch} = M_{ch}/M_c$, $\bar{Y}_{ch} = \Sigma_i Y_{chi}/M_{ch}$ and $\bar{Y}_c = \Sigma_h \Sigma_i Y_{chi}/M_c$.

fM_c is the expected value of the random sample size $m_c = \Sigma m_{ch}$. We can also express the variance in a slightly different form:

$$\text{var}(\bar{y}_c) = \frac{1-f}{fM_c} \{\sigma_c^2 - \Sigma \bar{M}_{ch} W_{ch} (\bar{Y}_{ch} - \bar{Y}_c)^2\}, \tag{2.1'}$$

where

$$\sigma_c^2 = \Sigma_h \Sigma_i (Y_{chi} - \bar{Y}_c)^2 / M_c \quad (i = 1, \ldots, M_{ch}).$$

Notice that the element variance in brackets approximately takes the place of σ_c^2 (or S_c^2) in the variance one would have from a simple random sample of m_c out of M_c elements in the subclass. On the other hand, a proportionate stratified sample with $m_{ch} = fM_{ch}$ in every stratum would have an element variance of

$$\sigma_c^2 - \Sigma W_{ch} (\bar{Y}_{ch} - \bar{Y}_c)^2.$$

The last term is the between-stratum variance, to be gained from proportionate stratification for the subclass itself. Note that for *means of crossclasses the gains of proportionate stratification*, from the between-stratum components $(\bar{Y}_{ch} - \bar{Y}_c)$, *tend to vanish* in proportion to $\bar{M}_{ch} = M_{ch}/N_h$, *and the variance approaches that of simple random sampling*. Durbin (1958) wrote: "...if the proportion in the domain of study is small most of the advantage of stratification has been lost, while only if the proportion is close to unity has the advantage been retained." For example, a gain of 12 per cent for the entire sample would be reduced to 1.2 per cent for a crossclass of 10 per cent.

The simple formulas above neglect only the factors $N_h/(N_h - 1)$ in the precise and general variance, without the assumption $(f_h = f)$ for proportionate selection:

$$\text{var}(\bar{y}_c) \doteq \sum_h \frac{1-f_h}{f_h M_{ch}} \frac{N_h}{N_h - 1} W_{ch}^2 \{T_{ch}^2 - \bar{M}_{ch} (\bar{Y}_{ch} - \bar{Y}_c)^2\} \tag{2.2}$$

$$= \sum_h \frac{1-f_h}{f_h M_{ch}} \frac{N_h}{N_h - 1} W_{ch}^2 \{\sigma_{ch}^2 + (1 - \bar{M}_{ch})(\bar{Y}_{ch} - \bar{Y}_h)^2\}, \tag{2.2'}$$

where

$$T_{ch}^2 = \sigma_{ch}^2 + (\bar{Y}_{ch} - \bar{Y}_c)^2 = \Sigma_i (Y_{chi} - \bar{Y}_c)^2 / M_{ch}.$$

The approximation in the above is due only to the use of a ratio mean. For derivations see Durbin (1958), Hartley (1959) or Kish (1961a, 1965a).

The approach to simple random sampling signalled by the T_{ch}^2 terms is even faster for the difference of the means of two subclasses. Computing the difference of two subclass means, $\bar{y}_c$ and $\bar{y}_b$, from the same sample is a most common technique for measuring relationships. The variance of the difference may be written for proportionate sampling as

$$\text{var}(\bar{y}_c - \bar{y}_b) \doteq \frac{1-f}{fM_c}\sigma_c^2 + \frac{1-f}{fM_b}\sigma_b^2 - (1-f)\sum_h \frac{1}{n_h}\{W_{ch}(\bar{Y}_{ch} - \bar{Y}_c) - W_{bh}(\bar{Y}_{bh} - \bar{Y}_b)\}^2. \tag{2.3}$$

This neglects factors of $N_h/(N_h-1)$. More precisely and generally the above is

$$\text{var}(\bar{y}_c - \bar{y}_b) \doteq$$

$$\sum_h (1-f_h)\frac{N_h}{N_h-1}\left[\frac{W_{ch}^2 T_{ch}^2}{f_h M_{ch}} + \frac{W_{bh}^2 T_{bh}^2}{f_h M_{bh}} - \frac{1}{n_h}\{W_{ch}(\bar{Y}_{ch} - \bar{Y}_c) - W_{bh}(\bar{Y}_{bh} - \bar{Y}_b)\}^2\right]. \tag{2.4}$$

The third term will tend to become relatively small because $n_h = f_h N_h$ is large compared to $f_h M_{ch}$ for small subclasses. Furthermore, since strata typically tend to have "additive" effects, it tends to become negligible due to similar, and therefore cancelling, stratum differentials. Hence, *for the difference of two crossclasses, the gains of proportionate stratification tend to vanish.* The two variances may be computed as if for unstratified random samples, except for weighting for unequal sampling rates.

Controlling the sample sizes, m_{ch} and m_{bh}, for crossclasses is difficult in practical surveys; but where feasible, it suggests optimal allocation for the differences of crossclass means (Sedransk 1957).

The formulas for computing sample variances reflect the above (2.2–2.4):

$$\text{var}(\bar{y}_c) = \Sigma(1-f_h)\frac{w_{ch}^2}{m'_{ch}}\{t_{ch}^2 - \bar{m}_{ch}(\bar{y}_{ch} - \bar{y}_c)^2\}, \tag{2.5}$$

where

$$w_{ch} = F_h m_{ch} / \Sigma F_h m_{ch}, \quad F_h = N_h / n_h, \quad \bar{m}_{ch} = m_{ch} / n_h,$$

$$m'_{ch} = m_{ch}(n_h - 1)/n_h \text{ and } t_{ch}^2 = \sum_j (y_{chj} - \bar{y}_c)^2 / m_{ch}.$$

For the difference of two means approximately

$$\text{var}(\bar{y}_c - \bar{y}_b) \doteq \Sigma(1-f_h)\left(\frac{w_{ch}^2 t_{ch}^2}{m'_{ch}} + \frac{w_{bh}^2 t_{bh}^2}{m'_{bh}}\right). \tag{2.6}$$

For proportionate samples $f_h = f$ and (2.6) becomes

$$\text{var}(\bar{y}_c - \bar{y}_b) = \frac{1-f}{fm_c}\hat{\sigma}_c^2 + \frac{1-f}{fm_b}\hat{\sigma}_b^2, \tag{2.7}$$

where

$$m_c = \Sigma m_{ch} \text{ and } \hat{\sigma}_c^2 \doteq \Sigma_h w_{ch} t_{ch}^2 n_h / (n_h - 1) = \Sigma_h n_h (n_h - 1)^{-1} \Sigma_j (\bar{y}_{chj} - \bar{y}_c)^2 / m_c.$$

Differences between means provide the most common bases for measuring relationships between variables in survey data. Furthermore, they also provide grounds for conjectures about the sampling fluctuations of other analytical statistics (cell B3 in Figure 1) used for measuring relations between variables in stratified samples. These conjectures are necessary for the more complex statistics for which standard theoretical analysis cannot provide measures of sampling fluctuations in accord with stratified designs.

Table 1 contains remarkable confirmation of these conjectures applied to a large and diverse group of chi-squared tests. Eight sets of data from stratified samples were involved, and on each, three sophisticated iterated techniques (Nathan 1972, 1973) were used to fit their stratified selections. Then Professor Nathan agreed to compute the same tests, but with "naïve" *SRS* assumptions. Finally, we computed the ratios of the sophisticated to the naïve results. Note that in the last iterations the ratios are all within 4, and mostly within 1 percent of 1·00. These values measure how close the naïve estimates are to the last iterations, hence may slightly overestimate.

TABLE 1. Ratios of Three Iterated Chi-Squared Tests to SRS Tests
Eight contingency tables based on proportionate stratified samples from Israel: Nos. 1–4 of savings, No. 5 of attitudes, No. 6 of hospital data, No. 7 of poultry medicament, No. 8 of perception experiments. Adapted from data of Nathan (1972)

Data set	*No. of strata*	*Row × columns*	*Sample size*	Nathan's three tests: *First iteration*			*Last iteration*		
				X^2	X_1^2	G	X^2	X_1^2	G
1	4	3 × 3	845	1·028	0·992	1·017	1·004	1·004	1·005
2	4	3 × 3	821	1·088	0·963	1·043	0·999	1·003	1·001
3	4	3 × 3	491	1·740	0·707	1·406	1·011	1·001	1·009
4	4	3 × 3	2581	1·095	0·959	1·049	1·003	1·005	1·003
5	6	2 × 4	500	1·079	0·967	1·040	1·004	1·003	1·003
6	3	2 × 4	120	1·013	0·967	1·009	1·008	0·969	1·007
7	5	2 × 4	269	1·076	0·989	1·043	1·011	1·015	1·011
8	2	2 × 4	81	1·368	0·889	1·186	1·029	1·037	1·029

We conjecture that similar results usually will be obtained on other data, and also on other analytical statistics based on stratified random selections (case B3 in Figure 1). It is a useful conjecture, because appropriate computations of sampling errors for analytical statistics will be difficult for the foreseeable future. Clearly we need more research, both theoretical and empirical. The accumulation of empirical evidence will be most useful, but alone it is slow, subject to sampling fluctuations, and not completely convincing. Theoretical foundations would strengthen and hasten understanding, but alone they cannot suffice. The conjecture involves parameters with empirical content, and it can be contradicted in rare situations.

What attitudes should we adopt concerning the results on crossclass comparisons (2.6, 2.7), and the analogous conjectures about analytical statistics? On the one hand, if justified, we should welcome the convenience of the many formulas available on simple random assumptions. On the other hand, we may be surprised and annoyed that the effects of proportionate stratification tend rapidly to vanish altogether. It is not a result that would suggest itself to intuition.

We may conveniently summarize this section in terms of "design effects," a concept we shall use repeatedly. "The *design effect* or *Deff* is the ratio of the actual variance of a sample to the variance of a simple random sample of the same number of elements" (Kish 1965a, p. 285); this may do here briefly for elements selected with equal probability. This concept, under diverse names, has been long and widely used. The design effect, in proportionate stratified element samples, (a) is less than 1 typically for means based on entire samples; (b) tends towards 1 from below for crossclasses, as these become small; (c) is close to 1 for differences between crossclass means; and (d), we conjecture, is close to 1 for analytical statistics in general.

3. SUBCLASSES AND COMPARISONS IN CLUSTER SAMPLING (C2)

When diverse subclasses are completely segregated in separate clusters and strata (as for regional estimates in area samples) their treatment needs no new methods, although problems of multipurpose design and allocations arise (Kish 1973). However, most frequently we must deal with new problems due to *crossclasses*, that is, subclasses that cut across clusters; examples of these are age, sex and social classes in area samples. Dealing with crossclasses instead of the entire sample produces two principal effects on the sizes of the sample clusters, (1) a decrease in average size by the factor $\overline{M}_c$, and (2) an increase in the coefficient of variation CV(x).

Control of sample size, either with stratification or with PPS (probability proportional to size), is typically imperfect even for the entire sample. Unequal cluster sizes lead to common use of the combined ratio mean, $\bar{y}_c = r = y / x$, and this estimator also serves subclasses. This estimator is rather robust, but not when the denominator x is subject to wild variation. A sufficiently small coefficient of variation CV(x) assures a low bias for the ratio; this is also assumed for deriving its variance (Hansen et al. 1953; Kish 1965a). For subclasses, if small or unevenly distributed, the loss of control over cluster sizes may permit CV(x) to become too large. Our computing programs (Kish et al. 1972) have monitoring features to catch cases when CV(x) is too large for comfort. Actual situations are generally comfortable, because the bias ratio, Bias(r)/Ste(r) = $-\rho_{rx}$CV(x), is small when either component is small; empirical investigations have been encouraging (Kish, Namboodiri, and Pillai 1962).

It is useful to consider a simple model for the variance of crossclass means as a function of the proportion $\bar{M}_c$ in the crossclass. We begin with the well-known $\text{var}(\bar{y}) = \{1+\text{Rho}(n/a - 1)\}S^2/n$, where brackets contain the design effect for clusters of size n/a. Then:

$$\text{var}(\bar{y}_c) = \{1 + \text{Rho}_c(\bar{M}_c n / a - 1)\}S_c^2 / m_c. \tag{3.1}$$

Here

n = the number of elements in the entire sample,

a = the number of clusters in the sample,

$\bar{M}_c$ = the proportion of crossclass elements in the sample,

m_c = $n\bar{M}_c$, the number of crossclass members expected in the sample,

Rho_c = intraclass correlation for the crossclass,

S_c^2 = element variance of crossclass members.

The formula fails to make separate allowance for the effects of unequal sizes of sample clusters, and for the effects of stratification. We may consider these either as having been ignored, or as having been implicitly included in the definitions of the parameters of the equation. It obviously breaks down when $n\bar{M}_c$ approaches a, and should not be taken seriously for such small clusters of crossclass members. Subject to these limitations, the design effect in brackets is viewed as a function of $\bar{M}_c$, and of Rho_c and S_c^2. To the extent that the latter two are relatively constant for a group of similar variables, we see the increase over 1 of the design effect in relation to $\bar{M}_c$: the design effect tends toward 1 for decreasing crossclasses.

The estimation of the variance of $\bar{y}_c$ proceeds according to standard formulas for the ratio mean (y/x) of two random variables. However, small and fluctuating sizes of sample clusters cause problems; but these may be countered with two types of averaging procedures. First, with "combined strata" (Kish 1965a) we can combine primary selections, chosen at random across strata, to form larger units for computing the variance. The procedure introduces no bias into the estimated variance, but increases its variance.

Second, and much more important, are procedures for averaging computed variances for a group of means or other variates. Survey results are produced for so many variates—for different survey variables, and each of these for many subclasses—that the process of computing and presenting sampling errors for all of them usually becomes too costly and cumbersome. But we may compute them for a subset, and then make inferences from their average to the entire set. Furthermore, such averages may be computed for several meaningful sets of results. Another reason for averaging is to produce more stable estimates of variances than sample designs usually yield; averaging should increase the accuracy (lower mean-square error) in spite of introducing some "bias" for the individual variances. To control and reduce that bias, averaging should be confined to groups of similar variates, and should be performed with methods that promise stability within those groups.

A common method for averaging is to plot a graph of the computed design effects deff (or $\sqrt{\text{deff}}$) against subsample sizes m_c. Here Deff and Rho refer to population values and deff and rho to sample values. Using deffs removes two obvious sources of disturbing factors, S_c^2 and m_c, from the averaging of variances computed for different variates. This method assumes a common Deff and Rho$_c$ for variates within a pooled set, as a function of m_c and due chiefly to the same design. The averaging may be done separately for more or less similar groups of the variates. If Rho is constant over values of m_c, then Deff approaches 1 linearly with decreasing m_c (Kalton and Blunden 1973).

Variances for differences between subclass means raise new issues. Formulas for computing $\text{var}(\bar{y}_c - \bar{y}_b) = \text{var}(\bar{y}_c) + \text{var}(\bar{y}_b) - 2\,\text{cov}(\bar{y}_c, \bar{y}_b)$ are merely extensions of variances for single means (Kish 1965a, Section 6.5). Because subclass comparisons are so basic and common in survey analysis, it is annoying to find computations of their variances so rarely even today. At the Survey Research Center they have been imbedded into our computing programs for variances since 1952, and from hundreds of computations we found bases for the inequalities:

$$\frac{S_c^2}{m_c} + \frac{S_b^2}{m_b} < \text{var}(\bar{y}_c - \bar{y}_b) < \text{var}(\bar{y}_c) + \text{var}(\bar{y}_b). \tag{3.2}$$

In words, the variance of the difference of two means from clustered samples shows the design effect of a positive intraclass correlation, but that effect is less than for the separate means. In other words, the covariance is positive, but not great enough to cancel the design effects of the separate means. See Kish (1965a, Section 14.1), and Kalton and Blunden (1973).

This is an empirical statement about the additive nature of positive clustering effects in crossclasses. In actual computations subject to large sampling variations, it has often been contradicted, but in our experience these exceptions were negated when recomputed on similar data. Although it cannot be logically perfect, it is a dependable and useful empirical law. It is clearly preferable to the common practice of assuming equality at either extreme. Most commonly, equality is assumed on the left, as if the samples were simple random. Less commonly, the equality on the right is assumed, with the "conservative" estimate that disregards the covariance of crossclasses selected from the same sample of clusters.

Subclass comparisons represent a basic measure of relations between variables. Our findings about them lead to conjectures about design effects for other statistics that measure relations, such as regression coefficients. When techniques were unavailable for computing variances for them, we conjectured that design effects were greater than 1, but less than for the means of the variables involved (Kish 1957). These conjectures have received empirical confirmations (Kish and Frankel 1970; Frankel 1971) as discussed in the following two sections.

4. COMPLEX STATISTICS FROM COMPLEX SAMPLES (C3)

Here we deal with clustered samples and with statistics more complex than the difference of subclass means. The following section describes how new techniques now make possible the computations of variances that incorporate the complexities of the sample. We shall justify the need for such computations with three broad propositions:

1. Statistics (means, regression coefficients, etc.) approach their population values as the sample size increases.
2. The approach is generally slowed by design effects.
3. The design effects differ for different statistics, for different variables, and for different sample designs.

The three propositions presuppose that we are concerned with finite and real measurements and populations. This philosophy, which should be assumed by anyone involved in the application of statistics, pervades survey sampling theory. Here we extend it from means to measures of relationships. Consider a realistic view of regression:

I. There exists a finite population of N elements. Associated with each of these elements is a vector of k+1 values $Y_i, X_{1i}, X_{2i}, \ldots, X_{ki}$;

$$P = \{(Y_i, X_{1i}, X_{2i}, \ldots, X_{ki}) \mid i = 1, \ldots, N\}.$$

II. Our parameters are numbers B_j such that $\Sigma_i^N (Y_i - \Sigma_j^k B_j X_{ji})^2$ is minimum subject to

$$\Sigma_i^N (Y_i - \Sigma_j^k B_j X_{ji}) = 0.$$

III. Given a sample of n vectors from the population of N vectors our desire is to estimate the parameters B_j.

The regression model stated in (II) does not in practice correspond exactly (or even closely) to the complex relationship among the actual population of vectors. The error term measures (usually in a least-square sense) the extent to which the model departs from the actual complex relations among the population of vectors.

The statistical theory of regression begins at the other end—the theoretical end. It first assumes a basic structure of relationships. Letting $\mathbf{X}_i = (X_{1i}, \ldots, X_{ki})^T$, and $\mathbf{B} = (B_1,\ldots, B_k)^T$, it uses the model $Y_i = \mathbf{B}^T\mathbf{X}_i + \varepsilon_i$. It then makes several strong assumptions:

(A) *linearity*: $E(\varepsilon_i \mid \mathbf{X}_i) = 0$, for all i;
(B) *homoscedasticity*: $\text{var}(\varepsilon_i \mid \mathbf{X}_i) = \sigma^2$, for all i;
(C) *independence between observations*: $\text{cov}(\varepsilon_i\varepsilon_j \mid \mathbf{X}_i,\mathbf{X}_j) = 0$, for all $i \neq j$;
(D) *normality* for the ε_i.

Assumptions (A), (B) and (D) concern the basic structure of the universe of the model, whereas (C) involves independent selections from it. This (or a similar) well-specified model yields several desirable results; the standard least-square estimates $\hat{b}$ are minimum variance, linear, unbiased, normal, etc. Literature and textbooks are written about this pretty model; this is what the research workers find in statistical textbooks, but they find very little about how to reconcile this model with the real population they are investigating.

Specifically, we need the real population model to describe the principal effect of a complex selection design: assumption (C) fails to hold. Clustered selection tends to introduce positive correlations between the errors of the model and, as will be shown later, these often have serious consequences.

The first proposition states that the correlation between elements *does not prevent* the approach of "first-order statistics" based on large samples to their respective population values (parameters). By *first-order statistics* we mean estimates of *parameters of the population distribution*; these parameters are (a) the substantive objectives of research, (b) based on *all* population elements taken *individually* and (c) unaffected by the sample design (for example, means, element variances, regression and correlation coefficients). On the other hand, by *second-order statistics* we mean *measures of variation* (variance, standard error, mean-square error) of first-order statistics. They estimate second order parameters, e.g., $\text{E}\{y - \text{E}(y)\}^2$, that are based on aggregate results of samples obtained under specific designs, and they are affected by the correlations between elements induced by the designs.

We need laws of large numbers and central limit theorems for samples in which elements are not independent. A difficult problem here is to specify what a "large" sample may be. It is not sufficient merely to say that it contains a large number of elements if these come highly correlated from a few clusters (i.e., the primary selections). To allow the clusters to become very numerous would suffice, but that would place unrealistic demands on many practical sample designs. Large numbers of primary selections are often neither possible nor necessary; a moderate number will suffice if the elements are numerous and the correlations between them not too great. We recognize that a rigorous theoretical formulation of the last statement, so needed for practical work, stands as a difficult challenge for theoretical statisticians. We shall merely sketch first what we know today about this topic in terms of the unbiased (or almost unbiased) nature of the expected values of the results of probability samples, without assuming independence of selections.

The weighted sum of sample observations is an unbiased estimator of the population aggregate: $E(\Sigma_j^n y_j / P_j) = \Sigma_i^N Y_i$, where P_j is the selection probability of the jth sample element. This is also true of vectors $\mathbf{y}_j$ and $\mathbf{Y}_i$ for several variables. (This expectation seems to be the basis for selections without replacement.) It is also true of moments of sample values, such as $\Sigma y_j^k / P_j$ or $\Sigma y_j^k x_j^m / P_j$, where k and m are real numbers, usually integers. Means based on these sums are also unbiased if the denominator is fixed; otherwise they are consistent and close (Kish 1965a, Section 2.8). Often the denominator is a correlated sum of sample observations as in the ratio mean $r = (\Sigma y_j / P_j) / (\Sigma x_j / P_j)$; empirical evidence is reassuring (Kish et al. 1962). Many statistics are complex versions of functions of ratio means, e.g.,

$$b = (\Sigma y_j x_j / P_j)/(\Sigma x_j^2 / P_j).$$

We conjecture that the biases in estimates of regression and correlation coefficients (as with ratio means) are functions of the departure of actual population relations from assumptions (A) – (D) above, and of sample sizes.

Analytical expressions would be complex and are not available now. Hence our inferences must rely heavily on accumulating empirical evidence. In Table 2 we summarize results on the biases of five types of estimators from a complex sample (Frankel 1971). The results from this large study, described in Section 5, are reassuring. Even for small sizes (12 units from 6 strata) the relative biases (ratio of bias to parameter) are small, and they decrease with the sample size for means and for three diverse types of regression coefficients. Because the standard errors also decrease the bias ratios (ratio of bias to standard error) fluctuate, but they remain small or moderate. The multiple correlation coefficient shows worse behaviour; but we think this is due to the basic defect of the estimator, rather than to the design's complexity.

In addition to small biases, in Table 3 the same study also gives us comforting news about the approaches to normality of complex statistics from complex samples. The approach seems to be good even for 12 units from 6 strata, and it improves markedly in moderate sample sizes (see also the tables in Section 5).

The approach to population values promised by the first proposition must be accompanied by the second proposition's warning about the slowing of the approach caused by positive correlations among selected elements. The extent of the slowing is simply expressed by the design effect on the variance of the mean: Deff $= \{1 + \rho(\bar{n} - 1)\}$. This expression is well known for random selections of equal clusters. It also has been extended for sample means obtained from probability samples in general, with ρ expressing the pairwise correlation of sampled elements, and $\bar{n}$ a parameter of the selection design (Tharakan 1969). The same fruitful approach was used by "Student" (1909) in a paper that, surprisingly, has been neglected in the literature of sampling.

TABLE 2. Biases in Complex Samples of Five Types of Estimators
Averages of relative biases = $bias(p_s)/P$, and of the bias ratios $= bias(p_s)/ste(p_s)$.
Data from Tables 5.1 and 5.2 of Frankel (1971)

Estimator	*No. in average*	6 *Strata design* 300 *samples*		12 *Strata design* 300 *samples*		30 *Strata design* 200 *samples*	
		Relative bias	*Bias ratio*	*Relative bias*	*Bias ratio*	*Relative bias*	*Bias ratio*
Ratio means	8	0·00425	0·04653	0·00216	0·03909	0·00295	0·08520
Simple correlations	12	0·06972	0·19847	0·05399	0·19013	0·01748	0·09896
Regression coefficients	8	0·04978	0·04429	0·03320	0·07108	0·02776	0·05558
Partial correlations	6	0·12333	0·21155	0·08365	0·20165	0·05863	0·17358
Multiple correlations	2	0·16002	0·72855	0·11115	0·52975	0·04670	0·29105

TABLE 3. Approach to Normality in Complex Samples of Five Types of Estimators
Averages of relative frequencies within stated intervals of statistics $(p_s - \bar{p})/\text{ste}(p_s)$.
Data from Table 5.3 of Frankel (1971)

Intervals	± 2·576	±1·960	± 1·645	± 1·282	± 1·000
P for normal deviate	0·9900	0·9500	0·9000	0·8000	0·6827
6-strata design (300 samples)					
8 Ratio means	0·9875	0·9533	0·9067	0·8075	0·6929
12 Simple correlations	0·9861	0·9533	0·9039	0·8061	0·6986
8 Regression coefficients	0·9742	0·9387	0·9067	0·8392	0·7425
6 Partial correlations	0·9822	0·9444	0·9050	0·8178	0·7039
2 Multiple correlations	0·9767	0·9467	0·9167	0·8417	0·7233
12-strata design (300 samples)					
8 Ratio means	0·9900	0·9550	0·8987	0·8804	0·6750
12 Simple correlations	0·9872	0·9461	0·9003	0·8064	0·6919
8 Regression coefficients	0·9808	0·9492	0·9092	0·8192	0·7079
6 Partial correlations	0·9872	0·9506	0·9050	0·8133	0·6861
2 Multiple correlations	0·9733	0·9383	0·9067	0·8350	0·7233
30-strata design (200 samples)					
8 Ratio means	0·9887	0·9544	0·9100	0·8069	0·6744
12 Simple correlations	0·9900	0·9567	0·9017	0·7929	0·6867
8 Regression coefficients	0·9869	0·9444	0·9019	0·8144	0·6912
6 Partial correlations	0·9875	0·9575	0·8958	0·8000	0·6942
2 Multiple correlations	0·9900	0·9525	0·9100	0·8275	0·6825

Similar analytical expressions, in a few useful parameters, are needed for the effects of design on the variances of complex statistics. The effective content of the expressions must be statements about the structure of population variables, and the effect of the selection design on the variates studied. Meanwhile we must accumulate evidence about the magnitudes of these effects. Some of this empirical evidence is shown in Table 4, which summarizes results of Kish and Frankel (1970) and Frankel (1971).

We would like to know how the design effects tend to differ for different statistics obtained from complex selection designs. In addition to scientific curiosity, we have practical needs to discover reasonable regularities.

Often it is difficult to compute standard errors for all statistics, or to compute them with adequate precision. Hence reasonable conjectures would be most useful to researchers. Theory will help eventually, but it will need to be buttressed with empirical content. Our present conjectures have a light theoretical framework and some empirical background. They are phrased in terms of design effects Deff(b) for complex statistics b from complex samples.

i. Deff(b) > 1. In general, design effects for complex statistics are greater than 1. Hence standard errors based on simple random assumptions tend to underestimate the standard errors of complex statistics.
ii. Deff(b) < Deff($\bar{y}$). The design effects for complex statistics tend to be less than those for means of the same variables. The latter, more easily computable than the former, tend to be "safe" overestimates. (We noted earlier the "pathology" of multiple R.)
iii. Deff(b) is related to Deff($\bar{y}$). For variates with high Deff($\bar{y}$), values of Deff(b) tend also to be high. See Kish and Frankel (1970, Section 7) for a set of striking results.
iv. Deff(b) tends to resemble the Deff for differences of means. The latter is a simple measure of relations for which values of deff are easily computed, and for which (i)–(iii) also hold.

TABLE 4. Values of √Deff for Five Types of Estimators from Three Complex Samples
Set A from Table 2, Set B from Table 3 of Kish and Frankel (1970); Set C from Table E-1 of Frankel (1971)

	Sample set		
	A	B	C
Ratio means	1·106	1·800	1·438
Simple correlations	1·096	1·262	1·355
Regression coefficients	1·015	1·295	1·106
Partial correlation coefficients	1·041	1·400	1·360
Multiple correlation coefficients	NA	1·465	1·894

v. Deff(b) tends to have observable regularities for different statistics. This is a hope based on theoretical considerations; confirming results would help us make useful conjectures.

A simple model of the above would be

$$\text{Deff}(b_g) = 1 + f_g \{\text{Deff}(\bar{y}) - 1\}, \tag{4.1}$$

with Deff($\bar{y}$) > 1, $0 < f_g < 1$ and f_g specific to the variables and statistic denoted by g.

5. THREE METHODS FOR COMPUTING SAMPLING ERRORS

We shall compare here three basic methods for computing sampling errors from stratified clustered sample designs: The Taylor expansion method (*TAYLOR*), the method of balanced repeated replication (*BRR*) and the method of jackknife replication (*JRR*). These names are convenient, but not unique. The comparisons are based on a large-scale empirical study which contains a fuller discussion of all three (Frankel 1971).

In that study for the sake of simplicity, we used sample designs with equal numbers A of primary units within all strata, and with two of those units selected from each stratum with random choice, without replacement and without subsampling. Thus, we have a clustered stratified sample design, where each population element has equal probability ($f = 2/A$) of appearing in the sample. However, all three methods for computing variances can deal with appropriate weighting to compensate for unequal probabilities of selection within and between strata. The extension to any number of primary selections per stratum is straightforward for the *TAYLOR*. With modifications and with the use of collapsed and combined strata techniques, methods *BRR* and *JRR* also can be applied to other sample designs (Kish and Frankel 1970, Section 12).

5.1. Taylor Expansion Method

The use of the Taylor expansion for computing variances of ratio means has been described in textbooks. Deming (1960), Kish (1965a) and Woodruff (1971) describe its use for estimating variance for other functions of the basic sample sums. The method is also known as the linearization or delta (δ) method. A detailed published extension of this method to more complex first-order estimates specific to survey sampling is due to Tepping (1968). This method produces an approximate estimate for the variance of a first-order statistic, based on variances of the linear terms of the Taylor expansion of the statistic (Brillinger and Tukey 1964).

Let $\mathbf{y} = (y_1,\ldots,y_i,\ldots,y_k)^{\mathrm{T}}$ be a vector of sample totals; the sample total y_i is the aggregate of primary selection totals y_{iha}, where the indexes h and a denote strata and primary selections:

$$y_i = \sum_h \sum_a y_{iha}, \text{ where } h = 1,\ldots,H \text{ and } a = 1,2. \tag{5.1}$$

The y_{iha} values are sums over primary selections of element values y_{ihaj}, weighted by the inverses of selection probabilities, so that $\mathrm{E}(y_i) = KY_i$, the corresponding population value, with K some convenient constant.

The y_i's are chosen so that $g(Y)$ is the parameter we wish to estimate with the first-order statistic $g(y)$. Using the linear terms of the Taylor expansion $g(y)$ near $g(Y)$, the estimator of $\mathrm{var}\{g(y)\}$ is given by

$$\mathrm{var}\{g(y)\} = (1-f)\sum_h \left\{ \sum_i \frac{\partial g(Y)}{\partial Y_i} y_{ih1} - \sum_i \frac{\partial g(Y)}{\partial Y_i} y_{ih2} \right\}^2. \tag{5.2}$$

$\partial g(Y)/\partial Y_i$ is the partial derivative of $g(Y_i)$ with respect to the variable Y_i, and taken at the expected value Y_i; we must use $\partial g(y)/\partial y_i$ as sample estimators of $\partial g(Y)/\partial Y_i$.

5.2 Balanced Repeated Replication (*BRR*) Methods

The approach of repeated replications was developed at the U.S. Census Bureau (Deming 1956) from basic replication concepts (Mahalanobis 1946) and orthogonal balancing was added later (McCarthy 1966; Kish and Frankel 1969). The *BRR* methods can be briefly described as follows. Assume that we have a stratified sample design with two primary selections from each stratum. Let S denote the entire sample; let H_i denote the ith half-sample formed by including one of the two primary selections from each of the strata; and let C_i denote the ith complement half-sample, formed by the primary selections in S not in H_i. The method we used for choosing the pattern of primary units that form the half-samples, H_i and C_i, is known as "full-orthogonal balance." If we form k half-samples $H_1,\ldots,H_k$, and corresponding complement half-samples $C_1,\ldots,C_k$, then we form *BRR* second-order estimators in one of two ways:

$$\mathrm{var}_{BRR\text{-}S}\{g(S)\} = \frac{1-f}{2k}\sum_{i=1}^{k}[\{g(H_i)-g(S)\}^2 + \{g((C_i)-g(S)\}^2], \tag{5.3}$$

or

$$\mathrm{var}_{BRR\text{-}D}\{g(S)\} = \frac{1-f}{4k}\sum_{i=1}^{k}\{g(H_i)-g(C_i)\}^2. \tag{5.4}$$

Each of the two components in the *BRR-S* form also may be used separately for a less costly but less precise second-order estimator (Kish and Frankel 1970; Frankel 1971).

5.3 Jackknife Repeated Replication (*JRR*) Methods

The term *JRR* refers to a set of second-order estimation methods motivated by jackknife estimation procedures (Brillinger 1964) and by *BRR*. With *BRR* methods, each of the k replications estimates the variance of the entire sample. With the *JRR* methods, each replication measures the variance contributed by a single stratum. The technique used to measure these contributions to the variance from the strata was suggested by the jackknife method for variances; it was formed by leaving out replicates from the sample. The specific procedures below were first used and described in Frankel (1971).

Assume again that we have two selections from each of H strata. Let S denote the entire sample; let J_i ($i = 1,\ldots,H$) denote the replicate formed by removing from S one selection in the ith stratum, but including twice the other selection in that stratum. Let CJ_i ($i = 1,\ldots,H$) denote the complement replicate formed from S by interchanging the eliminated and duplicated selections in the ith stratum.

Two *JRR* estimators of variance are defined as follows:

$$\text{var}_{JRR\text{-}S}\{g(S)\} = \frac{1-f}{2}\sum_{i=1}^{h}[\{g(J_i)-g(S)\}^2+\{g(CJ_i)-g(S)\}^2] \tag{5.5}$$

and

$$\text{var}_{JRR\text{-}D}\{g(S)\} = \frac{1-f}{4}\sum_{i=1}^{h}\{g(J_i)-g(CJ_i)\}^2. \tag{5.6}$$

5.4 Accuracy of the Three Methods

If we assume first-order estimates $g(y)$ or $g(S)$ that are linear functions of statistics, then a number of exact analytical results can be derived for all three variance estimation methods: *TAYLOR, BRR* and *JRR*. However, when we consider the first-order estimates actually used by survey analysts (e.g., ratios, correlation and regression coefficients) we find that usable methods for exact (non-approximate, non-asymptotic) results evade us. Since estimators of sampling errors are needed now, we follow a tradition among statisticians that goes back at least as far as 1907, when "Student" (1908) selected 750 simple random samples to evaluate his theoretical derivation of the distribution of the sample mean divided by its estimated standard error.

We empirically evaluated and compared all three variance estimation methods, using three clustered and stratified sample designs. These called for paired selections (approximately 14 elements each) from 6 strata (approximately 170 elements), 12 strata (approximately 340 elements) and 30 strata (approximately 847 elements). The coefficients of variation CV(x) of the sample sizes were 0·19, 0·13 and 0·074 respectively. Thus we imposed rather harsh, demanding tests on the empirical validity of these methods. For a more complete description of this study, which makes use of data from the Current Population Survey of the U.S. Bureau of the Census, the reader is directed to Frankel (1971).

The three methods were used to compute sampling errors of several statistics: ratio means, simple correlations and multiple regression coefficients. *BRR* and *JRR* methods also were used to compute sampling errors for partial and multiple correlation coefficients, but for the *TAYLOR* method we were unable to find tractable forms for the partial derivatives.

For standards of comparisons we used robust fundamentals based on the definitions of means and variances. The bias of first-order statistics was judged against population parameters, based on the entire population of 45,737 households in 3,240 primary sampling units. The statistics were based on 300 independent drawings for the 6-strata and the 12-strata samples, and 200 drawings for the 30-strata sample. We used 8 variables in 2 multiple regression equations, each with 3 predictor variables. Thus the 8 coefficients of regression, 12 of simple correlation, and 6 of partial correlation, represent averaging 300 × (8, 12 and 6) statistics in the 6 and the 12 strata, and 200 × (8, 12 and 6) in the 30 strata. The total of about 400,000 complex computations made good use of modern computers.

We could not afford to compute *all* the possible combinations for the second-order statistics such as $\mathrm{E}\{\bar{y}-\mathrm{E}(\bar{y})\}^2$ for the variance, and $\mathrm{E}\{\bar{y}-\bar{Y}\}^2$ for the mean-square error; these were averaged from the 300, 300 and 200 statistics. To these standards for second-order statistics the corresponding results of the three methods were compared.

Here we can only summarize a large set of results. In the original publication (Frankel 1971), the large volume of details for distinct statistics provides firmer bases for the tables here and for the conclusions derived from them. For first-order estimators the biases on the average were relatively small; this was true both in terms of the relative bias (bias/estimate) and of the bias ratio (the ratio of bias to standard error). These were in the neighbourhoods of 0·05 for means and regression coefficients, and of 0·1–0·2 for simple and partial coefficients; the multiple correlation coefficient was around 0·3–0·7 and clearly presented problems as noted above (Table 2). As for the variability of first-order estimators, the strong design effects were shown in Table 4.

We are chiefly concerned here with the performances of the three methods of computing variances. These are summarized in Tables 5 and 6. Table 5 summarizes the averages of relative biases for the mse's (three mean-square errors), and the averages of their dispersions, measured as mean-square errors of the mse's. The corresponding results for the computed variances (Frankel 1971, Tables 6.1 and 6.3) were close to these results shown for mse's, because the biases that would separate them ($\mathrm{mse} = \mathrm{var} + \mathrm{bias}^2$) were small or negligible.

Table 6 presents results for the criterion we consider most significant because it measures directly the inputs of the three methods into inference statements. Against the accepted standards of probability levels (for 6, 12 and 30 degrees of freedom), this table shows the levels actually attained on the average by the sample functions

$$t(s) = \frac{g(s)-\mathrm{E}\{g(s)\}}{\mathrm{ste}\{g(s)\}}. \qquad (5.7)$$

TABLE 5. Accuracy for Mean-Square Errors (MSE) for Three Methods of Error Computations
Adapted from Tables 6.2 and 6.4 of Frankel (1971)

Relative bias of mse *Bias (mse) / MSE*				*Relative MSE of mse* *MSE (mse) / MSE²*		
BRR	*JRR*	*TAYLOR*		*BRR*	*JRR*	*TAYLOR*
			6 Strata			
0·032	−0·019	−0·041	Means	0·543	0·501	0·483
0·188	−0·006	−0·075	Regression coefficients	4·207	2·803	2·437
−0·040	−0·163	−0·278	Simple correlations	0·772	0·678	0·431
0·029	−0·153	–	Partial correlations	0·989	0·852	–
−0·297	−0·426	–	Multiple correlations	1·168	1·079	–
			12 Strata			
0·064	0·035	0·022	Means	0·437	0·418	0·381
0·097	−0·010	−0·034	Regression coefficients	1·425	1·180	1·134
−0·072	−0·159	−0.243	Simple correlations	0·530	0·483	0·326
−0·013	−0·157	–	Partial correlations	0·686	0·603	–
−0·330	−0·439	–	Multiple correlations	0·993	0·906	–
			30 Strata			
0·004	−0·011	−0·014	Means	0·156	0·152	0·147
0·068	0·019	0·014	Regression coefficients	0·608	0·558	0·554
−0·036	−0·104	−0·159	Simple correlations	0·405	0·349	0·231
0·012	−0·101	–	Partial correlations	0·578	0·497	–
−0·161	−0·286	–	Multiple correlations	1·050	0·895	–

The proportion of times that the ratio $t(s)$, computed for each sample, fell within fixed symmetric intervals t_p (± 2·576, ± 1·960, ± 1·645, ± 1·282) are shown against the Student's P_t expected probabilities. Relative frequencies are shown for three methods: *BRR-S, JRR-S* and *TAYLOR* (from Frankel 1971, Tables 7.4, 7.8 and 7.1 respectively). We omitted data for *BRR-D, BRR-H, BRR-C* and for *JRR-D, JRR-H, JRR-C*; the differences of these from the results shown for *BRR-S* and *JRR-S* are less important and are discussed elsewhere (Kish and Frankel 1970; Frankel, 1971). The latter gives Tables (7.1–7.9) for all nine variations of the three methods and for asymmetric (one-sided) intervals, for which the performances were less satisfactory (because of skewed distributions) especially for the 6-strata design. Many tables of Frankel (1971), Appendices) give results for separate variables: 8 means, 8 regressions, 12 simple, 6 partial and 2 multiple correlation coefficients.

We derive from these tables several summary conclusions useful for survey sampling.

1. All three methods gave good results for several statistics: means, coefficients of regression and of correlation, simple and partial. The mse values

TABLE 6. Relative Frequencies of P_t Intervals for Three Methods of Error Computations
Value of $t=\{g - E(g)\}/\{\text{ste}(g)\}$ computed, then for each type of (statistic × design × method) the proportions that fall within $\pm t$ intervals. Adapted from Tables 7.1, 7.4, 7.8 of Frankel (1971)

BRR	*JRR*	*TAYLOR*		*BRR*	*JRR*	*TAYLOR*
	$t = \pm 2{\cdot}576$				$t = \pm 1{\cdot}960$	
			6 Strata			
	$P_t = 0{\cdot}9580$				$P_t = 0{\cdot}9023$	
0·956	0·951	0·948	Means	0·904	0·894	0·888
0·966	0·952	0·942	Regression coefficients	0·915	0·883	0·873
0·948	0·931	0·916	Simple correlations	0·886	0·863	0·837
0·957	0·937	–	Partial correlations	0·908	0·868	–
0·935	0·912	–	Multiple correlations	0·895	0·840	–
			12 Strata			
	$P_t = 0{\cdot}9757$				$P_t = 0{\cdot}9264$	
0·972	0·971	0·971	Means	0·922	0·920	0·919
0·973	0·968	0·966	Regression coefficients	0·934	0·916	0·912
0·955	0·944	0·933	Simple correlations	0·897	0·875	0·859
0·966	0·949	–	Partial correlations	0·912	0·888	–
0·920	0·895	–	Multiple correlations	0·850	0·813	–
			30 Strata			
	$P_t = 0{\cdot}9848$				$P_t = 0{\cdot}9407$	
0·983	0·982	0·982	Means	0·944	0·943	0·943
0·983	0·980	0·979	Regression coefficients	0·938	0·933	0·932
0·973	0·966	0·965	Simple correlations	0·911	0·902	0·898
0·955	0·946	–	Partial correlations	0·897	0·879	–
0·913	0·895	–	Multiple correlations	0·825	0·793	–
	$t = \pm 1{\cdot}645$				$t = \pm 1{\cdot}282$	
			6 Strata			
	$P_t = 0{\cdot}8489$				$P_t = 0{\cdot}7529$	
0·845	0·836	0·833	Means	0·756	0·742	0·738
0·860	0·830	0·815	Regression coefficients	0·768	0·731	0·717
0·836	0·805	0·774	Simple correlations	0·739	0·699	0·671
0·855	0·810	–	Partial correlations	0·766	0·705	–
0·823	0·780	–	Multiple correlations	0·738	0·660	–
			12 Strata			
	$P_t = 0{\cdot}8741$				$P_t = 0{\cdot}7760$	
0·870	0·866	0·865	Means	0·769	0·765	0·763
0·875	0·854	0·850	Regression coefficients	0·773	0·750	0·744
0·844	0·826	0·803	Simple correlations	0·758	0·731	0·705
0·869	0·826	–	Partial correlations	0·754	0·711	–
0·790	0·738	–	Multiple correlations	0·677	0·633	–
			30 Strata			
	$P_t = 0{\cdot}8896$				$P_t = 0{\cdot}7903$	
0·891	0.889	0·888	Means	0·789	0·786	0·784
0·890	0.884	0·884	Regression coefficients	0·789	0·779	0·778
0·862	0.847	0·836	Simple correlations	0·753	0·735	0·723
0·844	0.819	–	Partial correlations	0·753	0·725	–
0·735	0.703	–	Multiple correlations	0·638	0·595	–

have small relative biases (Table 5), and the proportions of $t(s)$ values conform well to P_t expectations (Table 6). We now have three good methods for these difficult tasks.

2. The relative biases and the $t(s)$ proportions improve as expected for increasing sample size, from 6 to 12 to 30 strata.
3. The results for coefficients of multiple correlation are poor on all criteria, and they fail to improve for larger samples. We conjecture that this pathological behaviour does not result from the complexity of the selection design, but from more basic faults of the statistic.
4. The *BRR* method was consistently the best when judged by the criterion we believe most significant: the closeness to expected P_t of the actual proportions of $t(s)$ values. The *BRR* performed consistently better than *JRR*, and *JRR* performed better than *TAYLOR*. The *BRR*'s better performance is particularly noticeable for simple and partial correlation coefficients, where *JRR* and *TAYLOR* are less satisfactory.

   The weaker performance of *JRR* and *TAYLOR* for correlation coefficients on the $t(s)$ criterion is probably associated with the negative relative biases of the mse measures of order –0·10 to –0·16 for *JRR* and –0·16 to –0·28 for *TAYLOR*. Moderate positive biases for betas with *BRR* may explain its "conservative" high proportion of $t(s)$ values for 6 and 12 strata.
5. The variability, measured with mean-square errors in Table 5, shows interesting and surprising results. The values generally are greater than we should expect. Also, the decrease (consistency) for larger sample size is weaker than we expected. These results contrast sharply with the much better (and, we believe, more significant) results for proportions of $t(s)$ values in Table 6. Perhaps large numerators (deviations) and denominators tend to occur jointly with strong positive correlations. This possibility deserves further investigation.

   The variability is consistently lowest for *TAYLOR* and highest for *BRR*. The differences are small, and apparently have less effect than the relative biases on the closeness of $t(s)$ values.

   Clear differences in variabilities appear for the five kinds of statistics. Relative variation is least for means, and consistently decreases with larger sample sizes. For regression coefficients it is much greater, but also consistent. For correlation coefficients, both simple and partial, variability is somewhat greater than for means, and decreases for larger samples are rather weak.
6. When judged by several criteria, none of the three methods showed up strongly and consistently better or worse. The choice among methods may depend in most cases on relative costs and simplicity, and these will vary with the situation and with the statistics. *TAYLOR* methods may be best for simple statistics like ratio means, and *BRR* and *JRR* for complex statistics like coefficients in multiple regressions.

6. COMPUTING SAMPLING ERRORS

For complex samples (Row C in Figure 1), computing sampling errors seems both necessary and difficult. These computing methods are the necessary tools for inference, because the alternatives perform poorly in many practical situations. The difficulties must be great because actual computations still occur only as rare exceptions, rather than as the normal complement they should be for probability samples. The failure to compute sampling errors is a widely known scandal among practitioners. What difficulties cause this widespread evasion of an admitted duty? The list of difficulties to be overcome can also serve as *criteria for good practical programs.*

i. *Complexity.* Computing second-order statistics is inherently more complicated than computing the first-order statistics they serve. This problem becomes more acute for complex multivariate statistics.
ii. *Approximations.* Computations of variance typically involve approximations, and strategy involves a choice among them for validity and utility. We compared three methods; references contain further discussions (also see Kish 1965a, Sections 6.5, 8.6, 12.11, 14.1, 14.2).
iii. *Data input.* This appears as the most important component in machine-time, because surveys are typically large-scale, involving thousands of cases. It weighs heavily in large-scale computations for multipurpose surveys, and especially for many subclasses.
iv. *Multipurpose.* Surveys typically concern many variables, and these require many separate computations. The input for thousands of cases is multiplied by the number of distinct survey variables.
v. *Subclasses.* Survey statistics, and errors, are needed not only for the entire sample, but typically also for many domains. This further increases and complicates the volume of computations.
vi. *Interface.* For computing sampling errors we often need a triangular interface involving the researcher, the sampling statistician and the computer specialist. This is expensive, but omitting a side of this triangle without adequate planning can be dangerous.

We cannot present here a comprehensive treatment of these problems. In general we believe that for complex statistics (cell C3 in Figure 1), the strongest emphasis should be placed on dealing with complexity and with valid approximations. Here the *TAYLOR* method becomes too complex for practical work and *BRR* or *JRR* is needed. But for simpler statistics (cells C1 and C2), we think that the *TAYLOR* method offers a better approach to dealing with the last four criteria.

These approaches are incorporated in a set of computing programs we have designed and used over the years; more recently they have been made available to others. *SEPP* (Sampling Error Program Package) is a set of three programs which we have used for routine computations of sampling errors. Manuals and descriptions appear in a book with that title (Kish, Frankel, and Van Eck 1972), and a *SEPP* package of tape plus manuals may also be purchased. For brief descriptions see Kish (1971).

7. DISCUSSION OF ALTERNATIVES

The approximations proposed here for sampling errors should be useful for research workers involved with applications. We are concerned here chiefly with analytical statistics (B3 and especially C3 in Figure 1) and somewhat less with subclass means and their differences (B2 and C2). We know that we have raised more questions than we have answered. There are important contributions to be made by both theoretical and empirical investigations; we think it preferable that they be performed jointly. We urge the importance of the task by contrasting the proposed methods with the alternatives below.

1. To restrain analysis of data to those statistics for which mathematics provides explicit distribution theory for complex samples. That poses difficult and distant goals. Meanwhile there exist irrepressible demands for the analysis of data provided by survey technology and facilitated by computer technology.
2. To restrain samples to independent selections for which distribution theory is adequate. This would be wise sometimes, but often it is not practical because it would be too expensive. Furthermore, analyses of relations are often secondary to the collection of descriptive data, for which complex selections are much more efficient.
3. To omit computing and presenting sampling errors. This is common practice. The "first-order statistics" seem to be reasonably well behaved, and rigorous proofs of that may be obtained easier and sooner than for "second-order statistics" of variability. We believe, however, that this proposal is less acceptable to most than method 4.
4. To compute sampling errors with the available formulas based on independent observations. This often gives bad underestimates; our evidence will be buttressed by many others. The magnitudes of these mistakes testify to the magnitude of this problem (in ironic contrast with many research papers).
5. To select simple replicated (interpenetrating) samples, and to compute sampling errors using simple replications or jackknife modifications. This fundamental idea has much (and many) to recommend it, and it is useful sometimes. But more often in practice it is unsatisfactory for numerical reasons. If the replications are simple and few, estimates of error are poor (perhaps worse than those of method 4). Even averaging may not rescue them sufficiently and practitioners have been disappointed. On the other hand, many replications sacrifice stratification, simplicity and perhaps validity. Here we think that *BRR* is a better answer (Kish and Frankel 1971).

9

METHODS FOR DESIGN EFFECTS

LESLIE KISH

1. INTRODUCTION

I aim here to provide a simple, practical manual on where, why, and how values of deft should be computed for the sampling errors of statistics from complex survey samples. Deft, or design effects, are "only" tools rather than a theory or even a method. However, they are based on concepts that are theoretical, perhaps even philosophical, based on a different paradigm from the random variables (IID) of prevailing mathematical statistics, as we shall discuss briefly.

Deft are used to express the effects of sample design beyond the elemental variability (S^2/n), removing both the units of measurement and sample size as "nuisance parameters." With the removal of s, the units, and the sample size n, the design effects in the sampling errors are made generalizable ("transferable") to other statistics and to other variables, within the same survey, and even to other surveys.

I shall restrict this exposition to basic, essential approximations, which are sufficient in most cases. I must also give simple advice that may hold only in most common situations, say 0.95 or 0.99 of practical situations. I shall note with (E.x) possible exceptions relegated to the Appendix; for example (E.4) denotes a remark relegated from Section 4 to the Appendix. Thus forward momentum on basic concepts can be retained.

Users are computing design effects more often and in greater volume now, since the arrival of several computing packages. This overview should help these users to compute and to use them correctly—if not always, more often than now.

The exposition is divided into the following brief sections.

1. Introduction
2. Definitions of Design Effects
3. When Deft are Unnecessary

Reprinted with permission from *Journal of Official Statistics* (1995), 11, 55–77.
I gratefully acknowledge careful and creative suggestions from James M. Lepkowski and Thomas Piazza, and this would be a better paper with more time to accept all of them.

4. Necessary Deft
5. Other Needed Sampling Errors 6. Deft for Subclasses and Differences
7. Deft for Complex Statistics
8. Weighting and Generalization
9. Computing Deft.

2. DEFINITIONS OF DESIGN EFFECTS

Definitions should be public servants. Servants rather than our masters: thus rather than arguing about their titles and what they "really" are, we can designate what they should do for us. Public rather than private: to avoid confusion we users must agree on how we can use them.

Design effects have been first defined and commonly used for sample means $(\bar{y})$, and based on *properly* computed actual variances, $\text{var}(\bar{y})$, as

$$\text{deff} = \frac{\text{var}(\bar{y})}{(1-f)(s^2/n)} . \tag{2.1}$$

But, nowadays I and many others prefer a slightly different definition

$$\text{deft} = \sqrt{\frac{\text{var}(\bar{y})}{s^2/n}} . \tag{2.2}$$

Here s^2 denotes computed element variances, based on an achieved sample size n, selected with *EPSEM*, i.e., equal (fixed) sampling rate (probability) of f. Unequal probabilities and weighted samples are postponed to Section 8. Furthermore, applications to other, more complex (analytical) statistics are postponed to Sections 6 and 7.

We must note some strategic choices made here. First, the technique must be made simple (after the complex computations of variances) in order to allow easy computations of $\text{deft}(\bar{y})$ values for *all*, or most, or many survey means. This is necessary, because of the large variations of $\text{deft}(\bar{y})$ we found for different variables *within* the same surveys (Section 4). Second, we propose methods for inferring deft for more complex statistics from the $\text{deft}(\bar{y})$ based on regularities found empirically (Sections 6 and 7). Third, computations of $\text{var}(\bar{y})$ include the effects of clustering, stratification and weighting, and separation of the effects would be difficult (Section 8). The methods for computing *proper* estimates $\text{var}(\bar{y})$ of the true variances of $\bar{y}$ and of other statistics are beyond the scope of this paper. This is a central topic of most textbooks and many articles on survey sampling.

These strategies indicate changes (by me and others), based on empirical results found during the past 40 years. Earlier we computed values of $\text{deft}(\bar{y}_i)$ for a few variables i, with the stated aim of generalizing from their averages to all other variables on the same survey. Then we had assumed that the average design effect would allow us to generalize to other variables. But decades of computations with ever increasing numbers of deft facilitated by faster computers convinced many of us that some of the variables can have much larger values of deft, and generalizing to them is not safe. Second, however, we also found that values of deft($\bar{y}$) were most useful for generalizing to deft(b) for other statistics (b): Sections 6 and 7. Third, we also found that weighting posed more difficult and more common problems than we had anticipated: Section 8.

We also found that (2.2) has four advantages over (2.1), although the numerical differences are commonly trivial.

1. Deft is expressed in the same units as the factors in the intervals $\bar{y} \pm t\text{ste}(\bar{y})$, which it must chiefly serve. Thus it can appear directly as a multiplier either in t or ste($\bar{y}$).
2. It is easier to type deft^2 than $\sqrt{\text{deff}}$, when one of these is needed; and deft is needed more often.
3. The factor $(1 - f)$ when computed for the numerator (not often) may be considered as part of the design effects; the bases are variances of "unrestricted" sampling; i.e., simple random with replacement (IID).
4. The factor $(1 - f)$ may be difficult to compute when the selection is not *EPSEM* (equal probability selection method) with f.

These minor differences should not be taken seriously here, because deft should be viewed as rough measures for large effects. Similarly, the factor $(n-1)/n$ for computing s^2 is neglected here; and pq/n may be more convenient than $pq/(n-1)$ for s^2/n for proportions when $\bar{y} = p$.

However, we should distinguish the population parameters Deft and Deff from the statistics (2.2 and 2.1) based on sample results, hence subject to sampling variability; often very large variability (E.2). Similarly we distinguish $\text{Ste}(\bar{y}) = \sqrt{\text{Var}(\bar{y})}$ from $\text{ste}(\bar{y}) = \sqrt{\text{var}(\bar{y})}$. Also Deff should refer to the concept and theory of "design effects," not to specific cases. Most of the time here I use deft both for the singular and the plural, as in many deft values or many deft.

We shall also discuss deft(b) for various other statistics (b)

$$\text{deft}(b) = \sqrt{[\text{var}(b) / SRS\ \text{var}(b)]}. \tag{2.3}$$

I wish to alert you to two other decisions in defining Deff which have philosophical natures. First, we chose *SRS* variance S^2 / n for the standardizing denominator, because of its fundamental, classic character. Fisher's "efficiency" uses the "optimal" design for the denominator, but this concept is situationally restricted, like the choice of zero on the Fahrenheit scale, whereas our choice of *SRS* resembles the zero of the Celsius scale from the fundamental freezing point of water. Second,

$$\text{Deff} - 1 = \frac{\text{Var}(b) - SRS\ \text{Var}(b)}{SRS\ \text{Var}(b)}$$

would be often a more convenient base concept (as we shall see). However, these could result in negative values; e.g., for proportionate stratified random sampling. Deft^2 has a minimum of zero for $\text{Var}(b) = 0$, and this resembles the absolute zero of the Kelvin scale. (E.2)

3. WHEN deft ARE UNNECESSARY

It is not necessary to compute deft when *any one* of the following conditions holds:

(A) When the *population* closely resembles a well-mixed urn, its distribution is random; any portion can be regarded as IID, identically and independently distributed. Examples are a well-mixed urn, as in a good lottery, or a well-mixed deck of cards. But real populations, whether astronomical, physical, biological, or social, are never thoroughly mixed but are "grainy"; clustered—in my experience, model, and philosophy. This clustered status is measured by deft. This paradigm is not shared or is ignored by many, including "model-dependent" statisticians, mathematical statisticians, and econometricians; also opportunistic nonstatisticians in academic or market research. It is true that some populations come close enough to random for practical purposes, for example, the last four digits of the U.S. Social Security numbers; or the output of some of the better randomization programs on computers. Nevertheless, advice from a sampling expert should be sought to judge those populations.

(B) The *sample design* may be *SRS* or close enough for practical purposes. Here again, consulting a sampling expert (not any statistician) may help. For example, a systematic sample from a good list of elements (individuals) may yield a sample with deft only slightly lower than 1, due to mild stratification. Stronger stratifications are possible, but then they should be foreseen or detected with deft. However, even a telephone list may result in diverse "frame problems" (Kish 1965a, 2.7). Furthermore, selecting single adults from household telephones may result in nonnegligible deft, due to weighting (Section 8).

(C) Accepting either populations or samples as "approximately random" should be easier for *small samples*, where both the demands and resources are more restricted than for large samples. We are balancing possible biases of mild non-randomness against sampling errors, which are larger for smaller samples. Sampling errors also increase in decreasing sized subclasses of larger samples; and these subclasses are often important objectives of large samples.

(D) When only *descriptive statistics* are needed, and inferential (second-order) statistics will not be computed or used for the survey; that is, when standard errors and confidence intervals are ignored and only "point estimates" are made.

(E) *Sampling errors* and inferential statistics are needed only for one or *a few statistics*. For these few intervals $\bar{y} \pm t\,\text{ste}(\bar{y})$ may be sufficient without transformations through deft values. This means that there is no need for averaging with other deft values from other statistics, from either the same survey or other surveys.

These five cases represent situations when sampling errors are not needed (D); or can be computed with classic IID formulas (A, B, C); or with complex $\text{ste}(\bar{y}_i)$ survey formulas, but without conversions into deft, (E).

4. NECESSARY deft

Most populations are far from random, and many survey samples are based on selections that are far from *SRS*, and considerably clustered. Thus their sampling errors suffer considerable design effects from clustering. Therefore computing sampling variances that reflect appropriately the restrictions of the sample design is *necessary* for appropriate statistical (confidence) intervals and inference. These are the basis of "measurability" for survey samples (Kish 1987a, 7.1E). Proper variances without deft would suffice for computing and constructing probability intervals for one or a few statistics. However, variances and standard errors are not *sufficient* when averaging and generalizing are needed for many statistics; and deft are widely used for these purposes.

The principal reasons for computing deft from standard errors can be listed briefly:

(a) Averaging sampling error for different survey variables from the same survey. Averaging the standard errors would be meaningless for variables with diverse units of measurement. The averages of deft values over survey variables are considerably better and may be most popular now. However, I urge caution about these averages, because recent experience has shown great variations of deft between variables (Table 4a and 4b). Variations from 1.0 to 3.0 of deft are common, but note that standard errors can easily vary by factors of 100 or more. (E.4)

(b) Averaging over periodic surveys for the same variables. Deft values are better than standard errors when the effects of changes in sample sizes need to be removed.

(c) Relating the errors for different statistics on the same survey. Thus errors for complex statistics (b) such as regression coefficients, may be inferred by generalization from simpler statistics ($\bar{y}$); as in Sections 6 and 7.

(d) Generalizations from past surveys for designing other surveys from the same frame. Clearly these generalizations involve increasing risks with distancing of either survey variables or of sample designs for the new surveys. These warnings should be strengthened when other survey organizations try to "borrow" deft values.

TABLE 4a. Eight Fertility Surveys, Six Classes of Variables, High and Low Values of Deft and Mean Roh Values for Each Class

		Korea 1971	Korea 1973	Taiwan 1973	Peru 1969	USA 1970	Malaysia 1970 Rural	Urban	Metr.	All
Socio-economic	H	6.05	2.67	1.67	2.67	–	5.07	1.57	1.03	
	L	1.89	1.76	1.50	1.72	–	2.13	1.23	0.93	
	roh	.128	.081	.016	.126	–	–	–	–	.045
Demographic	H	1.78	1.73	1.86	1.35	2.02	3.21	1.48	1.23	
	L	1.13	1.06	1.86	1.23	2.02	1.22	0.78	0.71	
	roh	.014	.025	.025	.024	.105	–	–	–	.010
Fertility experience	H	2.01	1.38	1.83	1.67	1.66	1.60	0.88	0.95	
	L	1.08	1.02	1.21	1.00	1.03	1.35	0.65	0.70	
	roh	.016	.009	.014	.034	.019	–	–	–	.025
Contraceptive practice	H	4.49	1.90	2.88	2.02	1.72	2.17	0.90	0.94	
	L	1.31	0.80	1.19	1.08	1.25	1.67	0.71	0.66	
	roh	.047	.021	.030	.054	.029	–	–	–	.022
Birth preference	H	1.96	1.90	5.41	–	1.56	2.06	0.92	1.37	
	L	1.46	1.03	1.56	–	1.09	1.03	0.69	0.78	
	roh	.023	.024	.072	–	.019	–	–	–	.028
Attitudes	H	2.48	1.73	5.28	2.12	2.04	1.83	1.08	1.48	
	L	1.22	1.16	1.19	2.12	1.34	1.66	1.02	1.11	
	roh	.028	.026	.145	.094	.051	–	–	–	.017
Mean deft ($\bar{y}$)		2.16	1.47	2.35	1.65	1.47	1.92	0.99	1.01	
Mean roh		.050	.033	.059	.062	.038	–	–	–	
$\text{roh}_S/\text{roh}_t$		1.19	1.36	1.33	1.15	1.37	–	–	–	1.15
Sample *n*		6284	1919	5588	3327	–	–	–	–	
PSUs *a*		62	42	56	88	126				

Source: Kish, Groves, Krotki 1976

(e) Checking for gross mistakes in variance computations is greatly facilitated by deft. Usually a quick look at the printouts of deft values reveals to experienced eyes anything unusual that needs further investigation.

We come now to our central question. For what variables and for what statistics must values of deft be computed?

Compute values of deft$(\bar{y}_t)$ *for overall means of all survey variables* (*i*). This advice differs from past practice when values of deft were computed for only a few variables, chosen to be the most "important" or the most "representative." This change from our past practice and past advice is based on three realistic considerations:

1. Computations of sampling errors have become *much* easier with available machines and programs. Where this is not true, some may retreat partially toward the older restrictions.

TABLE 4b. Twelve Fertility Surveys, Five Classes of Variables, High and Low Values of Deft and Median Roh for Each Class

		Nepal	Mexico	Thai- land	Indonesia	Colombia	Peru	Bangla- desh	Fiji	Sri Lanka	Guyana	Jamaica	Costa Rica
Nuptiality	H	2.49	1.68	1.59	1.79	1.69	1.25	1.48	1.65	1.27	1.31	1.34	1.48
	L	1.14	1.12	0.98	1.14	1.07	1.02	1.04	0.97	1.06	1.09	1.06	0.99
All .02	roh	.02	.03	.01	.02	.01	.01	.01	.02	.02	.02	.04	.01
Fertility	H	3.02	2.08	1.98	1.56	1.63	1.50	1.20	1.30	1.29	1.07	1.21	1.12
	L	1.26	1.06	0.98	1.28	0.86	0.87	1.00	0.89	1.07	0.94	0.99	1.00
All .025	roh	.02	.06	.01	.02	.00	.04	.01	.00	.03	.05	.03	.02
Fertility	H	3.76	1.96	1.63	1.86	1.46	1.42	1.28	1.63	1.30	1.31	1.18	1.12
preference	L	2.13	1.06	1.15	1.31	1.06	0.97	1.10	0.94	1.04	0.93	1.00	0.91
All .03	roh	.05	.03	.02	.05	.01	.05	.02	.02	.03	.03	.07	.00
Contraceptive	H	4.19	3.20	2.77	2.60	2.84	2.10	1.80	1.66	1.47	1.61	1.35	1.17
knowledge	L	2.44	2.50	1.94	2.24	2.30	1.74	1.42	1.66	1.31	1.37	1.18	0.85
All .08	roh	.05	.17	.08	.14	.05	.14	.06	.04	.07	.08	.09	.01
Contraceptive	H	2.23	2.19	2.35	1.95	2.35	1.84	1.56	1.44	1.42	1.14	1.14	1.10
use	L	2.23	1.50	1.78	1.30	1.13	1.06	0.99	1.02	1.19	1.02	1.00	1.03
All .05	roh	.03	.11	.06	.06	.04	.09	.03	.02	.07	.03	.04	.02
Average .04	roh	.03	.07	.04	.05	.02	.05	.02	.02	.05	.03	.05	.01
Sample *n*		5940	6255	3820	9136	3302	5640	6513	4928	6810	3616	2765	3935
PSUs *a*		40	182	70	376	405	410	240	100	606	196	410	288

Source: Verma, Scott, O'Muircheartaigh 1980

2. Much empirical work has shown us that there are large variations in values of Deft for diverse survey variables (Tables 4a and b). These variations are of course much greater for values of deft^2, hence of "effective sample size," (n/deft), also of probability (p_α) values, also of standard errors. For example, a deft = 3 denotes $\text{deft}^2 = 9$, and instead of probability statements of $t = 2$ and $\alpha = .05$, you get $t = 2/3$ and $\alpha = 0.50$!
3. Empirical results have also revealed reasonably dependent relations from deft values for means to deft values of more complex statistics (Sections 6 and 7).

The $\text{deft}(\bar{y}_t)$ refers to overall means for the entire sample. Less often the simple expansion totals $\hat{Y}$ may be of greater interest, as in census counts, but totals are often computed from means $\bar{y}$.

Proportions p (and percentages $100p$) are the most common forms of survey means. Whether for proportions or otherwise, the means from survey samples are commonly ratio means $\bar{y} = y/x$, the ratio of random variables, hence they are often denoted as $r = y/x$. The denominator x is the sample size. When proportions come from dichotomies, like gender, with $p_1 = 1 - p_0$, then deft (p_1) = deft (p_0) and we compute only one deft. Often, however, proportions come from polytomies with $k > 2$ categories, such as religion, occupation, etc., which are unordered. But ordered categories like number of children born, years of education, or income classes are often presented also as proportions, although these may also be presented as means.

Compute values of deft(p_t) *for all k categories of categorical variables.* This advice goes beyond the common practice of computing and presenting sampling errors and deft(p) values only for one category, chosen as most "important" or "representative." But empirical data have convinced me recently that there occur sometimes (not always) large variations of deft(p) values for different categories of the same variable. Where computers and programs are readily available, they permit printing vast amounts of data, including deft for all variables, and all categories (E.6B).

The printout of the values of deft and sampling errors needs some interested and knowledgeable expert to examine the entire output. The values of deft should range mostly from 1 to about 3 or 5 in most situations, hence yield clues about mistakes, alarms, and outliers more readily than variances or standard errors.

However, the display and presentation of data printed in research reports, articles, and books should be much more restricted, and pointed to a less specialized audience. There may not be space for all categories, perhaps not even all variables, in many large survey reports, because including each ste($\bar{y}$) along with each $\bar{y}$ would unduly complicate them for most readers. The deft values may be presented in technical appendixes. One example presents all deft in decreasing order, with a code for variable classes, which the readers' eyes can examine (see Table 4a) (Kish, Groves, and Krotki 1976). In another publication, the classes have been formed by experts who present average deft to the reader (Table 4b) (Verma, O'Muircheartaigh, and Scott 1980). Even further averaging has been done for survey means represented mostly by proportions: tables of sampling errors are shown for a few proportions (0.5, 0.2, 0.1) with the sampling errors including average design effects, 2(ave deft)$\sqrt{(pq/n)}$. In addition to total n, some major limits of n for subclasses may be added. Problems arise due to differences of deft values between variables (Kish 1965a, Section 14.1). A new paper makes abundantly clear, with three variables from 56 countries, the great differences of deft values between variables, and the similarities across countries (Verma and Lê 1996).

In addition to deft values for the entire sample, separate deft values may also be computed for major regions. But this topic is better discussed with other subclasses, for which deft may also be computed but not necessarily published (Section 6). Deft for differences of means and of subclass means will also be discussed there. Deft for more complex, analytical statistics are discussed in Section 7.

5. OTHER NEEDED SAMPLING ERRORS

In addition to deft, some other functions of variances, also some other statistics related to them may be computed and presented at the same time. Most of these are needed for and computed as factors for deft; and others are easily available. Furthermore, these auxiliary statistics can also serve for interpreting deft, and sampling variability in general.

It is common to print out the mean ($\bar{y}$, or $r = y/x$, or p); and sometimes the sample total y or the estimated population total $\hat{Y}$. Also worthwhile is printing the values of ste($\bar{y}$) and *SRS* ste($\bar{y}$), whose ratio becomes deft($\bar{y}$), which is also printed.

But it is less clear that printing $\text{var}(\bar{y}) = \text{ste}^2(\bar{y})$ is needed; nor 2ste($\bar{y}$) and the two values $\bar{y} \pm 2\text{ste}(\bar{y})$ which have also been printed sometimes. This may be useful in some reports (perhaps medical or pharmaceutical) that publish only one or a few dozen statistics. Perhaps making graphs of the intervals $\bar{y} \pm 2\text{ste}(\bar{y})$ may be useful, for graphical interpretation.

It is usual to print n, usually the simple count of sample elements for *EPSEM* selections. This would not be necessary if the values of n for all variables were similar, because they come from the same *EPSEM* sample, with only small differences due to differential nonresponse. Then n and a, the number of PSUs, can be stated in the introduction, which may come from $a/2$ strata for paired selections. However the values of n can vary greatly if some variables have smaller bases, because other cases are not relevant; e.g., only homeowners or only registered (or intending) voters; only males, or females, etc., when subclasses are tabulated. Therefore, it is good to have the program print out n. I wish to point to some problems here, for which I cannot currently offer satisfactory solutions. Would it be better to print $n' = n/\text{deft}^2$, the "effective sample size?" This becomes more difficult for weighted samples. Sometimes roh = $(\text{deft}^2 - 1)/(n/a - 1)$, a synthetic but useful "ratio of homogeneity" is also printed. This practice is justified in Section 6 on subclasses. But its interpretation is more difficult for weighted samples (Section 8).

The coefficient of variation of the statistic $\bar{y}$ is printed because it is available as $\text{cv}(\bar{y}) = \text{ste}(\bar{y}) / \bar{y}$, or perhaps a percentage error $100\text{cv}(\bar{y})$. We must be careful with denominators near zero, because that would make the cv unstable. For example, for differences the $\text{cv}(\bar{y}_a - \bar{y}_b)$ varies around zero.

Our own OSIRIS programs have also included computing and printing $\text{cv}(x) = \text{ste}(x)/x$, the coefficient of variation of the denominator x of the ratio estimate y/x. This precaution is necessary to guard against estimates that may be unstable, when $\text{cv}(x) > 0.2$ (Hansen, Hurwitz, and Madow 1953, Vol. II, Section 4.12). But printing the values of cv(x) proved insufficient safety once, when nobody looked at them. Thereafter we had the program compare cv(x) against the floor level 0.1, and whenever the program found $\text{cv}(x) > 0.1$ it printed out in red letters STOP, LOOK, AND DO SOMETHING. We did not stop the printout, but the warning brought human help to the problem (E.5).

6. SUBCLASSES AND DIFFERENCES

Here I join the treatment for differences (comparisons) of means to those for subclasses for practical, empirical reasons, not because of theoretical considerations. The most common reasons for the frequent use of subclass statistics come from their comparisons. These take most commonly the form of differences $(\bar{y}_c - \bar{y}_b)$ between subclasses c and b, though ratios $(\bar{y}_c / \bar{y}_b)$ are also used, and sometimes other forms, such as $[p_b / (1 - p_b)] / [p_c / (1 - p_c)]$.

Before dealing with subclasses, we may look at differences $(\bar{y}_1 - \bar{y}_2)$ between two entire samples, and we may distinguish two common types from among many possible types. First, we may compare two independent samples, such as two regions, or two countries, then: $\text{Deft}^2(\bar{y}_1 - \bar{y}_2) = [\text{Var}(\bar{y}_1) + \text{Var}(\bar{y}_2)] / [SRS\ \text{Var}(\bar{y}_1) + SRS\ \text{Var}(\bar{y}_2)]$. If Deft and n are similar for both means, the Deft^2 of the difference will also be similar to the two. If the two differ, then the Deft^2 of the comparison will be the average of the two Deft^2, each weighted by $1/n_i$. For comparing regions from the same survey, it may be common and safest to use for the regions $(n_t / n_c)\text{Deft}^2(\bar{y}_t)$, where n_t and n_c are sample sizes for the total and the region.

However, for differences of two time periods (1 and 2) of the same survey we use

$$\text{Deft}^2(\bar{y}_1 - \bar{y}_2) = [\text{Var}(\bar{y}_1) + \text{Var}(\bar{y}_2) - 2\text{Cov}(\bar{y}_1, \bar{y}_2)] / [SRS\ \text{Var}(\bar{y}_1) + SRS\ \text{Var}(\bar{y}_2)]$$

because there are appreciable covariances that reduce significantly the value of Deft^2 for the difference (Kish 1965a, Table 14.1. IV). The covariances and reductions of Deft^2 have been found often from using the same clusters (primary, secondary, and lower), even if the elements (and final segments) differ. The covariances can also be computed and presented as correlations R_{12}.

One type of subclass, called a "proper" or "domain" subclass contains independent samples and resembles the first type above. Two examples are regions, also urban and rural subclasses and their comparisons, which are often presented. The computations of variances, hence deft, may differ between subclasses in these two examples. Methods of selection and clustering for urban samples may differ greatly from those for the rural sample, and the deft for the two may be quite distinct. Also there are usually enough primary selections, so that separate estimation of urban and rural variances and deft can be justified. However, for regions the situation may be quite different: when regions are small (and numerous) the number of primary selections (PSUs, "ultimate clusters") are few, the degrees of freedom fewer; and the variances and deft highly unstable. It is then preferable to use the overall $\text{deft}^2(\bar{y}_t)$ as an average and infer $(n_t / n_c)\text{deft}^2(\bar{y}_t)$ as the regional deft^2 with subsample size n_c.

The situation is quite different for the second type of subclasses, called "crossclasses," which are much more common: age and other demographic classes; education and other social classes; income, occupation and other economic classes; behavioral, attitudinal, and psychological classes, including those created by the survey questions, etc. These could not be and were not part of the clustering and stratification effect. They "cut across" the sample design more or less evenly, or randomly, and tend to be found in all or many of the primary selections. Therefore, crossclasses are based on (almost) the same number of primary selections and have as much (or little) stability as the entire sample. Thus the sizes of sample clusters n_c / a decrease on the average with sample size n_c, hence the "design effect" (also) decreases proportionately (almost).

This follows from $\text{deft}^2 = 1 + \text{roh}(n_c/a - 1)$, to the extent that roh remains the same for the crossclasses, and that is an empirical question. It does not follow necessarily or mathematically. However, it has been shown empirically for many and very diverse situations and survey designs and for many survey variables, that generally Deff decreases toward 1 with the decrease in the cluster sizes n_c/a. The decrease is not entirely smooth nor complete, due partly to increasing relative variances of cluster sizes. These irregularities are greater for subclasses that are not true "crossclasses"; instead their cluster sizes have greater variations than random. Socioeconomic subclasses were indeed found to have greater variation and greater rohs than demographic subclasses (Kalton and Blunden 1973; Kish, Groves, and Krotki 1976). Hence we recommend using 1.2 roh or 1.3 roh; see below and Table 4a.

With the experience of vast amounts of empirical data one can use a reasonably good model for inferring subclass variances $\text{var}(\bar{y}_c) = \text{ste}^2(\bar{y}_c)$ from $\text{var}(\bar{y}_t)$ for the entire sample for the same variable.

a. The variance is increased by (n_t/n_c) inversely proportional to the sample sizes. But this *SRS* adjustment needs modification when $\text{deft}(\bar{y}_t)$ is not close to 1, because $\text{deft}(\bar{y}_c)$ should then be increased, as below.
b. The size of the sample cluster is changed from n_t/a to n_c/a and the value of roh_t is increased by $k_c > 1$, so that

$$\text{deft}^2(\bar{y}_c) = 1 + k_c \text{roh}_t (n_c / a - 1) \tag{6.1}$$

Strict proportionality would imply k_c = 1 but a better value of k_c seems to point to k_c = 1.2 for some subclasses, but to k_c = 1.3 for socio-economic subclasses, which are less evenly distributed because they are clustered. Another way to compute (6.1) would be

$$\text{deft}^2(\bar{y}_c) = 1 + k_c \frac{p_c n/a - 1}{n/a - 1}[\text{deft}^2(\bar{y}_t) - 1] \tag{6.2}$$

where $p_c = n_c / n$, the proportion size of subclass c. Somewhat simpler is still another approach (E.6b)

$$\text{deft}^2(\bar{y}_c) = 1 + p_c[\text{deft}(\bar{y}_t) - 1]. \tag{6.3}$$

Differences between crossclass means $(\bar{y}_c - \bar{y}_b)$ are often principal objectives of sample surveys and the problems of design effects are somewhat different than for the crossclass means themselves. The following generalization has been found in many and diverse computations

$$SRS \text{ var}(\bar{y}_c) + SRS \text{ var}(\bar{y}_b) < \text{var}(\bar{y}_c - \bar{y}_b) < \text{var}(\bar{y}_c) + \text{var}(\bar{y}_b).$$

The left term is essentially $s^2/n_c + s^2/n_b$. The model behind this empirical generalization is similar to "additivity" in ANOVA: the primary clusters (within strata) that are high/low on variable y for crossclass c are also high/low for crossclass b. The wealth of empirical evidence is convincing, and the covariances tend to be positive and large so that most of the variances tend to fall near the lower *SRS* limits of Deft^2, near to 1. For most data even the upper limits are low because for small crossclasses they are mostly not much above 1. Thus the Deft^2 are squeezed to slightly above 1. Variances below the lower *SRS* limits also occur frequently, denoting deft^2 below 1, but both theory and experience teach us to attribute these to random variations and curtail deft^2 at 1. Instability of $\text{var}(\bar{y}_c - \bar{y}_b)$ is high, because it is the sum of three components, often each unstable.

In spite of my strong confidence in variances and deft for crossclasses from models based on the $\text{deft}^2(\bar{y}_t)$ computed for the entire sample, I strongly urge computation from actual data. Thus knowledge and confidence can be built. If meaningful contradictions are found, the results and their causes should discover better, even if more complex, models. Good programs exist which facilitate computations of variances, deft, and other sampling errors. Computing them for all subclasses may not be feasible, but perhaps we can find those that are most "important," and for the survey variables with the highest deft values. These would provide the strongest tests for our models.

7. Deft FOR COMPLEX STATISTICS

I must begin this section with two personal statements which I need as necessary cautions to the reader, and necessary defenses for myself. Other practicing samplers support these views, although it is difficult to find clear, unequivocal, written support. First, the existence of Deft and the need for probability selections for complex statistics are closely and causally linked. I cannot imagine a world where probability selection is irrelevant but which would also yield the empirical evidence of deft, as we have argued for four decades (Kish 1987a, 1.4–1.8) (E.7).

Second, the needs are widespread, as witnessed by computing programs for regressions, etc., with automatic standard errors based on IID, whereas probably most data going into them come from clustered samples, especially in the social sciences. The needs are greater than the reasonable conjectures we can offer; these outpace sufficient empirical results and are way ahead of solid mathematical theory. (Needs > conjecture > data base ≫ mathematics.) Let us view current conjectures with those *caveats*.

What are complex, analytical statistics? Let us go beyond subclasses and differences, already seen in Section 6. Instead of attempting a definition, let us view major examples and how we treat them. We deal here with conjectures for upper and lower limits for values of deft(b) for complex statistics based on deft($\bar{y}_t$) available for the same variables from the same survey. There are also computing programs for some complex statistics and the researchers should be encouraged to use them (Section 9). But the researchers may not be using these because they:

(a) either have no access to the program;
(b) or need features (weighting?) that the program lacks;
(c) or lack time, ability, assistance to use them;
(d) or have and need statistics for which no program of variances and deft exists;
(e) or have a sample design that baffles the programs.

Now on to conjectures from deft($\bar{y}_t$) to deft($\bar{y}_b$) for several statistics.

(a) *Ratios of ratio means and index numbers:* $r_1/r_0 = (y_1/x_1)/(y_0/x_0)$ may be used by researchers, also "odds ratios," sometimes in addition and sometimes instead of differences $(r_1 - r_0)$ of the ratio means of two surveys; these may be periodic surveys and r_0 the "base year." This may be somewhat difficult (though not impossible) to feed into programs of sampling errors. However, the variances for (r_1 / r_0) and $(r_1 - r_0)$ are similar, except that the former have "relvariance" terms like $\text{var}(r)/r^2$ instead of variances. Therefore, we conjectured and found that the deft for (r_1 / r_0) is similar to the deft for $(r_1 - r_0)$, which are easier to compute. Similarities may also be measured for linear combinations of these double ratios, such as $(r_2 / r_0 - r_1 / r_0)$ and *indexes* $\Sigma(r_{i1} / r_{i0})$ (Kish 1965a, 12.11; Kish 1968).

(b) *Medians and other quantiles* are often used by economists, sociologists and other researchers (in addition or in preference to means) for skewed distributions like income, wealth, time spent (in hospitals, prisons, on welfare, in queues), money spent, etc. Computing variances and deft may be difficult (even for the *SRS* variances), though possible (Woodruff 1952;

Kish 1965a, 12.9). However, it has been conjectured and found that deft for medians should be similar to deft for proportions near 0.5, and these are easy to compute (similarly for other quantiles). Also similarly, the difference of medians (e.g., between two subclasses, or two years) should have deft similar to those for differences between proportions from the two samples, also easy to compute.

(c) *Differences* $(p_i - p_j)$ *of the proportions of pairs of categories* (i, j) from the same variable, with $k > 2$ categories have been investigated recently. For example, difference of preferences between two candidates, automobiles, religions, contraceptives, etc. It was found regularly on diverse surveys from several countries that to a good approximation in each situation $\text{deft}(p_i - p_j) = \frac{1}{2}[\text{deft}(p_i) + \text{deft}(p_j)]$. This holds even when the deft(p_i) and deft(p_j) are not close. Furthermore the $(p_i - p_j)$ may represent the net change (+ − minus − +) from two waves of a panel (Kish, Frankel, Verma, and Kaciroti 1995).

(d) *Coefficients for linear regressions* are the best known and most important of methods for multivariate statistical analysis. Deft have been computed and theory developed recently by a few investigators, since the development of computing programs and of resampling methods, as we shall see. For forty years, however, I had to argue with econometricians, mathematical statisticians and others, for the very *existence* and validity of Deft for regression. Their arguments were "model-based," or based on Bayesian and likelihood arguments, but happily some changed their minds since then. I believe that in a regression $\Sigma \pm b_i x_i$ $(i = 0, 1, 2\ldots)$ the choice of the variables x_i, their exponents (1, or 0.5 or 2, or −1), their signs (+ or −) all come from the model of the researcher (economist, etc.) with little help from the statistician, although some statistical tests may help with the choices. However, the values of the b_i must come from empirical data, hence from specified samples from specified populations. The values of the b_i are conditional on the populations from which they are sampled, they are subject to sampling errors, and subject to deft, which must be measured (Kish 1987a, 1.4 – 1.8).

Here follow several suggestions for estimating deft for coefficients in linear regressions:

1. The usual programs for computing linear regressions display estimates for the diverse coefficients (regression, partial, and simple bivariate) that are acceptable for complex samples. The standard errors, however, based on IID assumptions, serve as denominators for the deft values [E.2].
2. Computing programs exist for linear regressions from complex samples, with either the Taylor (delta, linear) approximations or one of three resampling methods BRR, JRR, or Bootstrap (but for bootstrap no programs seem yet to be useful for complex surveys). The relative advantages are discussed in Section 9.

3. Two cautions should be sounded here. First, weighting may be difficult (or impossible?) with some programs. Some model-dependent econometricians denied the need for weighting, but I do not share those views (E.7). Second, long multivariate equations may overtax some computing programs; perhaps an abbreviated program of the most important predictors will yield enough deft for reasonable conjectures.
4. Conjectures from deft($\bar{y}$) to deft(b) of diverse coefficients have been made for three decades (Kish and Frankel 1974).

A simple model of the results in Table 7 would be

$$\text{Deff}(b_g) = 1 + f_g\{\text{Deff}(\bar{y}) - 1\}$$

with $\text{Deff}(\bar{y}) > 1$, $0 < f_g < 1$ and f_g specific to the variables and statistic denoted by g.

Notice that the lowest deft appear for regression coefficients in all three studies, though we do not know why. Nevertheless, even these deft of 1.106 and 1.295 are not negligible for statistical inference. We should like to see further research that would link deft(b_i) to deft($\bar{y}_i$) for specific variables, to achieve tighter inference.

TABLE 7. Values of √deff for Five Types of Estimators From Three Complex Samples. Set A from Table 2, Set B from Table 3 of Kish and Frankel (1970), Set C from Table E-1 of Frankel (1971)

	Sample set		
	A	B	C
Ratio means	1.106	1.800	1.438
Simple correlations	1.096	1.262	1.355
Regression coefficients	1.015	1.295	1.106
Partial correlation coefficients	1.041	1.400	1.360
Multiple correlation coefficients	NA	1.465	1.894

i. Deff(b) > 1. In general, design effects for complex statistics are greater than 1. Hence standard errors based on simple random assumptions tend to underestimate the standard errors of complex statistics.
ii. Deff(b) < Deff($\bar{y}$). The design effects for complex statistics tend to be less than those for means of the same variables. The latter, more easily computable than the former, tend to be "safe" overestimates. (We noted earlier the "pathology" of multiple R.)
iii. Deff(b) is related to Deff($\bar{y}$). For variants with high Deff($\bar{y}$) values of Deff(b) tend also to be high. See Kish and Frankel (1970, Section 7) for a set of striking results.
iv. Deff(b) tends to resemble the Deff for differences of means. The latter is a simple measure of relations for which values of deff are easily computed, and for which (i)–(iii) also hold.
v. Deff(b) tends to have observable regularities for different statistics. This is a hope based on theoretical considerations; confirming results would help us make useful conjectures.
From Kish and Frankel (1974).

(e) *For dummy variables and for categorical data regressions*, the above results and conjectures should also be helpful. With LISREL and similar sophisticated programs I have no experience and no guidance to offer.

(f) *For Chi square tests*—such as $k \times m$ tests—from survey data I also have little experience or interest, but nevertheless useful conjectures, based on the essential similarity of 2×2 tests to differences of two proportions. For proportionate stratified element sampling Deff goes to 1 (Kish and Frankel 1974). For clustered samples the Deft are much reduced and can be computed. For $k > 2$ and $m > 2$ we can argue by analogy from the deft values of the pairs of differences (Nathan, Rao, and Scott 1987).

8. WEIGHTING AND GENERALIZATION

All statistics involve generalizations; and the overall means $\bar{y}_t$, variances $\text{var}(\bar{y}_t)$, and also $\text{deft}^2(\bar{y}_t)$ for the entire sample ignore and average variations between its separate domains. For example, in a country's sample, the urban domain of metropolitan and large cities may have very different values of deft^2 than the rural, because both the cluster sizes b and the homogeneity roh can be different in $[1+\text{roh}(\bar{b}-1)]$. The design can be further complicated if a different (higher or lower) sampling rate was used in the urban domain than in the rural. The capital's area in a developing country may contain a minor fraction of the population for which a larger sample is desired; but in developed countries the small rural areas may need increased sampling. Perhaps separate values of the deft^2 should be computed and used.

The needs for generalizations from computed values of deft lead to conflicts that are especially difficult when complicated by weighting. *Within* the same survey Deff carries the effects of clustering and stratification, often in several stages. After standardizing for S^2/n the $\text{Deft}^2 = \text{Var}(\bar{y})/(S^2/n)$, we must recognize for each variable distinct Deft values, due to varying factors of homogeneity *roh*. However, we must distinguish four major *sources* of unequal selection probabilities (P_i) and hence of weighting ($w_i \propto 1/P_i$), because they need distinct treatments for computing deft (Kish 1992).

(a) *Nonresponses* may be compensated with differential weighting in classes. These differences should be either small or rare, or both, if nonresponses are under reasonable control.

(b) *Frame problems* result in unequal selection probabilities, because these could not be measured or controlled before selection. These also are (or should be) relatively small or rare. An example is weighting with number N_i of adults when one is selected at random from each sample household; we have seen values of deft from 1.05 to 1.20.

(c) *Allocation to separate domains* of different sampling fractions occurs often, either to increase sample sizes in some entire domains, or to reduce costs in others; and sometimes to decrease variances in the total sample.

These distinct domains may be regions, or urban/rural strata, large/small units, etc.

(d) *Disproportionate sampling fractions* may be introduced into crossclasses, which are not domains, deliberately for "optimal allocation," or in order to increase the sample sizes of some subclass. For example, households found to have an ethnic or age group may be oversampled.

(e) *Post-stratification and ratio estimates* for population control and adjustments face us with difficulties. Often the range of weights is not great, mostly within $w(max) / w(min) < 2$; for these the unweighted deft values may suffice. I do not have a simple, general solution.

Classes (a) and (b) may often be treated simply, because the effects of weighting on deft may be small. The effects on the descriptive (first-order) statistics, like $\bar{y}$, may be larger than negligible (or weights would not need to be used). Yet their effect on the variances may be small. And beyond that their effects on deft should be similar on both sides of the ratio $\mathrm{Var}(\bar{y})/(S^2/n)$, hence even smaller.

Class (c) needs a different treatment and separate values of deft^2 can be calculated for two or a few domains. If the deft^2 are approximately similar they can be combined for simpler joint presentation.

Class (d) faces us with the most difficult problem and confronts us with the conflict for which I know of no satisfactory simple solution. The treatment we offer here applies also to classes (a), (b), and (c), when the simple treatments seem unsatisfactory, and must be treated, like (d).

(d1) For *internal* use and inference, the standard definition, $\mathrm{var}(\bar{y})/(s^2/n)$ yields $\mathrm{deft}^2(\bar{y})$ that combines and confounds the effects of the specific weighting used with those of clustering and stratification. The numerator $\mathrm{var}(\bar{y})$ contains the weighted variance from the input data. In the denominator the *weighted* s^2 merely estimates S^2 in the population; thus s^2/n estimates the variance of an *SRS sample of size n*. Since all statistics and variables have the same weights over the entire sample these $\mathrm{deft}(\bar{y})$ values yield the proper Deff corrections for standard errors of the sample for *these* weights and these weights only.

(d2) *Internal generalizations within the survey*, such as conjectures from $\mathrm{deft}(\bar{y}_t)$ to other statistics are possible, but only with caution. For example, the effects of weighting on "crossclasses," like age classes, will be similar ("inherited"); that is, they will be about the same in the subclasses as in the total sample, unlike cluster effects which decrease in subclasses. However, for subclasses correlated with the selection/allocation probabilities, the effects may be either increased or decreased. The effects of allocations may also differ for diverse statistics. For example, for some of the means the unequal selection rates may produce reductions of the variance; these may overcome the clustering effects and thus result in deft < 1. However for some statistics, especially proportions, the same allocation may result in losses (increases of variances). Thus, an "optimal allocation" for mean incomes resulted in losses for median incomes. These results came from an investigation that used a simplified analysis, which treated the first phase ratings of dwellings from a multistage sample, as if it were a stratified element sample (Kish 1961).

(d3) Deft for *external* use are difficult to fashion, though they would be highly desirable, since generalization is the chief reason for computing deft. First, one may wish to design an *EPSEM* (equal f rates) or a different allocation using the same sampling frame. Second, one may wish to plan a survey for a different sampling frame. Separation of the effects of weighting from the effects of clustering is needed.

One approach is to compute values of unweighted

$$\text{deft}^2 = \frac{\text{var}_u(\bar{y})}{s_u^2 / n} \tag{8.1}$$

without weights. This estimates the values of deft^2 in a population in which the frequency distribution has been biased by the selection probabilities of the selection factors of the sample. The values of Deff in this distribution should differ some from those in the actual population.

This population can be approached with a subsample selection that reduces the oversampled portions in order to produce an *EPSEM* selection. The (unweighted) computation of values of deft^2 estimate Deff in the actual population. These can be compared both to the deft^2 above and to the weighted estimates. This approach requires a separate research project that will seldom be undertaken.

(d4) *Haphazard or Random Weights* can be dealt with more easily. Weights due to frame problems (b) and to nonresponses (a) may often be considered approximately random. In these situations I assume that the variances are increased by a factor $1+L = n\Sigma k_j^2 / (\Sigma k_j)^2$, where L represents the "loss" due to the element weights k_j. This loss L is easily shown to equal the "relative variance" of the weights k_j (Kish 1992). Then in $\text{deft}^2 = \text{var}(\bar{y}) / [(1+L)s^2 / n]$, both terms contain losses for random weights, and we have estimates of deft without them.

9. COMPUTING deft

A. The most basic concept of "measurability," for sampling errors for variances and for deft, is *random replication*. In complex clustered samples, *paired* selections of paired primary selections (i.e., ultimate clusters) from strata form most commonly the bases for computations. But sometimes the $a/2$ pairs are actually "collapsed" from single selections from a strata. Or a systematic selections are treated as $a/2$ pairs; or as $a - 1$ nonindependent pairs. Larger strata with $a = \Sigma a_h$ and $a_h > 2$ are used less commonly. And "interpenetrating samples" of k independent selections are rare in practice, I believe. When they occur, the variance computations should be easy, but they may be highly unstable (variable) when k is not large.

Paired selections are not necessary (as has been stated sometimes in the past) but they are convenient and also often efficient. They satisfy approximately three conflicting needs in cluster sampling:

1. to restrict the spread of the sample to a primary selections, because these are costly;
2. to use more strata for greater efficiency and because information is available; and
3. two random replicates are needed for computing variances.

The sizes of clustered samples should be judged not only in terms of the numbers n of elements, but also in numbers of clusters at all stages. The numbers of primary clusters is particularly important, not only for reducing the standard errors and Deff of statistics, but also for reducing the instability of standard errors and of deft.

There are many articles and books dealing with aspects of unbiased estimators ("exact" and "best") of variances. Under practical conditions we generally must use only approximately unbiased estimates, and we must avoid large biases, such as *SRS* estimates of variances that ignore Deff. But, alas, too little is written about the precision or stability of estimates of variances, standard errors, and deft. The single basic fact is that the coefficient of variation of deft is no less than $1/\sqrt{2d}$, where d is the number of degrees of freedom; a little more in practice because the clusters are unequal, and the *SRS* denominators s^2/n also contribute a little to the instability (E.5). The famous jackknife design of ten replicates has 9 degrees of freedom, hence cv(deft) $> 1/\sqrt{18} \approx 0.25$. But this 25% variation, or 50% for two sigmas, is of little use for evaluating deft. Statisticians then fall back on averaging deft, but we saw that deft vary greatly between variables. With $a = 60$ systematic selections, 30 paired selections yield $d = 30$ and cv(deft) $> 1/\sqrt{60} \approx 0.13$ and 2cv(deft) of 26%. If all $(60-1)$ differences are used, d is about $(4/3)(a/2) = 40$, and cv(deft) $> 1/\sqrt{80} \approx 0.11$ and 2cv(deft) of 22% (Dumouchel, Govindarajulu, and Rothman 1973).

For each primary selection (ultimate cluster) for valid computation of variances, hence of deft, we need the sample totals $y_{h\alpha i} = \Sigma_j y_{h\alpha ij}$ for every variable i; for each primary selection α (often only two, $\alpha = a$ or b), within each stratum h; j is the element case count, which is commonly on the tape. However, the stratum and primary selection identification numbers are carelessly omitted often from survey data and for those surveys the computation of sampling errors and deft are not feasible. (Only sampling code numbers are needed, and deliberate omission of names of units for confidentiality can be pursued.) We may consider the existence on data tapes of data for the covariance matrix of primary selections as necessary and sufficient for computing var($\bar{y}$) for clustered *EPSEM* samples. For weighted data the element case weight w_j is needed also.

For the denominator s^2/n, and generally for *SRS* var(b), the element values y_j and w_j are needed; but the identifications of the primary selections are not. This is also true for the mean $\bar{y}$ or p or $r = y/x$ and for other descriptive statistics (E.9). The computation of s^2/n should cause no problems. Furthermore, for complex statistics, like regression coefficients, I believe that computing programs automatically print useable estimates of *SRS* standard errors.

Several good computing programs are available for computing sampling errors that also compute deft values. I cannot attempt to be comprehensive; that would be futile and also would soon become obsolescent. There are also many more being used privately but not publicly available. In the Appendix the few best known in the U.S.A. are checklisted [E.9] (Francis 1981; Cohen, Burt, and Jones 1986; Cohen, Xanthopoulos, and Jones 1988).

10. APPENDIX

(E.2A) For the *SRS* variance s^2 or $\hat{s}^2 = s^2(n-1)/n$ in the denominator of deft^2, I am not concerned with small factors like $(n-1)/n$, or whether pq/n or $pq(n-1)$ should be used, or about sampling with/without replacement. But we should be concerned about using $\hat{s}^2 = \Sigma y_j^2/n - (y/n)^2$, for samples that are clustered and stratified, and $\hat{s}^2 = \Sigma w_j y_j^2/\Sigma w_j - (\Sigma w_j y_j/\Sigma w_j)^2$ when they are weighted also. This is obvious to a few, but unforeseen or surprising for most but in either case it is most important and convenient, that for all data from probability samples, we have simply

$$E(\hat{s}^2) = \sigma^2 - \text{Var}(\bar{y})$$

that s^2 is an almost unbiased estimate of $\sigma^2 = \Sigma Y_i^2/N - \bar{Y}$, the element variance in the population. It is only a slight overestimate, since $\text{Var}(\bar{y})$ is smaller than σ^2 by n^{-1}. For *any EPSEM* we have (Kish 1965a, 2.8)

$$E(\bar{y}) = E(\Sigma y_j/n) = \Sigma Y_i/N = \bar{Y}$$

$$E(\Sigma y_j^2/n) = \Sigma Y_i^2/N.$$

By definition

$$\text{Var}(\bar{y}) = E(\bar{y} - \bar{Y})^2 = E(\bar{y}^2) - \bar{Y}^2.$$

Then

$$E(\hat{s}^2) = E(\Sigma y_j^2/n) - E(\bar{y}^2) = \Sigma Y_i^2/N - E(\bar{y}^2)$$

$$= \Sigma Y_i^2/N - \bar{Y}^2 - \text{Var}(\bar{y}) = \sigma^2 - \text{Var}(\bar{y}).$$

(a) Dividing by random variables (not fixed) n results in approximate or conditional expectations.

(b) The same derivation holds for P_j weighted samples that are not *EPSEM* (Kish 1965a, 2.8).

(c) The same derivations hold for $\Sigma y_i x_i$ terms in the covariance matrix; also for higher moments. Therefore, also for estimates of b and *SRS* var(b) of analytical statistics in Section 7. From the above we deduce that $\hat{s}^2 + \text{var}(\bar{y}) = \hat{s}^2(1 + \text{deft}^2/n)$ will give adequate estimates of σ^2. Also that $s^2 = \hat{s}^2(1 + 1/n)$ will suffice for the denominator when deft^2 is near 1.

(E.2B) A *brief history* of the background of Deff may be useful, before this name was introduced (Kish 1965a). Since then high speed computers and the spread of probability sampling have made Deffs well known, widely computed, and used. The earliest reference to a similar concept I found in earlier textbooks (Yule and Kendall 1965) under the name "Lexis ratio" (Kendall and Buckland 1982), traced to the publications of W. Lexis in German at the end of the 19th century. (This dictionary still lacks "design effects.") The concepts of "intraclass correlation" by Fisher (1950) are related through the within/between components in the analysis of variance. From the late 40s and through the 50s there were a few of us computing "ratios of variances" (Hansen, Hurwitz, and Madow 1953, 12D); and values of true var / *srs* var are given for five large scale samples (Kish 1957). The U.S. Census Bureau had computing programs in the 1950s, but the first published program with deft appeared in 1972 (Kish, Frankel, and Van Eck 1972).

(E.4A) An older alternative to deft for generalizing sampling errors are the coefficients of variation $\text{CV}(\bar{y}) = \text{Ste}(\bar{y})/\bar{Y}$ and $\text{CV}^2(\bar{y}) = \text{Var}(\bar{y})/\bar{Y}^2$, called relvariances. They also remove the units of measurement, and have the advantage of being easily understood, as $100\ \text{CV}(\bar{y})$ = percent variation in $\bar{y}$. Thus CVs for means and totals of quantities (people, money, acres) can be compared and the technique has been used for proportions. The name Generalized Variance Function (GVF) has been coined for fitting curves for generalizing to diverse statistics within a survey (Hansen, Hurwitz, and Madow 1953, Vol. I, 12.B.15; Wolter 1985).

However, CVs have several drawbacks compared to deft for generalizing and inference.

1. CVs are functions of $\sqrt{n}$, but these are removed from deft.
2. CVs are also subject to Deff, but without explicit expression.
3. They are unstable when the denominator is small; e.g., net change or difference, small or rare values.
4. $\text{CV}(p) \gg \text{CV}(1-p)$ for small p, although $\text{Ste}(p) = \text{Ste}(1-p)$ (Kish 1965a, 2.5).

(E.4B) The two tables are abstracted from two frequently quoted publications on sampling errors of fertility-related variables. Each contained the broadest set of sampling errors at its time (but a new "champion" will be presented by Verma and Le to the ISI in August 1995). Each table represents tens of thousands of cases (n) from hundreds of PSUs (a). Note the large variation between the highest and lowest deft in each class of variables. The highest deft in each table 6.05 and 4.19 mean increases of deff of 36.6 and 17.6 in actual/*srs* variances, and similar decreases in the effective sizes of n. Even modest average values of 2^2 have drastic effect on variances and effective sizes. Note some reasonable regularities by classes of variables, and by countries. Nevertheless, substantial differences are revealcd between variables, classes, and countries. Note also the rather regular ratios from 1.2 to 1.3 for roh, of the total sample over roh for subclasses (Table 4a).

(E.5) CV(x) = σ_x / a is directly proportional to the CV of the primary unit sizes X_α and inversely proportional to the number a of those units in the sample. Therefore, it can be reduced with better control of units size X_α, e.g., with stratification, PPS selection, redefinitions; or by taking more primary units a.

Reduction of CV(x) is also needed to control the technical bias of ratio estimates, and these estimates are common in surveys—though happily not the biases (Hansen, Hurwitz, and Madow 1953, Vol. II; Kish 1965a, 6.6B).

(E.6A) This (6.3) has been proposed as not only simpler, but also as giving a better fit (Skinner, Holt, and Smith 1989, Chap. 3). Skinner et al. (1989) present interesting discussions about Deff in chapters 2 and 3, and show that their formula (3.12) (our 6.3) is better than my (6.2) with $k = 1.0$, and their more complex (3.13) even a little better. But with $k = 1.2$ my (6.2) would perform as well, and seems preferable theoretically, because it will take Deft($\overline{y}$) to 1, when $n_c\ / a = 1$, I believe.

Another technique for imputing deft $(\overline{y}_c)$ from deft(y_t) presents a novel and sophisticated method for modeling deft for subclasses (Verma, Scott, and O'Muircheartaigh 1980). All three techniques have in common some model that roh$_c$ for subclasses will be somewhat higher than roh$_t$ for the entire sample, but also that the two are strongly related. I believe that they all lead to useable values of deft($\overline{y}_c$). Also that deciding between their relative values will require some empirical investigation based on several diverse surveys, because we lack a strong theoretical model for deciding between them.

(E.6B) The need for computing deft(p_i) separately, for each of the k categories (i = 1, 2, ... k) of categorical variables with $k > 2$, is illustrated in a recent article (Kish, Frankel, Verma, and Kaciroti 1995). They found in eight surveys from five countries that the values of deft(p_i) as well as the ste(p_i) = deft(p_i) $\sqrt{[p_i(1-p_i)]}$ varied a great deal. Therefore, that choosing only one of the categories to represent as "typical" the diverse categories of that variable was not sufficiently accurate. This result has been found by others before, I believe, but not emphasized. The variation between the variables was even greater than between categories of the same variable. These are empirical results for which simple models would be difficult to construct.

In these eight surveys it was also found, surprisingly for most of us, that deft for the differences ($p_i - p_j$) of two correlated proportions from the same variable is similar to the average of the two deft(p_i) and deft(p_j). That is, deft($p_i - p_j$) ≈ (1/2)[deft(p_i) + deft(p_j)] approximately.

(E.7A) "A great need exists for mathematical bases of analytical statistics to deal with data originating in complex sample designs. At present, these analytical statistics are not computed or they are computed incorrectly under *SRS* assumptions. The latter results in gross mistakes chiefly because of the effects of "clustering" on the sample. Because of these mistakes the researcher often may be actually using confidence coefficients which are distorted (unknown to him or her) from $P = .99$ or $P = .95$ to $P = .50$! These problems require the urgent attention of mathematical statisticians, particularly to provide formulas—valid under complex clustered designs—for some of the most important statistics. An example of this is: The coefficients of multivariate analysis and their variances:..." From my talk Sept. 7, 1956, to a joint meeting of ASA and IMS.

Although right in calling attention to this important neglected problem (and in some other aspects) I was wrong and naïve in several others:

1. "First, mathematical statistics has not and *will not* give us complete distribution theories that will be useful directly, because there are too many parameters in the double complexity of analytical statistics from complex surveys. Second, model builders cannot make those complexities vanish." (Kish 1984).
2. The problems of *inferential* statistics should be more clearly separated from those of *descriptive* statistics. Inferential (second order) statistics depend on pairwise probabilities P_{ij} of selection, hence need ultimate cluster and stratum identifications of "measurability"; but descriptive (first order) statistics depend only on element probabilities P_i and the usual estimates suffice (E.2) (Kish and Frankel 1974).
3. Several methods exist now for computing useful estimates of sampling errors and deft. These depend on Taylor and repeated replication methods without mathematical distribution theories, and on computing developments (Section 9).
4. I failed to foresee that 40 years later only a score or two of mathematical statisticians will pay any attention to the theoretical problems of survey sampling. Their names are in 200 plus References (Skinner, Holt, and Smith 1989). There are many more names of statisticians designing, operating, and analyzing the growing body of probability samples around the world.

(E.7B) In Table 4 we related the average deft(b) for multivariate coefficients to the average deft($\bar{x}$) of means. But after much empirical evidence about large differences between the deft(x_i) for different means it would be better to find more specific relations of the deft(b_i) to the respective deft($\bar{x}_i$).

(E.9) These are but four outstanding examples of available programs from the United States. There will be others in the future, and some that we have missed. In other countries there are still others. Also many more that have been and will be prepared for internal institutional use but not easily available to the outside.

Each of the programs in the table below can deal with case weights, more or less easily. Each of them can handle means, proportions, and ratio estimates.

	Clusters	SUPERCARP	SUDAAN	OSIRIS
Relative Cost	Low	Low	High	Moderate
Relative Simplicity	Easy?	Complex	Complex	Easy?
Data Input	ASCII	ASCII or SAS	ASCII or SAS	ASCII or OSIRIS
S.E. Methods	Taylor	Taylor	Taylor	Taylor or BRR, JRR
Analytical Stats		Regression Chi Square	Regression Log Regression Chi Square Survival	Regression

10

WEIGHTING FOR UNEQUAL P_i

LESLIE KISH

Abstract: Four distinct sources for unequal selection probabilities P_i of elements are distinguished concerning their origins, their effects, and their need for weights $k_i \propto 1/P_i$. Three other types of weighting for estimation are also identified. Survey sampling theory is for unbiased estimation with weights k_i but model based theory is against. The main disadvantage of weighting is the increase in variances from S^2/n to $S^2(1+C_k^2)/n$ for weighted estimates $\bar{y}_w$, where C_k^2 is the relvariance of the k_i. This is balanced against the increase of the mean square error of the unweighted estimate $\bar{y}_u$ from S^2/n to $(S^2/n + R_{ky}^2 C_k^2 S^2)$, where $R_{ky} C_k S$ is the bias = $\bar{y}_u - \bar{y}_w$ of $\bar{y}_u$. This comparison of the mean square errors is explored for reasonable choices between $\bar{y}_w$ and $\bar{y}_u$. Very recently some compromises are being suggested, especially "trimming" extreme weights, and "shrinkage" estimators (Potter 1990; Spencer and Cohen 1991). The problem becomes difficult for multipurpose surveys, which are much more common than a single purpose $\bar{y}_w$.

Key words: Selection probabilities; unequal selections; selection biases; self-weighting.

1. INTRODUCTION

Fundamental questions about weighting seem to be *the* most common during the analysis of survey data and I encounter them almost every week, requiring prompt and practical actions. The requests come from social researchers of all kinds. But I cannot find textbooks or references for them, because we "lack a single, reasonably comprehensive, introductory explanation of the process of weighting" (Sharot 1986), readily available to and usable by survey practitioners, who are looking for simple guid-

---

Reprinted with permission from the *Journal of Official Statistics*, Statistics Sweden (1992), 183–200.
Acknowledgment: This version was considerably improved by critical remarks from Stuart Scott, Paul Flyer, Frank Potter, Vijay Verma, and an anonymous referee.

ance, and this paper aims chiefly to meet some of that need. Some partial treatments have appeared in the survey literature (e.g., Bailar, Bailey, and Corby 1978; Kish 1965a, Section 11.7, 1987a, Section 7.4, 1989), but the topic seldom appears even in the indexes. However, we can expect growing interest, as witnessed by many publications since 1987 listed in the references. With their concentrations and with their style they aim at technical statisticians, whereas I address social researchers and statisticians who want advice for applied problems.

This paper aims to help researchers to find reasonable solutions to practical problems of weighting their data. Here follow some typical questions posed to sampling consultants by client researchers concerned with their data of sample cases.

(a) Now that you (or we) have found that the cases had different selection probabilities (P_i), should these data be weighted by $w_i \propto P_i^{-1}$?
(b) In which situations is it "proper," or "necessary," or "important" to weight data?
(c) Should we weight for nonresponse rates $(1-r_h)$ which differ between classes h in the data?
(d) When does weighting make a considerable difference in estimates?
(e) How can we calculate proper and accurate weights?
(f) How can we apply weights to cases, tapes, and estimates?
(g) How do we apply weights in formulas and software?

The long neglect of weighting as a distinct topic in statistics textbooks reveals an interesting schism in our literature, I propose. Most of statistics deals with *identically and independently* distributed (IID) random variables, where differential weighting need not occur as a topic. On the other hand, sampling methods deal mostly with *selection* procedures; and this concentration is realistic, because in practice the statistical analysis of survey data is often removed in time and personnel from the selection process. But definitions of sample design often include both selection and estimation, and the two aspects cannot be completely separated. Weighting clearly pertains to estimation but it is also related to selection probabilities. Nevertheless, and despite their emphasis on "unbiased estimators," most sampling books refer to weighting only separately in connection with two or three distinct problems.

Some kind of weighting is frequently involved in the analysis of many survey reports, and *ad hoc* explanations appear sometimes, usually hidden in appendices behind project reports. On the other hand, we can also find articles with theoretical discussions that are concentrated only on some single specific aspect of weighting, such as stratification, or post-stratification, or nonresponses, or variance reductions.

We can also encounter misleading statements, even among some theoretical discussions based on diverse models, which are opposed to weighting. Researchers with sample data based on unequal selection probability must face the question whether to use weighted estimates like $\bar{y}_w = \Sigma w_i y_i / \Sigma w_i$ or simple unweighted (equal weighted) estimates like $\bar{y}_u = \Sigma y_i / n$. They are often confused by misleading statements resembling those below, though these are extreme forms of common misconceptions.

(a) Weighting data by $w_i \propto P_i^{-1}$ is a "simple" process that should be "always" applied to samples with unequal P_is (according to "design-based" theory).
(b) We cannot find any justification for weighting in "model-based" theory.
(c) Weighting is needed for means (like $\bar{y}_w$) but not for testing hypotheses or for regressions, because these are model-based.
(d) It is unethical to weight sample cases, because the process can be misused to produce biased results.

We cannot fully explore all the deep implications of such misleading statements. Rather we address the practitioners who want to know WHY, WHEN and HOW to weight their data.

I must avoid those arguments in this simple and brief treatment, which aims to be both general and useful (but see Brewer and Mellor 1973; Hansen, Madow, and Tepping 1983). In order to satisfy those two criteria, to be both simple and general, I had to forsake any attempt at profundity and precision. Anybody who tries to satisfy all three criteria of simplicity, generality, and profundity is bound to fail, probably on all three, I believe. Greater length would be especially needed to also treat the technical subjects of "optimal" weights for estimation. But here, as in the references cited, we are concerned with questions of whether and how to compensate with inverse weights for unequal selection probabilities, as clarified in Section 2. The basic problem is most simply stated by Spencer and Cohen (1991) in introducing "shrinking" for a compromise (but their unweighted Z_m is my $\bar{y}_u$ and their unbiased Z_u is our $\bar{y}_w$):

> A longstanding question in making inferences from unequal probability samples is whether to use an unweighted or other model-based estimator, say Z_m, or whether to use an approximately unbiased estimator Z_u that uses sampling weights reflecting the unequal selection probabilities. (Moments are defined with respect to the sampling design unless otherwise noted.) An unweighted estimator of a population mean often will have smaller variance than a weighted estimator but it will have a bias proportional to the correlation between the characteristic of interest and the sample weights (Rao 1966). For many sampling strategies, the variances of Z_m and Z_u alike decrease to zero as the sample size increases, but although the bias of Z_u is zero or approaches zero the bias of Z_m does not. In such cases, for sufficiently large samples Z_u will have smaller mean square error. On the other hand, for small samples Z_m may have a smaller mean square error (Cochran 1977, pp. 296–297). If one could know the mean square errors of Z_m and Z_u one could easily choose the optimal one. Fortunately, it is possible to use the sample itself to estimate the mean square errors, as DuMouchel and Duncan (1983) proposed in a different context.

2. REASONS FOR WEIGHTING

I distinguish here seven separate main sources of weighting, because they usually have very different effects, and also because they need different strategies and treatments. Of these seven the first three arise from different selection probabilities P_i for the sample cases. Compensatory weights ($w_i \propto 1/P_i$) are the main concern of this and many papers, and for much applied research.

The general and most useful form of weighting is to assign the weights w_i to the sample cases i, with $w_i = 1/P_i$. The selection probabilities P_i for all sample cases must be known for all probability samples by definition. (Obtaining the actual numbers is often a nontrivial but necessary task.) Then the weighted mean is computed as $\bar{y}_w = \Sigma w_i y_i / \Sigma w_i$, and similar "consistent" estimates are discussed in Section 3, as well as the use of convenient weights $w_i \propto 1/P_i$ permitted in averages by the "normalizing" sum of weights $(\Sigma w_i)^{-1}$. The probabilities P_i may have to be computed from complex multistage (or even multiphase) processes. The weights may also be used to compensate for nonresponses, so that $w_i = 1/(P_i r_i)$, where r_i is a response rate often calculated for classes of response.

1. *Disproportional sampling fractions* f_h can be introduced deliberately to decrease either variances or costs. Often these are made with "optimal allocations" to distinct strata h, according to the well known allocation formula $f_h \propto S_h / \sqrt{C_h}$. But they may also result from two (or multi) phase selections. Samplers often achieve spectacular gains in variances (and costs) with these methods, especially applied to surveys of establishments. Gains in household surveys are less spectacular and frequent, but possible (Kish 1961a). These deliberate differences in the sampling fractions f_h should be large to be effective, by factors from 2 to 10 and even greater; smaller differences seldom produce large enough effects to be worthwhile (Kish 1987a, Section 4.5; 1965a, Section 11.7). The differences among f_h should also be highly related to survey variables. The f_h may be simple integral multiples of a basic sampling rate f, like $2f$ or $10f$. These should (always) be compensated with inverse weights (e.g., 1/2 or 1/10) in order to avoid bad biases in combined statistics.
2. *Allocation to domains* of different sampling fractions f_h happens commonly and for distinct reasons. It is common to increase the sampling fraction from f to kf ($k > 1$) in order to reduce sampling errors in one or more domains, especially for small provinces. Or the rates in one or two provinces may be reduced to f/k to save overall costs. Sometimes equal sample sizes n^* are designed for unequal domain sizes N_h, so that the sampling rates are $f_h = n^*/N_h$. These inequalities in f_h are commonly compensated with unequal weights $w_h \propto 1/f_h = N_h/n_h$. However, the need for weights may not be quite as compelling as in 1, because the weights may be less extreme, and the relations of weights for domains to survey variables less strong. Estimates for means may be weighted but not for regressions perhaps (Section 6).
3. *Frame problems* may induce inequalities in selection probabilities P_i that may need compensating with $w_i \propto 1/P_i$. There are four basic classes of frame problems (Kish 1967, Section 2.7).

(1) *Small clusters* of unequal sizes N_i are common; for example, dwellings are commonly selected with equal f and then a single adult from the N_i adults in the dwelling, then $P_i = f/N_i$. Since the numbers of adults are mostly few (N_i = 1, 2, 3, 4 mostly), the biases may be moderate for unweighted means, and the weighted means have variance increases of 1.05 to 1.20 mostly. In one case buildings were selected with equal f, then a single dwelling with $1/N_i$, so that for the dwellings $P_i = f/N_i$ and the N_i ranged from 1 to 62; the biases for unweighted estimates were large, and the variance increase for weighted estimates was 2.6 (Kish 1977).

(2) *Duplicate (replicate) listings* may result in $P_i = d_i f$ when elements with d_i listings are selected with f applied to listings. If replications are common and uncorrected, considerable bias may result, and compensation with $w_i \propto 1/d_i$ is needed.

(3) *Blanks and foreign elements* among listings selected with f cause no inequalities if they are simply disregarded. But a common mistake of substituting the "next valid" listing often causes $P_i = L_i f$ where $(L_i - 1)$ represents the invalid blanks before the valid listing.

(4) *Missing units* (incomplete frame, noncoverage) refer to elements (units) missing from the sampling frame, hence $P_i = 0$. Since this would mean that w_i is not defined ($1/P_i = 1/0$), obviously other measures are needed, and they receive much attention, though never satisfaction.

4. *Nonresponses* present problems that differ from those of 1, 2, and 3, which can be compensated with "inverse P_i" weighting. Weights for nonresponses must involve models or assumptions of some kind, explicit or implicit. It is common practice to assume implicitly that nonresponses arise randomly within response subclasses, though they differ between these subclasses. Thus differential response rates r_h are computed within those subclasses. Thus the sample cases receive weights $w_i = 1/(P_i r_h)$. The subclasses h of the sample are formed with auxiliary variables, such as age, gender, geography etc. These variables must be

(a) available for response cases,
(b) somehow also for the nonresponse cases, and
(c) also related to the survey variables.

It is difficult and rare to obtain data either from the sample or from check statistics that closely satisfy the last two requirements safely and to a high degree. When nonresponses are not high, the differences between subclasses tend to be small, and then small differences in weights will not have large effects on combined results.

For *item nonresponses* compensations seem more often justified, and they are usually made with imputation (replication) of responses (Kalton 1983a; Rubin 1987;

Little and Rubin 1987). Corrections and weighting for *noncoverage* are much more difficult than for nonresponses, because coverage rates cannot be obtained from the sample itself, but only from outside sources. These may be done with "post-stratification," discussed below, where they more properly belong.

The sources above concern unequal probabilities of selection P_i, known from determinate selection operations. They are the chief subject of controversies about weighted versus unweighted estimates. However, the next three sources and types are motivated by estimation, rather than selection, and use models and auxiliary data sources. Some may question whether "weighting" is an appropriate term for these methods. Nevertheless the procedure can be summarized with factors c_i so that $w_i = c_i / (P_i r_i)$ can be used for the ith case. However, there are so many possible methods that we must limit our discussion to a few examples.

5. *Statistical adjustments* for improved estimates have diverse names: post-stratification, ratio estimators, and regression estimators are all described in the sampling literature, for reducing variances with controls that were not used in the selection process. In practice, however, post-stratification may denote a ratio estimator for reducing the biases of nonresponse and especially of noncoverage. Thus the ratio estimator $\Sigma X_h y_h / x_h$ with the auxiliary variable X is also $\Sigma N_h y_h / n_h$ in post-stratification for the population size N; these aggregates become means when divided by ΣX_h or ΣN_h. For example, check data from the census may be used for correcting for age-sex-race biases of surveys. As an extreme it adjusts for a noncoverage of young black males of 13% (U.S. Bureau of the Census 1978, chap. 5; Kish 1987a, Section 4.7). For those methods the data and the software must both be appraised for integrity; large biases may be introduced with inadequate models or data. Other technical methods also appear in the literature, such as weighting cases proportional to their precision, $w_i = 1/\sigma_i^2$ (Kalton 1968).
6. *Adjustments to match controls* can have a variety of motivations. Whereas the reasons under 5 for post-stratification and ratio estimation concerned mostly sampling variations, here we refer mostly to adjustments of samples from one frame population to some other target (standard) population(s), often known as *standardization*. For example, a sample from one province (state) may be reweighted to the national population (Kish 1987, Section 4.5). Or we may reweight samples from one country or period to another country or period. Generally, the subclasses of the sample are reweighted to the domains of the target population, and these controls must be available both for the sample and for the target population. If there are too many cells, the control data may be unavailable and the sample cases too few for stability, and then marginal adjustments may be used, with iterated fitting.

   Reweighting may also be used to examine differences, like $(\bar{y}_a - \bar{y}_b)$ of two (sub)populations (a) and (b) free from the effects of the different "compositions" (N_{ha} and N_{hb}) of the two populations in strata h (Kish 1987a, Section 4.5D).

Adjustments of nonprobability samples to fit check data in subclass cells, (e.g., age, sex, province) are also made. These adjustments can hardly overcome the biases of nonprobability selections within the subclasses. They may be viewed as similar to "quota" sampling, but with weighting substituted for selection.

7. *Combining samples* is becoming more popular, more important, and more feasible because of increasing numbers of samples that are available for combinations. All combinations concern weighting in some form, and one should always be explicit about the weights; and also careful about possible differences in measurements. We note also that any national sample combines diverse domains, some like provinces, some like diverse social or demographic classes; and all those domains differ in the distributions of the survey variables. Nowadays, one may also combine standardized national samples from several countries, e.g., African birth rates from separate national samples of the World Fertility Surveys. Similarly to spatial integration, we may also combine periodic samples into *rolling samples* integrated over a longer time span; e.g., annual averages of influenza, or cancer rates, or unemployment, or incomes from weekly or monthly surveys (Kish 1990b). *Meta-analysis* is a growing field for combining statistics, and already foreshadowed in 1924 by Yates and Cochran (1938). A special and simple method of combining can be the *cumulations* of individual cases (Kish 1987a, Section 6.6).

3. METHODS FOR IMPLEMENTING WEIGHTING

Four alternative procedures for weighting need individual attention because they require different techniques and also because they can have different effects on the variances.

1. *Individual case weights* (ICW) yield the most common, simple, practical, and flexible procedures, especially with modern computers and programs that can handle them. (Not all programs are equally adept.) The other procedures may be compared with and based on ICW, and they may increase variances more than ICW. The weights w_j for sample elements ($j = 1, \ldots, n$) may reflect a product $p_j r_j$ of the element probabilities p_j from complex statistical multistage selections with the response rates r_j, which may also include coverage rates. The weights $w_j = 1/(p_j r_j)$ are inversely proportional to these products. Both p_j and r_j should be available for all elements of probability samples, and newcomers to surveys must be made aware that those values must be obtained with careful bookkeeping. It is also possible to incorporate weights W_h for post-stratification (ratio) estimators so that $w_j'' = W_h/(p_j r_j)$.

The basic statistic is the weighted sum of sample values $\Sigma w_j y_j = \Sigma y_j / p_j$. This is a desired unbiased estimator of the population sum $\Sigma Y_i = Y = N\bar{Y}$.

For equal probability selections, or *epsem*, of n from N elements (whether simple random or more complex), $f = n/N$, we have the expected value $E(\Sigma y_{ji}/f) = \Sigma E(y_j N/n) = \Sigma \bar{Y} N / n = N\bar{Y} = Y$. For weighted estimates we also have $E(\Sigma y_j / p_j) = N\bar{Y} = Y$. This is shown in all sampling books, sometimes as a "Horvitz-Thompson" estimator (Cochran 1977, Section 9A.7; Kish 1965a, Section 2.8C). This simple expansion estimator is basic to probability sampling and should perhaps be called an "expectation estimator." In practice it is seldom used in this simple form and it needs adjustment for nonresponses, so that $w_j = 1/(p_j r_j)$.

The most common statistic is the weighted mean $\bar{y}_w = \Sigma w_j y_j / \Sigma w_j$. This is not "unbiased" technically (because it is a ratio estimator) but it is "consistent" as are the other similar statistics. Thus with $\Sigma w_j y_j^2 / \Sigma w_j - \bar{y}_w^2$ for the element variance; also with $\Sigma w_j y_j x_j / \Sigma w_j$ replacing $\Sigma y_j x_j / n$ from *epsem* selections. Because they are normalized (standardized) with Σw_j, the weights can be any positive numbers proportional to the expansion weights $1/(p_j r_j)$. It may help to note that for *epsem* selections we have the expansion weights $w_j = 1/f = N/n$ and $\Sigma w_j = N$; whereas in common formulas $w_j' = f / f = 1$ and $\Sigma w_j' = n$.

When we can find appropriate, unbiased, and dependable values of N_h for the population, the sampling fractions n_h/N_h in domains h can be used sometimes for $w_j = N_h/n_h$, and this is justifiable when the elements j are selected with actual equal probabilities within the domains h. On the other hand, in many situations the selection probabilities f_h must be applied, because reliable N_h are not available. However, it is misleading to confuse a mere fraction of elements in the sample with a true sampling probability; e.g., that a sample of n_h is selected from a population of N_h one may perhaps refer to a sampling "fraction" of n_h/N_h. But probabilities of selection must be justified with probability operations: otherwise we are faced with judgment samples, "quota" samples, and other model dependent sampling.

TABLE 1. Disproportionate Allocation to Illustrate Weights

W_h	N_h	f_h	r_h	n_h	$1/w_j$	w_j	$w_j n_h$
.90	90,000	.01	.90	810	.009	111.111	90,000
.09	9,000	.10	.85	765	.085	11.765	9,000
.01	1,000	1.0	.8	800	.80	1.25	1,000
1.00	100,000 = N						$\Sigma w_j n_h$ = 100,000

A population of N = 100,000 elements was divided into three highly unequal strata of N_h elements. The disproportionate selection rates f_h applied and the different response rates r_h obtained result in Σn_h = 2375 observations. From $p_j r_j = 1/w_j$ in the three strata the weights $w_j = 1/(p_j r_j)$ are obtained. Note that $\Sigma w_j = w_j n_h = N_h$ exactly for each stratum, though in practice minor irregularities would cause small variations. For random variables Y_i we do not get the Y_h exactly, but we get the expectation $E(\Sigma w_{hj} y_{hj}) = Y_h$.

2. *Weighted statistics*, e.g., $\bar{y}_w = \Sigma W_h \bar{y}_h$, combine separate subpopulation statistics $\bar{y}_h$ with appropriate relative weights W_h, with $\Sigma W_h = 1$. This method may be preferred over ICW for:

   (a) combining published statistics when individual cases are not available;
   (b) combining a few strata based on disparate selection procedures; and for
   (c) relatively simple statistics, like means or totals.

   But they are not as useful for complex analyses of single surveys. Dependable weights W_h are needed from justifiable sources. These can also be used as $w_j = W_h/(p_j r_j)$ with the ICW, as above.
3. *Duplication of cases* may be used instead of ICW in order to prepare self-weighting tapes for convenience in some situations. It is especially convenient for item nonresponses, and particularly for complex analytical statistics, because both reasons hinder individual weights (Kalton 1983a). Some compromise between random selection and "closest" matching to reduce bias is generally used for duplication within subclass cells. If the response rate is r_h in cell h, $(1 - r_h)$ cases can be duplicated to fabricate $(1 - r_h)$ pseudo cases; either randomly selecting with probability $(1-r_h)/r_h$ or by finding the "closest" matching of that fraction of cases. Duplication increases variances over individual weighting ICW, but those increases are not great for duplicating only a small proportion of the samples. Furthermore, these increases of variances can be almost eliminated with procedures of "multiple replications" (Little and Rubin 1987; Rubin 1987).

   We must caution against the crude mistake of accepting from the computing programs the tape counts (or card counts) m, which contain $(m - n)$ replicates as well as n genuine cases. The n genuine cases can be "tagged" for counting. But the "effective number" may be further diminished by duplication to $n' = n/(1 + L)$, as noted in point 2 in Section 4.
4. *Elimination of cases* can be justified in some situations, although throwing away information may appear statistically criminal and is seldom practiced. Nevertheless, consider three justifiable situations.

   (a) Large samples have been selected with different sampling rates for a nation's several provinces; then a self-weighting sample is designated for complex national analysis, with rates suited to the lowest provincial rates; "microtape" samples can be made self-weighting.
   (b) A small domain has been greatly oversampled for separate analysis, but a proportionate sample has been "tagged" from it for joint complex analysis, which could be difficult and not much more precise with the extra cases from the small domain.
   (c) Eliminating a small proportion of cases (say $<.05$) increases the variance only little more than duplication of a similar proportion. This counterintuitive

result can be used for compromise adjustment for differential nonresponses between strata (Kish 1965a, Section 11.7B).

4. REASONS AGAINST WEIGHTING

1. *Complications* often arise from weighting, even when good computing programs are available, and this factor is often neglected and difficult to quantify. That is why I put it first, though it should not be the most important. Mistakes arise in the man-machine system and they tend to increase for more complex analyses. Other complications are more basic: for complex, analytical statistics, and for inferential statistics, such as tests of significance, adequate methods may not be available for weighted estimators or for their sampling errors. Theoretical contributions are now fast developing, but they are not useful, general and simple enough to be "available" for many survey practitioners (Kott 1991a, b; Rao and Scott 1981).
2. *Increased variances* can result from weighting for random, or haphazard, or irregular differences in selection probabilities, when these are not "optimal." For example, the inequalities due to frame problems or to nonresponses are generally of this kind. Furthermore, these increases of variances (unlike those due to clustering) tend to persist undiminished for most such subclasses and for all statistics, as if they were to increase the element variances from σ^2 to $(1+L)\sigma^2$, or to decrease the "effective" number of elements from n to $n/(1 + L)$. Here L denotes relative loss, so that $L = 0.8$ means a factor of $1 + L = 1.8$ increase in the variances, explained below.

   "Haphazard" sources of weights are most common in survey work, but they are counterintuitive to minds attuned to "optimal allocation," source 1 in Section 2. Those weights are highly related to stratum values $\bar{y}_h$ and σ_h. But $(1 + L)$ refers to the other sources (2, 3, and 4), small domains, frame problems, and non-responses, which are hardly related (negatively or positively) to most survey variables. Therefore the best summary measure of their effect is a relative increase of $(1 + L)$ in variances, a statement based on much experience in multipurpose surveys. For example, even a sample with optimal allocation for *mean* incomes and assets turns out to be less efficient than proportional allocation for buying behavior and even for *median* income and assets in the same sample (Kish 1961a; Verma, Scott, and O'Muircheartaigh 1980).

   Three simple formulas yield adequate estimates of the increases $(1 + L)$ in element variances. The choice between these three alternatives depends on the situation and the data available for the weights to be used. In these formulas N_h represent population sizes and n_h sample sizes for strata h, and $W_h = N_h/\Sigma N_h$ and $w_h = n_h/\Sigma n_h$ denote relative sizes, so that $\Sigma W_h = \Sigma w_h = 1$. The weights are represented by k_h, which can be $1/f_h$, the inverse of the selection rates, but they may be only *relative* values proportional to them, $k_h \propto$

$1/f_h$. I use k_h or k_j for w_j here for easy comparison with references, where 4.1 to 4.5 are derived (Kish 1965, Section 11.C; 1976, 1988a).

(a) In the design stage one may consider using sampling fractions and weights in the proportions k_h in strata with relative sizes W_h. If the element variances are roughly equal ($\sigma_h^2 \simeq \sigma^2$ approximately) then the variance of the mean (and many other statistics) will increase approximately by the factor

$$(1+L) = (\Sigma W_h k_h)(\Sigma W_h / k_h). \tag{4.1}$$

(b) In the analysis stage, if n_h cases have weights k_h, the increase in the variance (with conditions as above) is approximately

$$(1+L) = \Sigma n_h \Sigma n_h k_h^2 / (\Sigma n_h k_h)^2. \tag{4.2}$$

For individual element weights k_j when $n_j = 1$ for all n cases (4.2) becomes simply

$$(1+L) = n\Sigma k_j^2 / (\Sigma k_j)^2. \tag{4.3}$$

Note also that the relative increase or loss L may be viewed as the *relvariance* cv^2 = variance/mean^2 of the relative weights k_j, because

$$\begin{aligned} cv^2 &= n\Sigma k_j^2 / (\Sigma k_j)^2 - (\Sigma k_j)^2 / (\Sigma k_j)^2 \\ &= (1+L) - 1 = L. \end{aligned} \tag{4.4}$$

Thus the factor $1 + cv^2 = 1 + L$ depends on the relative variances of case weights k_j. It serves as good precaution to compute the cv^2 or $1 + L$, or the frequency distribution of the weights to estimate what the increase may be.

(c) When the population sizes N_h and sample sizes n_h are both directly available they can be used directly without the relative weights k_h to compute

$$1 + L = (\Sigma N_h^2 / n_h) n / N^2. \tag{4.5}$$

3. *Lower mean square errors* (MSE) may be achieved by unweighted, biased estimators, such as means $\bar{y}_u$. Comparisons with weighted means $\bar{y}_w$ can be based on $\text{MSE}(\bar{y}_u) = S^2(n^{-1} + B^2/S^2)$ versus $\text{Var}(\bar{y}_w) = (1+L)S^2/n$. The bias ratio B/S for $\bar{y}_u$ can be estimated from the ratios $(\bar{y}_u - \bar{y}_w)/\text{ste}(\bar{y}_u)$ computed from survey data for several (many) survey variables.

Similarly, the factor $(1 + L)$ can be estimated with $(1 + C_k^2)$ from the sample with (4.3) or anticipated in the design with (4.1), and this increase in the variance is rather constant for most statistics. The factor C_k^2 is also important for B^2/S^2 (equation 5.3), but these bias ratios differ greatly for diverse survey variables of the same survey. Furthermore for subclasses and their comparisons the variances are much higher, hence the ratios much lower (Section 7).

4. *Model dependent arguments* have been advanced that weighting corrections for selection biases are not needed for regressions from surveys (Brewer and Mellor 1973; Hansen, Madow, and Tepping 1978).
5. *Public relations or ethics* may also hinder overt and differential weighting, because it is possible to misuse it to produce subjectively desired, prejudiced results (Sharot 1986). For example, the combined mean $\bar{y}_w = \Sigma W_h \bar{y}_h$ could be made to approach any of the components $\bar{y}_h$ with extreme weights W_h. Journalists, alas, do this commonly, by using the cost of either automobiles and TV sets, or housing and health care to contrast the cost of living in economies with contrasting price systems. The naively prejudiced weights tend to escape the public's scrutiny, but any explicit weighting system suffers from exposure, unfortunately.

5. BALANCING VARIANCE INCREASES AGAINST BIASES

We saw that that ratio of increase of variances due to (haphazard) weighting can be computed as

$$1 + L = 1 + C_k^2 = 1 + \sigma_k^2 / \bar{k}^2 \tag{5.1}$$

from the data (4.3) or anticipated in the design (4.1). This has been shown to be true generally for departures from optimal allocating in linear sample designs (Kish 1976). The relative bias, $-B = (\bar{Y}_w - \bar{Y}_u)/\bar{Y}_u$, of unweighted samples can also be estimated from the sample data. But it is remarkable and useful that these biases for means can also be shown to depend on the same C_k^2, and on the correlation R_{ky} between the weights and the survey variables (Kish 1987a, equation 7.4.12). Thus:

$$\begin{aligned} -\text{Bias} &= \bar{Y}_w - \bar{Y}_u = \Sigma y_i k_i / \Sigma k_i - \bar{Y}_u \\ &= \bar{k}^{-1}[N^{-1}\Sigma y_i k_i - \bar{k}\,\bar{Y}_u] \\ &= \bar{k}^{-1}\text{Cov}(k_i y_i) \\ &= \bar{k}^{-1} R_{ky}\sigma_k \sigma_y, \end{aligned} \tag{5.2}$$

where $\bar{k} = \Sigma k_i / N$. These summations are to the population N, but they can be made to the sample n in sample estimates, which concern us most. The sample difference $(\bar{y}_w - \bar{y}_u)$ estimates the population difference $(\bar{Y}_w - \bar{Y}_u)$, with the expectations: $\text{E}(\bar{y}_w) = \Sigma(P_i Y_i / w_i)/N = \Sigma Y_i / N = \bar{Y}$, the population mean; but

$E(\bar{y}_u) = \Sigma P_i Y_i / N = \bar{Y}_u = \bar{Y}$ + Bias, with the entire population exposed to unequal selection.

Thus $\bar{Y}_u - \bar{Y}_w =$ Bias, because $\bar{Y}_w = \bar{Y}$ is unbiased, but $\bar{Y}_u$ is biased; but I used (5.2) for convenience, because in the sample we have $\bar{y}_u$ most conveniently. Similarly relative values based on $\bar{Y}_w$ are preferable on theoretical grounds, for B, S, and C_y, as in (5.5). But I used $\bar{Y}_u$ for everyday convenience and because it has lower variances.

$$-B = (\bar{Y}_w - \bar{Y}_u)/\bar{Y}_u = R_{ky} C_k C_y$$

and

$$B^2 = R_{ky}^2 C_y^2 C_k^2. \tag{5.3}$$

In order to contrast the increase in variance $(1 + C_k^2)$ with the effects of biases, let us consider a mean with the effective sample size $n_d = n/\text{Deff}$, where Deff is the "design effect," often appreciably greater than 1; these effects, Deff = $\text{Var}(\bar{y})/(S_y^2/n) > 1$, have been computed and used in many studies (Kish 1976). Then the relative mean square errors (RMSE) for $\bar{y}_w$ and $\bar{y}_u$, respectively, are:

$$S^2(1 + C_k^2) = C_y^2(1/n_d + C_k^2/n_d) \tag{5.4}$$

and

$$B^2 + S^2 = C_y^2(1/n_d + R_{ky}^2 C_k^2). \tag{5.5}$$

These relations were expressed in relative terms, with the biases and variances divided by $\bar{Y}_u^2$. I preferred these in order to make comparisons easier for many variables within the same survey and also between surveys. However, some may prefer to avoid that, especially for situations where division by $\bar{Y}_u^2$ is inappropriate, as when $\bar{Y}_u$ is near zero or is a proportion when P or $(1 - P)$ may be confused. But the same relation may be found in

$$\text{Var}(\bar{y}_w) = S_y^2(1/n_d + C_k^2/n_d) \tag{5.6}$$

$$\text{Bias}^2 + \text{Var}(\bar{y}_u) = S_y^2(1/n_d + R_{ky}^2 C_k^2). \tag{5.7}$$

Thus the relative increase C_k^2/n_d for the variance of $\bar{y}_w$ decreases along with the variance itself. But the effects of the biases in $\bar{y}_u$ do not decrease, hence come to dominate for large samples; and these functions of R_{ky}^2 tend to vary greatly between variables.

From the relations above we construct some useful guidelines for practical work.

1. Values of $(1 + C_k^2)$ should be estimated in the design (4.1) or from the sample (4.3). When C_k^2 is moderate its effects on both biases and variances will be small, except for very large n_d and large R_{ky}^2. For

TABLE 2. Relative Losses (L) for Six Models of Population Weights (U_i); for Discrete (L_d) and Continuous (L_c) Weights: for Relative Departures (k_i) in the Range from 1 to K[a,b]

Models	K	1.3	1.5	2	3	4	5	10	20	50	100
Dichotomous $U(1-U)$											
(0.5)(0.5)		0.017	0.042	0.125	0.333	0.562	0.800	2.025	4.512	12.005	24.50
(0.2)(0.8)		0.011	0.027	0.080	0.213	0.360	0.512	1.296	2.888	7.683	15.68
(0.1)(0.9)		0.006	0.015	0.045	0.120	0.202	0.288	0.729	1.624	4.322	8.82
Rectangular	L_d	0.017*	0.042*	0.125*	0.222	0.302	0.370	0.611	0.889	1.295	1.620
$U_i \propto 1/K$	L_c	0.006	0.014	0.040	0.099	0.155	0.207	0.407	0.656	1.036	1.349
Linear decrease	L_d	0.017*	0.040*	0.111*	0.203	0.283	0.353	0.616	0.940	1.437	1.917
$U_i \propto K+1-K_i$	L_c	0.006	0.014	0.040	0.097	0.153	0.205	0.409	0.680	1.127	1.514
Hyperbolic decrease	L_d	0.017*	0.040	0.111*	0.215	0.312	0.404	0.807	1.466	3.014	5.076
$U_i \propto 1/k_i$	L_c	0.006	0.014	0.041	0.103	0.171	0.235	0.528	1.011	2.138	3.621
Quadratic decrease	L_d	0.016*	0.036	0.080	0.150	0.211	0.264	0.460	0.696	1.048	1.333
$U_i \propto 1/k_i^2$	L_c	0.006	0.014	0.040	0.099	0.155	0.207	0.407	0.656	1.036	1.349
Linear increase	L_d	0.017*	0.040*	0.111*	0.167	0.200	0.222	0.273	0.302	0.320	0.327
$U_i \propto k_i$	L_c	0.006	0.013	0.037	0.088	0.120	0.148	0.223	0.273	0.308	0.320

[a]From Kish (1976, 1987a).

[b]Dichotomous, $1 + L = 1 + U(1-U)(K-1)^2/K$. Also all*. Discrete, $1 + L_d = (\Sigma U_i k_i)(\Sigma U_i / k_i)$, with $k_i = i = 1, 2, 3, \ldots, K$.
Continuous, $1 + L_c = \int Uk\, dk \int (U/k)\, dk$, with $1 \le k \le K$. Only two values, 1 and K, were used for L_d for $K = 1.3$, 1.5, and 2.

example, when $C_k^2 < 1$, $C_k^2/100 < 0.01$ and $R_{ky}^2 < 0.01$ if $R_{ky} < 0.1$. Furthermore $(1 + L)$ can be often guessed well enough from Table 2 and values of the range $K = k_{max}/k_{min}$ of the relative k_i. For example, for $K = 1.3$, L is between 0.01 and 0.02; and this may be true for many nonresponse weights; for $K = 1.5$, L is between 0.015 and 0.04. For $K = 2$, L is between 0.04 and 0.125 and may be worth computing; and this is true for $K \geq 3$. For example, the value of $L = 0.33$ I computed for the Current Population Surveys would explain why only mild differences $(\bar{y}_w - \bar{y}_u)$ were found by Bloom and Idson (1991).

2. Note that C_k^2 depends on both the range of the k_i and the frequency distribution represented by W_h or $n_h = W_h/k_h$. Hence C_k^2 will be large only when k_{max} is large for large portions, W_h or n_h. Large values of L in Table 2 occur for large K, and for dichotomies, especially for W_h (or U) = 0.5.
3. C_k^2 is stable for the sample, but C_k^2/n_d is much increased for small n_d in subclasses. On the contrary, the values of R_{ky}^2 can vary by orders of magnitude between variables of the same survey. Therefore the effect of biases can be large for some variables, especially for the entire sample with large n_d; but it may be small for subclasses with small n_c, or small R_{ky}^2. Weighted estimates should be preferred when $R_{ky}^2 > 1/n_d$. Perceptible biases are seldom found, because the squared correlations have small effects.
4. We have disregarded here

   (a) the extra costs (troubles) of weighting;
   (b) the advantages of self-weighting samples;
   (c) the advantages of compromises between weighted and unweighted estimates, noted below in Section 8;
   (d) implications also for more complex statistics, such as regression coefficients.

6. EPSEM SELECTIONS FOR SELF-WEIGHTING SAMPLES

Self-weighting samples are often preferred, because they possess considerable advantages in reduced variances, in simplicity, and in robustness. Statistical theory also, from the lowest to the highest, overwhelmingly assumes self-weighting samples in one form or another. Furthermore, selections of elements with equal probabilities, *epsem* for short, often seems desirable and reasonable when the survey variables are more or less evenly distributed over the population. Voting by all adults springs readily to mind, but there are other behaviors, attitudes, and opinions which are also democratically, evenly distributed, or at least roughly so.

Self-weighting samples for analysis is a goal for many surveys and epsem selection is the principal means toward that end. Because we often confuse the two let me clarify how they exist in common practice. Disproportionate "optimal allocations" (Source 1 in Section 2) clearly are not epsem, and hardly anybody would use them for self-weighted (i.e., unweighted) analysis. Samples with oversampled (or undersampled) domains (source 2) are not epsem and they also are not self-weighting. When frame problems (source 3) result in unequal P_i of selection the question of weighting becomes quantitative (Sections 4, 5, and 7). For example, fertility studies may find that 1 to 5 % of an epsem sample of households have two women of childbearing age; both can be selected to maintain the epsem f. If one is selected at random her probability is reduced to $f/2$, but in most cases the analysis may disregard the small factor $(1+B^2/S^2)$ and proceed with self-weighting. On the contrary the example in Section 2 of single dwellings selected from buildings with N_i dwellings ($N_i = 1, 2, ..., 62$) presented gross violations with $P_i = f/N_i$; the variance factor was $(1 + L) = 2.6$, but the bias factors $(1 + B^2/S^2)$ were highly variable and much worse. It is best to avoid them, an important part of the sampler's art.

Nonresponses are impossible to avoid, but much can be done to reduce them and their effects (too much to detail here). Because of their omnipresence they should not cancel the label epsem, but if weights are used for compensation the analysis is not self-weighting, and we saw above that epsem is not necessary in cases of trivial frame problems. The two are closely related but keeping the two concepts distinct clarifies both.

Similarly we must avoid the common confusion of epsem with simple random sampling (srs). Probably most survey samples are epsem, but very few are srs (outside academic writing). That popularity of epsem and self-weighting samples is due to their "robustness," I claim (Kish 1977). Three reasons were given in Section 4: avoiding complexity, the variance factor $(1+L)$, and the bias factors $(1+ B^2/S^2)$. The fourth reason comes from the multipurpose nature of most surveys, and that self-weighting "satisfices" most purposes and comes close to optimizing ("satisficing") many or even most purposes (Kish 1976, 1988a).

Unequal probabilities can be justified by "optimal" allocation if the purposes (all, or preponderant) of the survey justify it. In other situations, however, the sampler should search for ways to avoid the dilemma between the biases of unweighted and the increased variances of weighted samples. Achieving epsem samples during the selection operation is a fundamental skill in the art of survey sampling. This includes complex multistage selections with probabilities proportional (first directly, then inversely) to measures of sizes. Often it also requires clever handling of imperfections in the sampling frame. After selection, achieving acceptable response rates often needs skillful and devoted care.

7. DIVERSE EFFECTS FOR DIFFERENT STATISTICS

Users must be warned about these wide differences between effects, because many may be misled by the mere phrasing of "THE Bias." The biases can be estimated from the sample by $(\bar{y}_u - \bar{y}_w)$, and should be. They vary greatly between the variables and depend strongly on the correlations between the survey variables y_j and the weights w_j; the bias may be negligible for most variables but large for others. The standard errors can differ also, increasing for small subclasses with small $\sqrt{n}$. However, they tend to differ less than the biases, and especially for proportions when $\sigma = \sqrt{[P(1-P)]}$ remains rather constant for all but extreme values of P.

1. *Expansion totals* $\hat{Y} = y / f$ are most sensitive to biases from weights; for example, even uniform and random nonresponses can result in bad underestimates, if not adjusted. Also expansions like $\hat{Y} = \Sigma N_h (y_h / n_h)$ can be very sensitive to biases in the borrowed values of the N_h. But differences or ratios of such totals from periodic studies would be less sensitive, therefore biases may be more tolerated in such comparisons.
2. *Means* are usually less affected than totals. Sample surveys survive the terrible nonresponse rates now prevailing in the USA (and elsewhere) only because nonrespondents do not differ drastically from respondents for most survey variables. Large biases result only from combinations of differences in *both* weights and survey variables within subclasses. If *either* of these is uniform over subclasses the net bias tends to be small (Kish 1965a, Section 13.4B).
3. For *subclass means* the variances increase in proportion to the decrease of the sample bases, roughly in a ratio that may be denoted by $S_c = \sqrt{(\text{Deff}_c \sigma_c^2 / n_c)}$. The n_c, σ_c, and Deff_c all refer to the subclasses c; the "design effects" Deff_c in cluster samples tend to decrease slightly from Deff > 1 toward 1 with decreasing subclass size, especially for "crossclasses," but much more slowly than n_c. Crossclasses refer to the majority of subclasses which cut across the clusters of sample designs; thus the Deff = $[1 + \text{roh}(\bar{b} - 1)]$ of clustering tends to be reduced toward 1 as the average size of clusters reduces from $\bar{b}$ to $\bar{b}_c$ along with n to n_c (Kish 1987a, Section 2.3; Verma, Scott and O'Muircheartaigh 1980). Since the biases B_c tend to be the same, either generally or on the average, the bias ratio B_c / S_c tends to decrease as S_c increases with decreasing sizes of subclasses.
4. For *subclass differences* $(\bar{y}_c - \bar{y}_b)$ the above process is greatly enhanced, because the standard errors $S = \sqrt{(S_c^2 + S_b^2)}$ are greater than for one subclass, and even more because biases often tend to be in the same direction, and thus tend to cancel in the differences. Then the bias ratio B/S with a drastically reduced numerator, and increased denominator, becomes greatly reduced (Kish 1987a, Sections 2.4–2.6). Hence the mean-square-errors $S^2(1 + B^2/S^2)$ become dominated by the variances S^2. "Model dependent" inference may go further and claim that weights can be disregarded in estimating subclass differences. However, I reject that null limit for B on philosophical grounds (Kish 1987a, Sections 2.4 and 1.8).

5. *Analytical statistics* can be of many kinds, and a general statement about *B/S* seems difficult. I share the "population bound" view of inference that weighting matters and data have shown this for such analytical statistics as regression coefficients (Holt, Smith, and Winter 1980). When considerable differences are found between weighted and unweighted statistics (e.g., regression coefficients), I trust the former. It is also true that computational, methodological, and mathematical problems may pose formidable problems for weighted estimates. *Sampling errors*, inferential statistics, and tests of significance can also pose severe problems of computation, methodology, and interpretation for weighted estimates.

8. TO WEIGHT OR NOT TO WEIGHT? COMPROMISES AND STRATEGIES

The literature on this topic, including my references, concern themselves with *the* bias of some statistic, often a regression equation, as if this were the single purpose of the survey. But most surveys, all in my experience, have many purposes; they are multipurpose and in several dimensions: several variables, several statistics, subclasses, etc. (Kish 1988a). All these exhibit different relative effects $(1 + B^2/S^2)$ of biases. Is it possible or desirable to treat each statistic separately, perhaps using weighted estimates for means, but unweighted estimates for regressions, as implied by some?

Another paradox arises when we contrast academic literature with practice. In the former there are sharp contrasts made between the "population-bound" and the "model dependent" approaches (Brewer and Mellor 1973; Hansen, Madow, and Tepping 1978; Kish 1987a, Sections 1.4 and 1.8). Perhaps the contrasts are sharpened by exaggerations of the opponent's misstatements. Most important, actual practice often leaned toward compromise, but this needs the guidance of theory.

Some theory has been coming in the past few years as my references show, and we may confidently await growing interest; the trend is toward compromises and the criterion of mean-square-errors predominates. Though the terms of $(1 + L)$ and $(1 + B^2/S^2)$ for relative increases in variances and biases may be mine, most of the concepts are not contradictory to them. I find it interesting and worthwhile to distinguish four "levels" of compromises.

1. At the very least comes the recognition that choices should be made for specific situations rather than a blanket population-bound stand of always weighting for unequal P_i versus a blanket model-based position of never weighting, especially for relationships and regressions (Bloom and Idson 1991; DuMouchel and Duncan 1981; Graubard and Korn 1991; Iannacchione, Milne, and Folsom 1991; Kish 1965a, p. 400). I propose, however, that the MSE criterion of $(1 + B^2/S^2)$ is preferable to tests of significance for $|B/S| > 1.96$ at the 5% level.

2. At the next level we find some practice of *trimming* small percentages of extreme weights to accept small biases against large reductions in the $(1+C_k^2) = (1+L)$. It may be especially useful to trim the extreme "right tail" where a few cases of large k_j from small p_j (some of these outliers may be mistakes, others haphazard events) may greatly increase $(1+C_k^2)$ (Flyer, Rust, and Morganstein 1989; Hidiroglou and Srinath 1981; Kott 1991a; Lee 1991; Potter 1988). I wonder if some smoother transformation than trimming of the k_j may do even better.
3. The *shrinkage* of weights seems to be a natural development that holds promise (Spencer and Cohen 1991). Instead of the *ad hoc* nature of trimming decisions, it points to a generalized, uniform treatment with theoretical bases. However shrinkage weights are specific for each variable and they can vary widely on multipurpose surveys. Trimming is uniform, and it affects only a small portion of the sample.

   Briefly, a shrinkage mean $\theta \bar{y}_u + (1-\theta)\bar{y}_w$, with $0 \leq \theta \leq 1$, in practice implies transforming the weights to $g_j = \theta \bar{k} + (1-\theta)k_j = k_j - \theta(k_j - \bar{k})$. On the other hand, with trimming we get $g_j = k_j$ for $k_j \leq K$, but $g_j = K$ for $k_j > K$, where K is some strategically chosen constant; for two sided trimming $|k_j| > K$ may be used. However, some other compromise transformation between these two transformations may be even better. For example, a square root transformation $g_j = \sqrt{k_j}$, or $g_j = k_j^c$ with $0 \leq c \leq 1$. Empirical investigations would be welcome in situations where C_k^2 is large enough to distinguish between the diverse gains.
4. Nevertheless the *multipurpose* nature of most surveys raises problems not addressed by my references. When are specific answers for each statistic of a survey feasible and desirable? Or is a general overall answer for all statistics on a survey more acceptable, and how does one arrive at such a compromise? Having had to answer these questions in practice convinces me that they need more theory. The good news is that there is need for much research, both theoretical and empirical, and especially combined. I also hope for compromise average weights, adapted from those for allocations (Kish 1976, 1988a).

I end with a few pieces of advice.

(a) Always compute estimates of the factor $1+C_k^2$ (Section 5).
(b) If this $(1 + L)$ is large, see how much you can reduce it with trimming.
(c) Compute many (30 or 60?) estimates of $(\bar{y}_w - \bar{y}_u)$ for different variables and different statistics.
(d) Make comparisons with the factors $(1 + B^2/S^2)$ and justify your decision.

NONSAMPLING ERRORS

COLM O'MUIRCHEARTAIGH AND ROBERT M. GROVES[1]

Most of Kish's work was concerned with sample design and analysis of complex surveys. However, he did not focus narrowly on sampling and sampling errors, but rather he took a broader view of the total survey process. He was well aware that sampling errors are only one source of error in survey estimates and that, indeed, with large samples they are often far less important than the various sources of nonsampling error. Chapter 13 of Kish's (1965) text *Survey Sampling* provides an excellent presentation of total error models for surveys, nonresponse and noncoverage, response biases and variance, and measurement errors of other sources.

A practical limitation to probability sampling in survey research is that the sample actually achieved suffers deficiencies from noncoverage and nonresponse. Both of these can cause bias in the survey estimates. Noncoverage occurs when certain elements in the target population are not included on the sampling frame from which the sample is selected. With household area sampling, in principle there is no noncoverage of the residential noninstitutionalized population, but in practice some people are missed because of incomplete listings of households in sampled areas and others are missed because of incomplete listings of persons in selected households. Kish and Hess (1958) were concerned about an estimated 10 percent noncoverage rate in area samples conducted at the Survey Research Center in the 1950s, and carried out a study to examine the problem. That study led to improvements in field procedures that brought the rate down to 3 percent. This paper is one of only a few on noncoverage in area samples. Yet such noncoverage is a significant concern, and particularly so in developing countries, where it is often a much more serious threat to survey quality than nonresponse.

To deal with nonresponse from noncontacts, Kish and Hess (1959b) introduced an imaginative procedure in which noncontacts in one survey are replaced by noncontacts in a previous, similar, survey. Since the time that paper was written, willingness to participate in surveys has declined greatly and nonresponse is now a critical concern to survey researchers, at least in developed countries. A great deal of

[1] Colm O'Muircheartaigh, National Opinion Research Center, Chicago, IL 60637, and Irving B. Harris Graduate School of Public Policy Studies, University of Chicago. colm@uchicago.edu.
Robert Groves, Survey Research Center, University of Michigan, Ann Arbor, MI 48106–1248. bgroves@isr.umich.edu.

research has been conducted in recent years on the causes of nonresponse, on ways to avoid it, and on ways to make adjustments to compensate for it at the analysis stage. For an introduction to the current literature, see, for example, Groves and Couper (1998) and Groves et al. (2002).

Measurement errors are another important potential source of error in survey estimates that have received a great deal of attention in recent years (e.g., see Biemer et al. 1991; Lyberg et al. 1997). For this volume we have chosen two of Kish's papers that examine this source of error. The first is "Response Errors in Estimating the Value of Homes" (Kish and Lansing 1954). The paper presents an elegant contrast of measurement methods, with a focus on bias impacts on key survey statistics. The key indicator is the market value of the respondent's home. Respondents' reports are compared to reports from professional appraisers. To a contemporary survey methodologist, the paper is a careful dissection of many components of survey errors. What begins as a singular focus on response errors expands to revelations of how coder and interviewer errors might mask themselves as response errors. What begins with the default assumption that appraisers must be the undisputed "gold standard" reveals that sometimes appraisers used different definitions of the property than respondents. All of these are lessons followed by later generations of researchers who use validity study designs to examine response errors.

The second paper is "Studies of Interviewer Variance for Attitudinal Variables" (Kish 1962). The impact of interviewers on the responses to survey questions had from the earliest survey work been a topic of interest to social researchers. The initial concern was with interviewer bias, either for an individual interviewer or for a group of interviewers taken as a whole. Researchers had focused on the "extra-role" characteristics of interviewers—the interviewer's gender, age, social class, race—and there was a search for the optimum combination of interviewer characteristics for a particular target population and survey topic.

Among survey statisticians there were three principal approaches to studies of interviewer variance. First, Mahalanobis (1946) used replicates as a basic element of his sample designs and applied the same principles of replication to the allocation of field interviewers. Consequently his replicated estimates of variance included both interviewer variance and sampling variance. Second, Hansen, Hurwitz, and Madow (1953) put forward a model of survey error including both bias and variance; this model (the *Census Bureau model*) was refined and presented in Hansen, Hurwitz, and Bershad (1960). It is essentially a covariance model and the emphasis is on the way in which response errors are intercorrelated; Fellegi's (1964) paper on the same subject followed and extended the Census Bureau model and specified eight different forms of possible intercorrelation for repeated interviewing of the same respondents. Third, Kish's approach, though entirely compatible with the Census Bureau model, is a one-way ANOVA model in which correlated errors are seen as the net biases of individual agents (in this case interviewers) and the correlation between the errors is interpreted as the a function of the ratio of the between-interviewer variance to the within-interviewer variance of the observations.

The difference in conceptualization is important in its implications for the potential of the investigations to lead to progress in understanding the phenomenon and the practical implications of the results. Kish's interest was not so much in demonstrating that interviewer variance existed but in establishing how big a problem it was and whether the magnitude could be predicted for particular situations (classes of interviewer or classes of questions). Further, he was interested in estimators of interviewer variance components that might be transported over different survey situations.

Kish's ANOVA approach is intuitively more appealing (and easier to explain to nonstatisticians) than the covariance-based Census model. The ANOVA formulation was also particularly well suited to adaptation to multilevel modeling (hierarchical modeling), and in recent years there has been a steady development of such analyses, much of it referencing Kish's early work. O'Muircheartaigh and Wiggins (1981) presented a linear model with interviewer as a predictor; Anderson and Aitkin (1985) first presented a multilevel (variance components) approach. Subsequent developments in analytic tools and software allowed a combination of variance components and the inclusion of covariates in the models. This led to a group of papers (Hox, deLeeuw, and Kreft 1991, Pannekoek 1991, Wiggins, Longford, and O'Muircheartaigh 1992) that extended the estimation of possible interviewer impact from the impact on the mean to an impact on more complex statistics such as regression coefficients. The extension and development of software for such models (*MLWin*, *HLM*) made it possible to incorporate both complex analysis and covariates at different levels of the analysis; O'Muircheartaigh and Campanelli (1999, 2000) are examples of such applications.

In conclusion, it is instructive to note the considerable attention given to the experimental protocols employed in Kish's interviewer variance studies. The paper describes the practical steps taken to achieve randomization, the types of interviewers involved, and the mode of data collection. This care with method was a characteristic that Kish passed on to later generations of researchers at the Survey Research Center. For example, Groves and Magilavy (1986) devoted comparable care to the specification of appropriate procedures for random allocation of respondents to interviewers in a study of interviewer variance with telephone interviewing.

11

RESPONSE ERRORS IN ESTIMATING THE VALUE OF HOMES

LESLIE KISH AND JOHN B. LANSING

In the 1950 Survey of Consumer Finances home owners were asked to estimate the market value of their houses. Estimates for these same homes were later made by professional appraisers. These two estimates for each of 568 homes comprise the data analyzed here. The proportion of discrepancies between the two estimates is great: only 37 per cent of the estimates by respondents are within plus or minus 10 per cent of the appraisers' estimates. However, the errors tend to be offsetting, and in none of the ten price classes used is the difference in the relative frequencies for owners and appraisers statistically significant. Similarly, although the root-mean-square difference between the two measurements is high (an average of $3,100), the mean of the respondents' estimates is only $350 higher than the mean of $9,200 for the appraisers' estimates. The amount of variability is found to be rather similar for several sub-populations. However, for houses worth over $10,000 the mean-square difference between the measurements is found to increase with the value of the home. In the Appendix a model is developed for the statistical investigation of the data.

1. INTRODUCTION

Knowledge of the over-all financial position of consumers has been a primary objective of the Survey of Consumer Finances conducted annually since 1945 by the Board of Governors of the Federal Reserve System in cooperation with the Survey Research Center of the University of Michigan.

More than half of American families live in their own homes, and for the vast majority of these families, that home is their most valuable single asset. To be com-

Reprinted with permission from the *Journal of the American Statistical Association* (1954), 49, 520–538.

The authors are indebted to Clarke L. Fauver of the staff of the Federal Reserve Board, who initiated the research reported here while he was on the staff of the Board's division of Research and Statistics. They are also indebted to the American Institute of Real Estate Appraisers, the Federal Housing Administration, and the Society of Residential Appraisers for their participation in the field work.

plete, then, any analysis of the financial position of consumers must cover this asset.

In the 1950 Survey of Consumer Finances, respondents were asked to give their idea of what their house was worth (Katona, Kish, Lansing, and Dent 1950). The answers they gave have been tabulated and on the basis of those replies tables were published in the *Federal Reserve Bulletin* (Frechtling, Lorie and Schweiger 1950) showing distributions in class intervals of owners' estimates of the current value of their homes. The published distributions are for all owners, for owners having different incomes, owners with different occupations, and owners living in towns and cities of different sizes. This is important basic information for the student of housing economics.

The question naturally arises, how reliable are these data? How much does the average householder know about the going market price for his house? Assuming for the moment that he does know, is his answer to the interviewer's question likely to be seriously biased? Are recent buyers of homes more informed about current market conditions than owners who may have bought many years earlier?

It was in an effort to answer some of these questions that a special attempt was made to evaluate the responses given to questions concerning house values in the 1950 Survey of Consumer Finances. Respondents who reported they owned their own homes were asked in January and February 1950 whether they had purchased in 1949 or in some earlier year. Those who had purchased before 1949 were asked: "Could you tell me what the present value of this house is? I mean about what would it bring if you sold it today?" (A similar question was asked in the 1950 Census of Housing.) Those who had purchased their homes during the year 1949 were asked: "How much did the house and lot cost?"

Subsequent to the completion of interviewing, it was decided to check the estimates of respondents by obtaining estimates from qualified residential appraisers. Through the cooperation of the American Institute of Real Estate Appraisers, the Federal Housing Administration, and the Society of Residential Appraisers, arrangements were made to have professional appraisers visit a substantial number of the properties. The appraisers were not required to obtain access to the property; they were asked to look at it from the outside and to estimate its value in the light of their experience and familiarity with local real estate conditions.

From the sample of home owners found in the yearly survey a subsample was selected, including respondents who failed to answer the questions about home-ownership, but not including any potential respondents who had not been interviewed during the regular survey. (In the subselection a higher probability of selection was given to the more extreme house values.) The sample was distributed roughly evenly among the three participating organizations. The response rate in the follow-up study was 89 per cent. This high response rate, a result of the excellent cooperation on the part of these professional groups, made possible the analysis which follows.

The number of homes used in the first stage of the analysis is the 637 for which forms were returned. In 30 of the 637 cases the value of the property was not indicated on the completed form. In an additional 39 cases the respondent failed to give a usable answer to the question in the original survey. Hence there are 568 homes for which two estimates of value are available. (In calculating the response rate of 89 per cent mentioned above these 69 cases were treated as responses since some useful information is available about them. If the 69 were classified as non-responses, the response rate would be 79 per cent.)

Essentially, the analysis was divided into two stages. The first stage involved a simple comparison of the frequency distributions and cross tabulations obtained by the original survey and by the follow-up study. The second stage involved the statement of a mathematical model of the response error, and estimates of the terms of the basic equation of this model. Although the conclusions drawn in the second stage are described in the main body of this article, the model itself appears in the Appendix.

2. COMPARISONS OF CELL FREQUENCIES

The first step in the analysis was to compare the frequency distribution obtained from the survey of owners with that from the survey of appraisers. The results of the comparison appear as the first two columns of Table 1. Columns (3) and (4) show cumulative totals for columns (1) and (2), respectively.

The fifth column of Table 1 shows the distribution of appraisers' estimates for the 39 cases which were "Not Ascertained" in the survey. On seeing the "NAs" in any table one is led to wonder about their effect on the entire distribution. There is a mere suggestion of a concentration of several homes with very low values among these 39 cases. But anyone who assumed that the 39 cases should be distributed proportionally would not have been led far astray.

In column (6) the differences between the entries of columns (1) and (2) are given. These differences are subject to sampling variability. If we sent out both interviewers and appraisers to repeated samples of 600 cases under identical conditions, we would expect that the bracket distributions sometimes would show closer agreement than (1) and (2), and sometimes wider disagreement. The model in the Appendix permits us to estimate the probability that the proportions in any pair of cells will agree within a given range. That is, we can estimate how the differences shown in column (6) would fluctuate if the present study were repeated many times. The measures of this fluctuation, estimated from the data of this study, are presented in column (7) in terms of the standard errors of the differences. We may illustrate the interpretation of these columns as follows: the discrepancy between the proportion of homes placed in the bracket $2,500–4,999 by respondents and appraisers was 0.6 per cent in the present study; if the study were repeated many times, this difference would be less than 1.4 per cent in two studies out of three in the long run, and it would be less than 2.8 per cent in 19 studies out of 20.

The two distributions in (1) and (2) convey the same general impression about the proportion of owner-occupied homes of different values. The same would be true of other similar distributions from replications of the present study in view of the relatively small size of the error shown in (7). We make this

TABLE 1. Frequency Distributions of the Value of Owner-Occupied Homes Based on Estimates Reported by Owners and Appraisers (Uncorrected)[a]
(percentage distribution of homes)

Value of home	Respondents' estimates	Appraisers' estimates	Cumulative total of respondents' estimates	Cumulative total of appraisers' estimates	Appraisers' estimates where respondents' were not ascertained	Difference between proportions (1) − (2)	Standard error of the difference in (6)
	(1)	(2)	(3)	(4)	(5)	(6)	(7)
Under $2,500	2.9	2.3	2.9	2.3	14	+0.6%	0.7%
$2,500–4,999	13.1	13.7	16.0	16.0	14	−0.6%	1.4%
$5,000–7,499	19.6	19.3	35.6	35.3	20	+0.3%	1.9%
$7,500–9,999	21.5	24.3	57.1	59.6	18	−2.8%	1.9%
$10,000–12,499	19.1	16.8	76.2	76.4	7	+2.3%	1.8%
$12,500–14,999	6.5	8.8	82.7	85.2	10	−2.3%	1.2%
$15,000–19,999	7.2	6.3	89.9	91.5	3	+0.9%	1.1%
$20,000–29,999	2.8	2.2	92.7	93.7	3	+0.6%	0.7%
$30,000 and over	1.5	1.4	94.2	95.1	3	+0.1%	0.4%
Value not ascertained	5.6	4.7	99.8	99.8	8	+0.9%	
							1.2%
Total	99.8[b]	99.8[b]			100		
Number of homes	637	637	637	637	39		

[a]These "uncorrected" distributions contain clerical errors which were discovered and corrected in the course of comparing the data from respondents and appraisers. Later tables are based on corrected data except as indicated.
[b]Detail does not add to 100.0% owing to rounding.

judgment (and ask the reader to do likewise) within the general framework of the errors and requirements of surveys of this kind and size. It would be fruitless for us to raise here the question: for what kind of decisions are our results "reliable enough"? Our investigations do provide assurances against the existence in the procedures of large response errors. "Large" here is taken in the context of the actual sizes of the sample and of the sampling errors—but we must neglect the question of the relative cost of reducing the response error.

Although we find no reliable evidence of a net bias in any price class, it is possible (and even probable) that a large enough sample would uncover biases which escape detection in this sample. We have shown only that the differences between columns (1) and (2) *could* be the result of random response variation.

The second step in the analysis was to examine the discrepancies between the estimates of respondents and appraisers. The similarity between the first two columns in Table 1 could be the result either of few errors or of many off-setting errors. Table 2 compares the classification of the homes by respondents and appraisers. A sum of the proportions in the cells along the diagonal indicates that 43 per cent of the homes that were included in a given bracket by the respondents were also placed in that bracket by the appraisers. Errors were, in fact, frequent, but generally off-setting.

On close examination some of the differences shown in Table 2 seemed out of all reason—how could *any* house valued by a respondent at under $2,500 be valued by an appraiser at over $15,000? This question raises the possibility of errors in the survey process made by others than respondents and appraisers. The information in Table 2 was used to guide a special search for errors. All cases where the two estimates were in disagreement by more than one "bracket" (coded class of house value) up to a value of $15,000, and above that value all cases not in the same "bracket," were selected for study.

This search involved a comparison of the original interview, the appraiser's report, and the card on which the data had been punched. Such a search is unlikely to turn up errors by interviewers in recording the answers given by respondents, but it should disclose any errors in coding. An examination of 109 cases yielded 17 errors, all but two of them clerical errors by coders. Of the 17, four involved only errors in the conversion of a dollar amount (entered correctly) to a bracket (entered wrongly). There were 11 clerical errors made in coding the respondent's estimates, and two errors were made by interviewers. Ten of these 11 errors involved entries of one-tenth of the proper amount owing to the omission of a zero; in the one case, $11,000 was read as $77,000. In addition to these errors two exceptional cases were noted. In one, it seemed clear that the appraiser had included only part of the property which the respondent had in mind. In the other, the appraiser based his estimate on the commercial value of the property, while the respondent based his on the value for residential purposes.

The effect of these errors on the entries of a few cells in Table 2 is shown: certain cells which are emptied by the corrections or which contain only exceptional cases, have been indicated by being enclosed in "boxes." All of the most extreme discrepancies in Table 2 disappear but the marginal distributions are

TABLE 2. Relation Between Appraiser's Estimate and Respondent's Estimate (Uncorrected)[a]
(percentage distribution of homes)

	Respondent's estimate										
Appraiser's estimate	Under $2,500	$2,500 –4,999	$5,000 –7,499	$7,500 –9,999	$10,000 –12,499	$12,500 –14,999	$15,000 –19,999	$20,000 –29,999	$30,000 and over	Value not ascer- tained	Total
Under $2,500	1.0	0.4	0.2[b]							0.7	2.3
$2,500–4,999	0.7	7.2	3.8	1.1	0.2					0.7	13.7
$5,000–7,499		3.4	8.3	4.8	1.7					1.1	19.3
$7,500–9,999	0.2	0.5	5.0	11.1	5.7	0.2	0.6			1.0	24.3
$10,000–12,499	0.4		0.3	3.8	7.8	2.4	1.5	0.2		0.4	16.8
$12,500–14,999	0.2		0.7	0.1	2.3	3.0	1.2	0.6	0.1	0.6	8.8
$15,000–19,999	0.2				0.9	0.7	3.2	1.0	0.1	0.2	6.3
$20,000–29,999		0.2					0.6	0.8	0.4	0.2	2.2
$30,000 and over				0.2[b]				0.2	0.8	0.2	1.4
Value not ascertained	0.2	1.4	1.3	0.4	0.5	0.2	0.1		0.1	0.5	4.7
Total	2.9	13.1	19.6	21.5	19.1	6.5	7.2	2.8	1.5	5.6	99.8
Number of cases[c]	23	80	108	120	105	39	50	47	26	39	637

[a]For the difference between corrected and uncorrected data, see the discussion in the text. The principal effect of the corrections on this table is to empty the cells indicated by boxes, distributing the entries among other occupied cells. The table reads as follows: 1.0% of all houses in the sample were valued at under $2,500 by the respondent and also by the appraiser; 0.4% of all houses were valued at $2,500–$4,999 by the respondent, but at under $2,500 by the appraiser; etc.
[b]These two cells contain one case apiece. They are the exceptional cases noted in the text.
[c]Because there were three different weights used, the percentages are not simple ratios of the total of 637.

little changed by the corrections. It is interesting to note that the lowest class was composed in large part of errors.[1]

The comparison in Table 2 is supplemented by another approach in Table 3 below: this presents the distribution of each respondent's estimate divided by the appraiser's estimate on his home. This division was carried out for 568 homes. The respondents' estimates were within plus or minus 10 per cent of the appraisers' in 37 per cent of the cases. On the other hand, the discrepancy was more than plus or minus 30 per cent for 24 per cent. Of these 24 per cent, 18 per cent represent overestimates by respondents, suggesting a tendency for owners to overvalue their homes. This possibility can be better evaluated by comparing the means of the two distributions (after correction of the clerical errors).

3. COMPARISONS OF THE MEANS OF THE TWO DISTRIBUTIONS[2]

The difference between the means obtained by the two methods of measurements is \$9,560 − \$9,210 = \$350. That is: the mean of \$9,560 obtained from the responses of the home-owners seems to include a bias of \$350 (if we accept the appraisers' values as "true"). This bias is in the direction one would expect. The standard error of the difference was calculated (by a formula proper to the complexities of the sample design) to be \$170. Hence there appears to be a tendency (statistically

TABLE 3. Frequency Distribution of Respondent's Estimate Divided by Appraiser's[a]

Respondent's estimate divided by appraiser's	Proportion of homes
Under 70%	6
70 – 89%	20
90 – 109%	37
110 – 129%	19
130 – 149%	9
150% and over	9
Total	100
Number of homes	568

[a]Uncorrected for the clerical errors discovered only after comparison of the two estimates.

[1] This finding and the \$77,000 mistake have obvious implications for checking procedures. It should be noted that the checking procedure used in processing the 1950 Survey of Consumer Finances varied according to the nature and extent of the projected analysis of the data. The data on value of houses received the minimum amount of checking. For the type of distribution actually published in the *Federal Reserve Bulletin* these clerical errors were of little importance. The clerical errors do have a large effect on the errors of the estimated *mean* value; however, the mean was neither intended nor submitted for publication.

[2] We include this analysis because it may be of general interest. We repeat: the mean home value was not sought nor published in the original survey.

significant) for the home-owners to set higher values on their homes than do the professional appraisers. This tendency is small compared to the value of the home—about 4 per cent of the latter.

This net average bias may appear small also in comparison with the large discrepancies found in the two values obtained for individual homes. The mean square difference of the two measurements was estimated as 9,580,000 compared to the estimate of the squared bias of 100,000 (see equation (6.15) in Appendix). This result is consistent with the findings presented in Tables 1 and 2 which show also large discrepancies in individual estimates but small differences in the overall distribution of the two measurements.

The relative importance of a bias depends on the size of the survey to be taken. The sample mean of a simple random sample of n interviews with respondents may be expected to be subject to a total root-mean-square error of $\sqrt{[(V(r)/n)] + \bar{D}^2]}$ where the first term under the radical represents the total variability of the estimates from the survey of respondents about their own mean and the second, the square of the bias.[3] As the size of the sample increases the first term will decrease but the second will remain constant.

We may use the sample estimates obtained in our investigation to examine the effect of the bias on the total error. For $V(r)$ we have the estimate $v(r)$ = 32,650,000; and for $\bar{D}^2$ we have the estimate $\bar{d}^2 = 100,000$ (see equation (6.15) in Appendix). Now let us take the value of $\sqrt{[(V(r)/n) + \bar{D}^2]}$ for three different sample sizes, and under the two assumptions: that $\bar{D}^2$ = 100,000 and that $\bar{D}^2 = 0$.

The total (root-mean-square) error of the mean house value under six different conditions would then be as shown below. Note that the total error for a sample of 100 is not greatly increased by the bias term, but for a sample of 1,000, the effect of the bias term is large, while for a sample of 10,000 it is overwhelming. Where the facts are similar to those found in this investigation, an improvement in the accuracy of the estimate for surveys of a few hundred cases can probably be obtained most easily by increasing the size of the sample. For surveys of several thousand cases, however, it may be more efficient to allocate funds for a search for sources of bias and for development of techniques for reducing the bias than to allocate funds for an increase of the size of the sample.[4]

[3] The term $V(r)/n$ represents the variance of the sample mean as it is usually calculated, but it actually includes both the error resulting because not every member of the universe was interviewed—the sampling error proper—and any uncorrelated random response error which may be present in the methods used, such as random clerical errors (see equation (6.7) in Appendix). The *net* average of the response errors will be reflected in the squared bias term, $\bar{D}^2$. This expression shows that it is possible to increase the accuracy of the estimate of a mean from a simple random sample in one of three ways: by increasing the number of interviews (increasing n); by reducing the variability of the estimates, $V(r)$, (by reducing some of the errors of response); or by reducing the size of the bias $\bar{D}^2$ (for example, by more careful training of interviewers). The practical problem in the administration of surveys is to allocate resources among the three in such a way as to minimize the total error for a given outlay.

[4] From the data of Table 4 of the Appendix it seems that the contributions of the bias term are less for estimates of the proportions of houses falling into designated class intervals.

Value of D^2	Formula	Value if sample size (n) is 100	1,000	10,000
(1) Total error if $D^2 = 100{,}000$	$\sqrt{\frac{32,650,000}{n}+100,000}$	\$650	\$360	\$320
(2) Total error if $D^2 = 0$	$\sqrt{\frac{32,650,000}{n}}$	\$570	\$180	\$60

We have investigated the possibility that the discrepancy between the two measurements might prove to be a function of the value of the house. One can imagine, for example, that respondents might tend to overvalue low priced homes and to undervalue high priced homes. We divided the homes into groups based on the appraisers' value and estimated the mean value of the houses in each group, first on the basis of the respondents' value for the houses and then on the basis of the appraisers'. The difference between these two means has been plotted in Chart 1. (See solid line "A.") This graph indicates that the respondents tend to overvalue homes priced below about \$12,000. For homes priced above that amount no clear tendency to under- or over-valuation appears. We feel that the discrepancy below \$12,000 may be explained in part, but only in part, by errors in the estimates made by appraisers. Any such errors also would tend to give the graph below a general trend downward to the right.[5]

4. THE ROOT-MEAN SQUARE DIFFERENCE

As a measure of the average individual discrepancy between respondents and appraisers we use the root-mean-square difference, that is, the square root of the mean of the squared deviations between the pairs of estimates. We estimate this quantity at \$3,100 for the sample as a whole. In other words if we assume that the appraiser's estimate is the true value, the respondents are in error by an average of \$3,100 in their estimates. (From equation (6.15) in Appendix.) Actually there is no doubt that the appraisers also made errors, and the average discrepancy between the respondents' estimates and "true value" of the property would be less than \$3,100.

How does the average discrepancy vary with the value of the home? Is the discrepancy a constant amount, or a constant proportion of the house, or some

[5] For example, suppose the respondent estimated the value of this house correctly at \$11,000. The point on the chart to correspond would be X = \$11,000, Y = 0, if the appraiser made the same estimate. However, if the appraiser made an estimate of \$10,000, the corresponding point would be X = \$10,000, Y = \$1,000, which is above and to the left of the original point. An error by the appraiser in the other direction would lead to a point below and to the right of the original point.

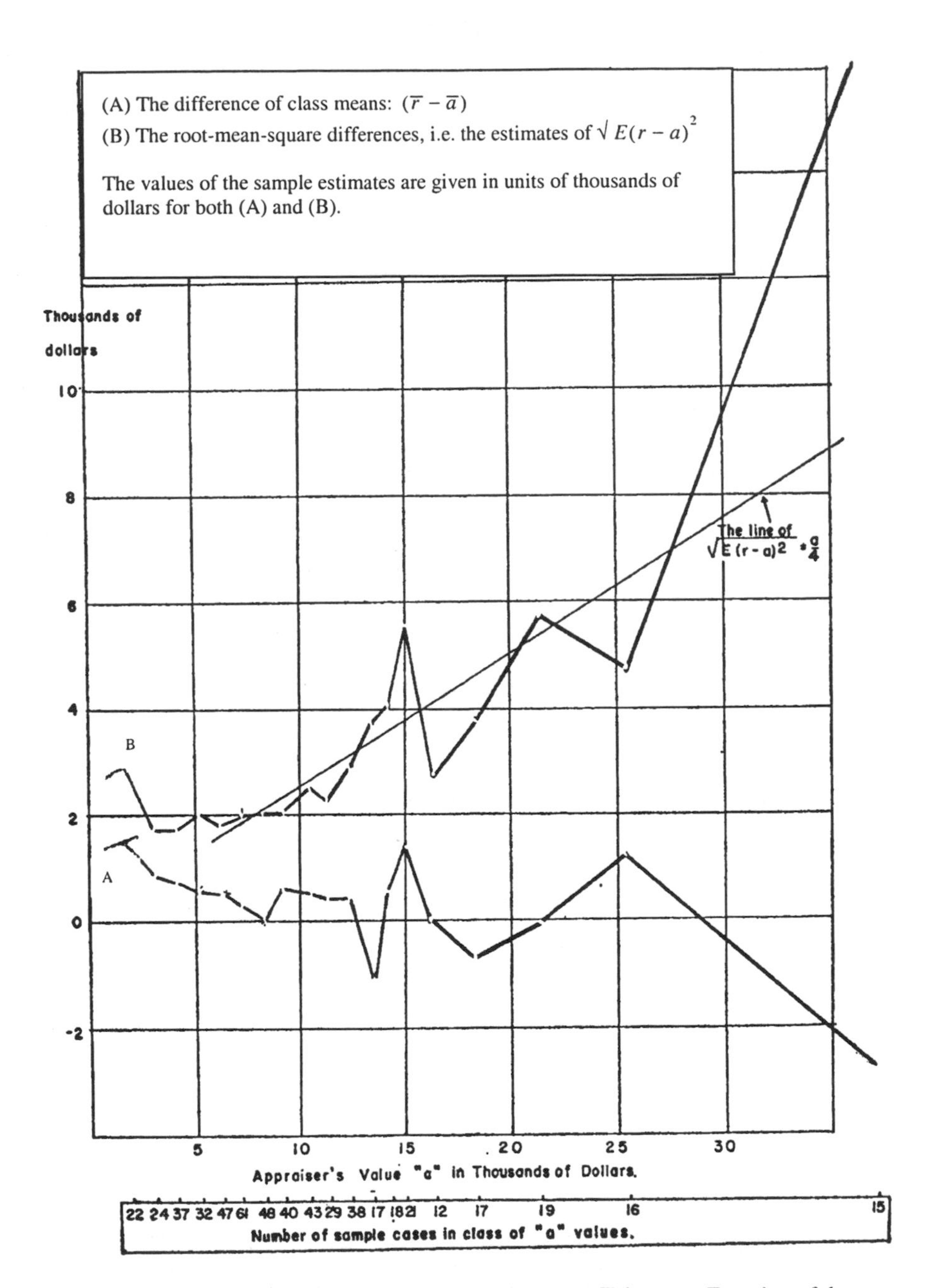

Chart 1. The Differences of Two Measurements (r–a) Taken as a Function of the Appraiser's Values

other function? On Chart 1 are plotted the root-mean-square differences—the rms(d) values—for each class of appraised values; the width of the intervals is $1,000 except at the ends, where classes were combined to obtain larger cells. (See the solid line "B.") For values below $10,000 the rms($d$) appears to be constant around $2,000. For values above $10,000 it is considerably more variable and larger, and it appears to be proportional to the estimated value of the home. The line which represents a root-mean-square difference of one-fourth of the appraised value is drawn in. It appears to the eye to fit the distribution above $10,000 fairly well. In other words in our data the expected absolute value of the difference between the respondents' and the appraisers' estimates is about $2,000 for a house worth less than $10,000; while for a house worth over $10,000, the expected value of the difference is one-fourth of the appraiser's estimate. For a $16,000 house, one would predict a respondent would differ from an appraiser by $4,000; for a $20,000 house, one would predict $5,000, and so forth.

5. ANALYSIS OF SOME SUBGROUPS

One aim of our investigation was to discover some of the variables which might be associated with response errors. For three cross-tabulations comparisons were made of the ratio of respondents' to appraisers' values. An attempt was made in the original survey to isolate those cases where the respondent seemed uncertain of his estimate. If this attempt were successful, it was thought that it might be possible to develop methods of analysis that would place more weight on the more reliable cases. The procedure tried was to instruct the interviewers as follows:

> Since some respondents have a very clear idea of the value of their house, based on such things as what the house next door just sold for, while others have only very vague notions, we have left space after question 31 in which you should note down any information he may give you about how he arrived at his estimate of the value of the house. Our objective is to distinguish between cases where we have the kind of accurate estimate we would prefer and cases where we have only vague information. In any case be sure to record the dollar value of the house.

The coders were then instructed to study the answer as recorded by the interviewer and attempt to assign a rating according to how sure the respondent seemed to be of his answer. This rating proved very difficult to make; coders disagreed frequently as to the proper point on the scale at which to place an answer. The relevant data (not given here) show that the assigned rating of the appearance of reliability had no validity: the errors were about equally large in the various classes of assigned reliability.

Secondly, occupation of the head of the family owning the house was selected as a measure of socio-economic status, on the hypothesis that people of higher status might be better informed. Thirdly, the population of the place (city) of residence of the respondent was selected on the hypothesis that knowledge of real estate values would be different in communities of different sizes. None of these hypotheses was substantiated; no sizeable differences were noted.

For four subgroups of the sample we calculated separately the estimates of our basic error equation (6.8). There exist *a priori* reasons why the accuracy of the estimates in each of these groups might turn out to be different than in the entire sample. The calculated equations are in the Appendix; here we shall summarize the results, using the root-mean-square difference— rms(d)—as the measure of accuracy. The conclusions we draw from these groups must be tempered by the knowledge that they were not properly selected subsamples of the entire sample; hence there may be other causes operating beyond that on which we focus our attention.

(a) In 65 cases the appraisers exceeded the minimum effort asked of them and went into the homes. We expected that their estimates would be more accurate, and that the rms(d) would be smaller. However, the rms(d) for these 65 cases turned out to be $2,700 compared with $3,100 for the entire sample. The appraisers' errors were not clearly increased by remaining outside the house.
(b) In homes purchased during the calendar year prior to the interview, the respondent was asked what he actually paid for his home. We expect that the reports of the respondents were fairly close approximations to the true value at the time of purchase. The rms(d) of $1,900 is reliably smaller than the $3,100 for the sample as a whole. One should not infer, however, that the entire $1,900 is the result of errors by appraisers. For one thing, real estate values change with time, and up to a year might elapse between the purchase and the original interview, with several months more passing before the visit of the appraiser.
(c) In the Surveys of Consumer Finances interviewers are instructed to make efforts to interview the head of the household rather than some other member. In the 91 cases where the interviews were taken from some other member of the household the rms(d) was not—contrary to expectations—larger than for the sample as a whole. In fact, it was $2,500 as against $3,100 for the entire sample.
(d) There were 59 cases where the head of the house was a female. For these the rms(d) was $3,900, which appears to be reliably higher than the entire sample.

The only important improvement in accuracy, then, was for respondents who purchased in the year prior to the survey. These respondents, as noted earlier, were asked what the property actually cost rather than their estimate of what it might be worth, hence it is not surprising that their responses are close to the appraisers' estimates.

6. APPENDIX

The Model. The symbol r_i denotes the value recorded at the ith home as a response in the interview survey; and a_i denotes the value assigned by the appraiser to the same home. The "true" (but unknown) value is y_i. Where there is little room for misunderstanding we shall drop the subscript i, and refer simply to r, a and y. The means over the entire population for the three sets of values may be designated by:

$$\bar{R} = E(r), \qquad \bar{A} = E(a), \qquad \bar{Y} = E(y). \tag{6.1}$$

The operator "E" denotes the "expected value of."[6] The variances of the three variables may be designated by:

$$V(r) = E(r - \bar{R})^2, \qquad V(a) = E(a - \bar{A})^2, \qquad V(y) = E(y - \bar{Y})^2. \tag{6.2}$$

The quantity $(r_i - y_i)$ denotes the individual error of the response in the interview survey for the ith home; and $(a_i - y_i)$ denotes the error in the appraiser's estimate for the same home. The difference between the two errors is equal to the difference of the two measurements:

$$d_i = (r_i - y_i) - (a_i - y_i) = (r_i - a_i). \tag{6.3}$$

Furthermore, let us call the mean value $(\bar{R} - \bar{Y})$ the response bias; $(\bar{A} - \bar{Y})$ the appraiser's bias; and the difference between the two biases is

$$\bar{D} = (\bar{R} - \bar{A}) = (\bar{R} - \bar{Y}) - (\bar{A} - \bar{Y}). \tag{6.4}$$

An important term in our model is the mean-square difference of the measurements:

$$\mathrm{MS}(d) = E(d^2) = E(r - a)^2. \tag{6.5}$$

We also need the expression for the covariance between the differences in measurements and the appraiser's values:

$$\begin{aligned} \mathrm{Cov}(da) &= E(d - \bar{D})(a - \bar{A}) = E(r - a - \bar{R} + \bar{A})(a - \bar{A}) \\ &= E(r - a)(a - \bar{A}), \end{aligned} \tag{6.6}$$

[6] The means of the measurements r_i and a_i over a finite population would be variables also due to the errors of measurements. But we may treat $\bar{R}$ and $\bar{A}$ as constants if we consider them as resulting from a large number of reported measurements, or as coming from a large population. By confining ourselves to large populations we may also disregard any "finite population corrections" in our variance formulas. The terms used here are generally in accord with those in Hansen, Hurwitz, and Madow (1953).

Another good treatment of the topic of errors of response may be found in Cochran (1953, chap. 13). However, none of the sources known to us develops the model we need in terms of the differences $(r - a)$ of two sets of measurements, both subject to error. To what extent these non-sampling errors may be considered to be random variables is a complex problem which we shall have to leave untreated.

also:

$$\begin{aligned}\text{Cov}(da) &= E(r - \bar{R} - a + \bar{A})(a - \bar{A}) \\ &= E(r - \bar{R})(a - \bar{A}) - E(a - \bar{A})^2 = \text{Cov}(ra) - V(a).\end{aligned} \qquad (6.6a)$$

With the above definitions, we may express the basic equation for our empirical investigations:

$$V(r) + \bar{D}^2 = V(a) + \text{MS}(d) + 2\text{Cov}(da). \qquad (6.7)$$

For proof express $E(r - \bar{A})^2$ in two different ways:

$$E(r - \bar{A})^2 = E(r - \bar{R} + \bar{R} - \bar{A})^2 = V(r) + \bar{D}^2$$

and

$$E(r - \bar{A})^2 = E[(r - a) + (a - \bar{A})]^2 = \text{MS}(d) + V(a) + 2\text{Cov}(da).$$

Our model would be simpler if the appraiser gave the "true" value for every home, so that $a_i = y_i$; and the error equation would become

$$V(r) + (\bar{R} - \bar{Y})^2 = V(y) + E(r - y)^2 + 2\text{Cov}(r - y)(y).$$

Here $V(y)$ is the "true" sampling variance, i.e., the variance among the y_i, which are the "true" values of the homes; and

$$V(r) + (\bar{R} - \bar{Y})^2 - V(y) = E(r - y)^2 + 2\text{Cov}(r - y)(y)$$

is the increase in the total mean-square error due to errors of measurement. Similarly, the increase in the total mean-square error, due to the lesser accuracy of the r_i than the a_i, may be measured as

$$V(r) + \bar{D}^2 - V(a) = E(r - a)^2 - 2\text{Cov}(r - a)(a).$$

It is also interesting to note the relationship

$$E(r - a)^2 = E(r - y)^2 + E(a - y)^2 - 2E(r - y)(a - y).$$

The covariance $E(r - y)(a - y)$ of the two measurements may be positive or zero, but it is not likely to be an important negative quantity in the present instance. Therefore, the term $E(r - a)^2$ available in this study is likely to be larger than the mean-square error of response $E(r - y)^2$, by a quantity no greater than (but perhaps almost equal to) the mean-square error $E(a - y)^2$ of the appraiser's measurements.[7]

Although the results were obtained from a complex multi-stage sample, the discussion is given in terms of the composition of the response error for the individual homes which are the ultimate elements comprising the population. The

[7] For the benefit of future researchers we should like to point out that an estimate of $E(a - y)^2$ could have been obtained had we assigned some of the homes to two appraisers each; we thought of this too late to carry out the necessary field work.

expressions of the relative effects of the bias and of the variable error are given in terms of simple random samples. It is hoped that in this form the data will be of greater general interest and usefulness in planning other surveys. The calculations are based on the "naïve" estimates from the pooled sample values; greater refinements did not seem to be warranted by the available data.[8]

The basic relationship shown in (6.7) may be expressed in terms of sample estimates as[9]

$$v(r) + \bar{d}^2 = v(a) + \text{ms}(d) + 2\,\text{cov}(da). \tag{6.8}$$

We have the following unbiased estimates:

$$\bar{r} = \frac{1}{n}\sum^{n} r, \qquad \bar{a} = \frac{1}{n}\sum^{n} a, \tag{6.9}$$

$$v(r) = \frac{1}{n-1}\sum^{n}(r - \bar{r})^2, \qquad v(a) = \frac{1}{n-1}\sum^{n}(a - \bar{a})^2, \tag{6.10}$$

$$\text{cov}(ra) = \frac{1}{n-1}\sum^{n}(r - \bar{r})(a - \bar{a}), \tag{6.11}$$

$$\text{cov}(da) = \text{cov}(ra) - v(a) \tag{6.12}$$

$$\text{ms}(d) = \frac{1}{n}\sum^{n} d^2 = \frac{1}{n}\sum^{n}(r - a)^2$$

$$= \frac{1}{n}\left[\sum^{n} r^2 + \sum^{n} a^2 - 2\sum^{n} ra\right]. \tag{6.13}$$

Although $(\bar{r} - \bar{a})$ is an unbiased estimate of $\bar{D}$, $(\bar{r} - \bar{a})^2$ is not an unbiased estimate of $\bar{D}^2$; but $\bar{d}^2$ is an unbiased estimate where

$$\bar{d}^2 = (\bar{r} - \bar{a})^2 - \frac{1}{n}[v(r) + v(a) - 2\,\text{cov}(ra)]. \tag{6.14}$$

This happens because

$$E(\bar{r} - \bar{a})^2 = E[(\bar{r} - \bar{R}) - (\bar{a} - \bar{A}) + (\bar{R} - \bar{A})]^2$$

$$= V(\bar{r}) + V(\bar{a}) - 2\,\text{cov}(\bar{r}\,\bar{a}) + \bar{D}^2.$$

---

[8] For the sake of simplicity and because we have no measure of it, we disregard the correlation among the errors of individual homes, such as may be caused by interviewer bias.

[9] Because the responses were "weighted" to correct for the use of different sampling rates, the actual sample calculations were somewhat different from those shown here. For example, $\bar{r}$ was calculated as $\Sigma^n w_j r_j / \Sigma^n w_j$, where r_j is the response, and w_j is the assigned weight of the jth home in the sample. The calculation of the variances may be illustrated by $V(r) = \frac{n}{n-1}\frac{1}{\Sigma^n w_j}\Sigma^n w_j (r_j - \bar{r})^2$.

The unbiased estimate $\overline{d}^2$ has some advantages: together with the other unbiased estimates from the sample, it yields values for our error equation (6.8) which balance out exactly. However, it also has some disadvantages: it is a residual of sample values and it turns out to be negative sometimes—an embarrassing situation for the square of a real quantity. (One may decide to truncate the distribution of $\overline{d}^2$ at zero by substituting the value zero for all negative sample estimates. Alternately, one may use simply $(\overline{r}-\overline{a})^2$ with the knowledge that it has a positive bias of known magnitude.)

6.1 Some Calculations on the Dollar Value of the House

The five terms of the basic equation (6.8) of the estimates of error components will be presented in this section for several situations. They will be given in units of \$1,000; since in these variance components the units are squared, a factor of 10^6 is needed to convert them to plain dollar values.

1. Our principal interest is in the components of the equation dealing with all the 568 cases:

$$32.65 + .10 = 26.69 + 9.58 - 3.52 \tag{6.15}$$

Note the relatively large ms(d) term which yields the $\sqrt{[9.58 \times 10^6]} = \$3,100$ estimate for the rms(d) between the two measure-ments on individual homes. But most of the discrepancies cancel leaving a much smaller net average error; the unbiased estimate of this bias is $\sqrt{[.10 \times 10^6]} = \$320$.

The total root-mean-square error of the responses is $\sqrt{v(r)} = \sqrt{[32.65 \times 10^6]} = \$5,700$ which is not much over the $\sqrt{v(a)} = \sqrt{[26.69 \times 10^6]} = \$5,200$ we would get from appraisers' estimates. Therefore, the practical surveyor may well be satisfied with the precision of the interview response—if the bias term is not too large.

The difference between the two variance terms is reduced by the sizeable negative covariance term (-3.52×10^6) between the difference in measurements ($r - a$) and the appraisers' values. This is probably due in part to overestimates among the appraisers' high values and underestimates among their low values. The negative covariance is in accord with the gentle negative slope of curve A on Chart 1.

2. If we allow the 13 gross coding errors (mentioned earlier) to stand uncorrected, the components are estimated as

$$37.35 + .04 = 26.69 + 15.81 - 5.11.$$

Thus over a third of the original ms(d) term of 15.81 was due to the 13 gross errors. However, the rise in the variance of the responses is more moderate (32 to 37). Moreover the estimate of the sample mean may be no worse off for these errors (ironically) because the bias term seems to be somewhat reduced. It seems that all these gross errors were in the direction of lowering the home-owners' estimates and, as noted above, home-owners suffer from a tendency to overestimate the value of their homes.

The effect of these coding errors on curve A in Chart 1 is to make the $(\overline{r}-\overline{a})$ values for the classes above $10,000 more depressed and more irregular. Curve B of the rms(d) values is also disturbed above $10,000: the curve becomes more irregular and the slope becomes greater (it seems to fit the line of $\sqrt{[E(r-a)^2]} = a/3$).

3. For the 65 cases where the appraiser went into the home the components are

$$37.07 - .09 = 34.31 + 7.29 - 4.62.$$

4. For the 61 cases where the response was in terms of the amount paid for a recently purchased home we have

$$23.14 - .05 = 21.76 + 3.67 - 2.34.$$

If we assume that the respondents gave the "true value" of their homes in these cases then we may accept this ms(d) term of 3.67 as a rough estimate of the appraisers' contribution to the discrepancy term.

5. For the 91 cases where the respondent was not the head of the household the values are

$$37.04 + .28 = 28.77 + 6.29 + 2.26.$$

6. For the 59 cases where the head of the household was a female the equation is

$$44.78 + .33 = 28.10 + 15.17 + 1.84.$$

6.2 Results on Proportions

When we deal with the proportion of cases which fall into any class interval our variables are binomial. The values of r_i and a_i are restricted to 0 and 1; and the value of $d_i = (r_i - a_i)$ is either 0, +1 or −1. The basic equation (6.8) of the estimated error components becomes:

$$[\frac{n}{n-1} p_r q_r] + [(p_r - p_a)^2 - \frac{1}{n-1}(p_r q_r + p_a q_a - 2p_{ra} + 2p_r p_a)] =$$

$$[\frac{n}{n-1} p_a q_a] + [p_r + p_a - 2p_{ra}] + [\frac{2n}{n-1}(p_{ra} - p_r p_a - p_a q_a)]. \qquad (6.16)$$

Here p_r is the proportion of the homes placed into a specific frequency group by the responses to the interviews, while p_a is the proportion placed into that group by the appraisers' estimates. Also p_{ra} is the proportion placed into the same group both by respondent and by appraiser. Furthermore, $q_r = 1 - p_r$ and $q_a = 1 - p_a$. The equation for the 5,000 − 7,499 group would be, as read from the values of Table 2:

$$[\frac{637}{636}(.196)(.804)]$$

$$+[(.196-.193)^2 - \frac{1}{636}\{(.196)(.804)+(.193)(.807)-2(.083)+2(.196)(.193)\}]$$

$$=[\frac{637}{636}(.193)(.807)]+[.196+.193-2(.083)]$$

$$+[2\frac{637}{636}\{(.083)-(.196)(.193)-(.193)(.807)\}].$$

In Table 4, columns (1) to (5), we present the estimates of the five components of equation (6.16) for each of the classes shown in Table 1. In column (6) we show the difference ($p_r - p_a$) between the proportion assigned to each bracket in the surveys of respondents and appraisers. In column (7) we show the standard error of each difference shown in column (6).

Note that the ms(d) terms, denoting the variability due to the difference of the two responses, in column (4) of Table 4 are large; generally they are as large as, or larger than, the $v(r)$ and $v(a)$ terms which ordinarily stand for sampling variability—shown in columns (1) and (3). One may be tempted to assume that this variability would be much less if larger groups were investigated; however, the two larger groups shown on the bottom two lines of Table 4, comprising respectively about 35 per cent and 60 per cent of the population, also have ms(d) terms almost as large as the $v(r)$ and $v(a)$ terms.

In spite of the large ms(d) the value of $[v(r)+\bar{d}^2]$ is hardly any larger than $v(a)$. This is due to the large negative covariance term. That is: there exists a large gross response variation but its net effect on variability is very small.

The net effect in terms of bias is even smaller. There is no bias term in column 2 which is reliable in terms of the standard error. If we average the ratios of the $\bar{d}^2$ values to the respective $v(r)$ values over the 10 classes we obtain .0005. In the calculations on the dollar mean the ratio of $\bar{d}^2$ term to the $v(r)$ term was .0030. Thus we may say that the bias term for the proportions remains undetected; and if it exists its effect on its total error is probably less than in the case of the dollar mean.

TABLE 4. Values of the Terms of the Error Equation (6.16) for the Proportion in Each of the Frequency Classes as Shown in Tables 1 and 2.

Frequency group	Values of the components of the error equation									(6) Difference between proportions found $(p_r - p_a)$	(7) Standard error of the difference $(p_r - p_a)$
	(1) $v(r)$	+	(2) $\bar{d}^2$	=	(3) $v(a)$	+	(4) ms(d)	+	(5) 2cov(da)		
$0–2,499	.0282	–	.0000	=	.0225	+	.0320	–	.0263	+0.6%	0.7%
$2,500–4,999	.1140	–	.0002	=	.1184	+	.1240	–	.1286	–0.6%	1.4%
$5,000–7,499	.1578	–	.0003	=	.1560	+	.2230	–	.2215	+0.3%	1.9%
$7,500–9,999	.1690	+	.0004	=	.1842	+	.2360	–	.2508	–2.8%	1.9%
$10,000–12,499	.1548	+	.0002	=	.1400	+	.2030	–	.1880	+2.3%	1.8%
$12,500–14,999	.0609	+	.0004	=	.0804	+	.0930	–	.1121	–2.3%	1.2%
$15,000–19,999	.0669	–	.0000	=	.0591	+	.0710	–	.0632	+0.9%	1.1%
$20,000–29,999	.0273	–	.0000	=	.0216	+	.0340	–	.0283	+0.6%	0.7%
$30,000 and over	.0148	–	.0000	=	.0138	+	.0130	–	.0120	+0.1%	0.4%
Not ascertained	.0530	–	.0001	=	.0449	+	.0930	–	.0850	+0.9%	1.2%
Two illustrative cumulated groups:											
$0–7,499	.2296	–	.0003	=	.2287	+	.2090	–	.2084	+0.3%	1.8%
$0–9,999	.2454	+	.0003	=	.2412	+	.2130	–	.2085	–2.5%	1.8%

12

STUDIES OF INTERVIEWER VARIANCE FOR ATTITUDINAL VARIABLES

LESLIE KISH

We studied the effects of the variable biases of interviewers on responses to a variety of items, mostly attitudinal, included in two surveys. Not satisfied with "testing statistical significance," we measured this interviewer variance as a component of the total variance per respondent; thus $s^2 = s_a^2 + s_b^2$, where s_a^2 is the "between interviewer" and s_b^2 the "within interviewer" component; $\rho = s_a^2 / s^2$ is the proportion of the interviewer effect. The contribution of the interviewer variance to total survey errors is examined for its implications for the planning and interpretation of survey statistics. Among these we included subclass means and comparisons among pairs of subclasses. The data yielded important results:

(1) We can obtain responses with low or moderate interviewer variance on highly "ambiguous" and "critical" attitudinal questions. The range of ρ's was mostly 0 to .07 in the first study, 0 to .05 in the second, and their average is about .01 or .02; these effects are not generally higher than for most items in a good Census.

(2) Nevertheless these seemingly small ρ's, combined with moderate workloads, can often increase the variance by factors as high as two or three, because the ratio of increase is $[1 + \rho(\bar{n} - 1)]$ where $\bar{n}$ is the average number of interviews per interviewer.

(3) The interviewer effects on subclass means were shown to be smaller in accord with $[1 + \rho(\bar{n}^* - 1)]$ where $\bar{n}^*$ is the average number of subclass members per interviewer.

(4) In the comparisons of subclass means, the interviewer effects tend to zero, in accord with an additive model of the effects.

Reprinted with permission from the *Journal of the American Statistical Association* (1962), 57, 92–115.

A brief version was presented at the Annual Meetings of the American Statistical Association, Palo Alto, August 1960, and appeared in Kish and Slater (1960). Dr. Carol W. Slater was a collaborator in much of this effort until drawn away by her other duties, much to my regret. N. Krishnan Namboodiri made several valuable contributions.

1. DESCRIPTION OF THE TWO STUDIES

Our data are based on analyses of two studies carried out by the Survey Research Center of the University of Michigan. Three basic features are common to both studies.

(a) Blue collar workers of an industrial plant were sampled with equal probability.
(b) The sample respondents were randomized among the interviewers with effective and practical procedures.
(c) The respondents, after strong assurances of anonymity, were asked many questions involving factual and attitudinal items about their jobs and related matters.

Their analyses formed the objectives of the surveys; our investigation is merely a methodological by-product.

One important difference between the two studies concerns the method of data collection. In the first study, the attitudes of respondents were obtained in personal interviews averaging an hour and a half and using open-ended questions. In the second study, after the personal interview averaging an hour, the respondents also filled out a paper-and-pencil questionnaire, for roughly another three-quarters of an hour, in the presence of the interviewer. In this second study, in contrast with the first, completely open-ended items were few: almost all questions included in the written questionnaire and many of those used in the interview involved asking the respondent to choose from a prepared list of fixed alternatives, the one coming closest to his own viewpoint. Pretesting of about one hundred cases had suggested that the overwhelming majority of respondents found such choices acceptable.

Another difference between the two studies existed in the background of the two sets of interviewers. In the first study, the 20 interviewers were selected and hired specifically for this study; initial interviews were used not only to evaluate potential performance but to eliminate applicants with extreme views on labor-management relations. All had some previous experience in interviewing, not necessarily in survey work; all were college graduates and about half had had some advanced training in the social sciences. A week of training was carried out before the study began and was augmented, as needed, by individual supervision and group sessions. In the second study, all nine interviewers were experienced members of the Center's field staff, including three regional supervisors. All had participated in other Center studies during periods ranging from two to 10 years and on a wide variety of topics; all had some college training.

Because of the high quality of the interviewers and the questionnaire we expected lower interviewer effects than is often feared on highly attitudinal items. We also expected somewhat higher effects on the first study than on the second because the interviewers had less training and because of the prevalence of open-ended questions. The results confirmed these expectations.

The first study, in 1948 at a large unionized auto plant in the Midwest, dealt mainly with laying the groundwork for a more detailed understanding of labor-management relationships in heavy industry. The interviews with the rank-and-file obtained attitudes towards foremen, stewards, the union, higher management and various aspects of their jobs. After stratification by sex, department, seniority and type of work the sample employees were selected with equal probability. Their names and addresses, typed on cards, were shuffled. Assignments were made randomly to the interviewers at the beginning of each day. The interviews were taken in the respondents' homes and with adequate call-backs a high rate of response was obtained. Altogether 462 employees were interviewed by 20 male interviewers.

The second study was conducted in 1958 at a large nonunionized oil refinery in Canada. This study covered a number of related content areas: the choice of reference figures in making wage comparisons, effects of shift work, relationships between experiences on and off the job. The male nonclerical hourly rated workers were selected with equal probability from a list stratified by type of work, department and pay rate. Interviews took place during or immediately before or after working hours, in offices provided by the company. In assigning respondents to interviewers we considered the fact that about 40 per cent of the men changed their working hours weekly, becoming unavailable for interviewing every third week. From the list of respondents available during each week, random assignments were allocated to the interviewers working during that period. In those few cases where it was impossible to keep the original appointment—because a respondent became ill or could not be spared from his unit—we tried, and usually managed, to ensure later assignment to the same interviewer. Altogether 489 respondents were obtained by nine female interviewers.

The first study also included an investigation by Cannell (1954) of the attitudes and expectations of the interviewers; generally these had no strong, if any, relationship to the response biases. (Hanson and Marks 1958 and Hyman et al. 1954 also present results on some kinds of relationships.) In both our studies we searched and reassured ourselves that generally the interviewer variance was fairly broadly spread and not confined to the idiosyncrasies of one or two interviewers.

2. THE MEASUREMENT OF INTERVIEWER VARIANCE

Besides sampling errors proper—those arising in selection or estimation procedures—survey results are also affected by errors which occur in the course of the observation (measurement), recording and processing of the data. These are treated in more or less detail in several books on sampling (especially in Hansen, Hurwitz and Madow 1953, Vol. 2, chap. 12). These errors may be divided into two types having very different effects on the summary results, such as means or totals, of a survey. The first type derives from "biases" or "systematic errors" imposed by the "essential survey conditions" (the average or "expected" deviations of sample estimates from their estimands, the population values); although important, these are not our present concern. The second type consists of *variable* errors: those not fixed by the "essential survey conditions." Some variable errors are uncorrelated among the elements, the individual respondents; these errors cannot be distinguished from sampling error among respondents unless replicate measurements are taken on the respondents. We are not here concerned with them; generally they can be regarded as random errors which increase the variance with contributions which enter automatically the computations of the variance. Some other variable errors, however, result from the *correlated* effect that each interviewer's bias can impose upon the respondents who make up his workload of interviews. Insofar as the individual interviewers have different average effects on their workloads, this *interviewer variance* will contribute to the variance of the sample mean: this contribution of the interviewer variance to the sampling variance is our present concern. Unlike the uncorrelated random errors, this contribution to the variance is not included in the usual computations of the variances.

Our model assumes the random selection of a sample of a interviewers from a very large pool of potential interviewers, the pool defined by the "essential survey conditions." Each interviewer has an individual average "interviewer bias" on the responses in his workload and we consider the effect of a random sample of these biases on the variances of sample means. This effect is expressed as an *interviewer variance* which decreases in proportion to the numbers of interviewers. Its contribution to the variances of sample means (s_a^2 / a) resembles other variance terms: it is proportional directly to the variance per interviewer and inversely to the number of interviewers. This contribution to the total variance may be substantial; by failing to take it into account (as when estimating the variance simply by s^2/n) one may be neglecting an important source of variation actually present in the design, having been introduced by the sampling of interviewers' biases. The interviewer variance s_a^2 should be viewed as a component of the total variance, denoted as

$$s^2 = s_b^2 + s_a^2 \tag{2.1}$$

where s_b^2 is the variance without any interviewer effect, all three terms being measured per element.

The next section presents the statistical results in terms of "ρ," essentially the statistical measure often called the "coefficient of intraclass correlation."

$$\rho = \frac{s_a^2}{s_a^2 + s_b^2} . \tag{2.2}$$

We listed in Table 1 those earlier investigations which permitted useful comparisons of the interviewer variance. To make these comparisons meaningful, the values of the effects should be made independent of the number of interviews and especially of interviewers; this can be done if the effects are expressed as variance components, preferably as ρ. Because in the literature we found the results expressed in terms of statistical tests (such as the F of the analysis of

TABLE 1. Comparisons with Some Other Investigations

	Range of ρ
Present investigation	
46 Variables in first study ($a = 20$)	0 to .07
48 Variables in second study ($a = 9$)	0 to .05
*Percy G. Gray** (1956) ($a = 20$)	
Eight "factual" items	0 to .02
Perception of and attitudes about neighbors' noises	0 to .08
Eight items about illnesses	0 to .11
*Gales and Kendall** (1957) ($a = 48$)	
Most semi-factual and attitudinal items about TV habits	0 to .05
1950 U.S. Census (1958) ($a = 705$)	
31 "age and sex" items	0 to .005
19 simple items	0 to .02
35 "difficult" items	.005 to .05
11 "not answered" entries	.01 to .07

*For these investigations we computed the above values from the published test statistics and this translation involves some uncertainties.

variance) that depend on sample size, we made approximate translations into ρ's—with the relationships shown in Section 6 and using the data available in the articles about average interviewer workloads and about the designs. Thus, these values of the ρ's are merely useful guesses, rather than proper estimates.

The attitudinal data in the Gray (1956) study come close to the type of variables we present, and his range of ρ's (0 to .08) is similar to ours, but perhaps on the high side. The Gales and Kendall (1957) study also shows rather similar ρ's (0 to .05) for comparable variables; they had one outstanding value of .19 for "Have you heard of any plans for commercial TV?" These comparisons are reassuring for the variables in our studies, which were perhaps more difficult and conceivably subject to greater interviewer influences.

That interviewer influences can be large indeed may be illustrated from several studies. Veroff et al. (1960) found values of .08–.29 in an investigation of motivation scores in Thematic Apperception Tests applied by 46 interviewers; and for length of protocol they found ρ's of .28 and .73! These values overestimate the actual effects, because the sample addresses were not randomized among the interviewers; but we think the overestimation was small and the given values useable. Stock and Hochstim (1951) found some large values for commodity pricing and for judging the "dilapidation" of homes; for several other items they found values ranging from 0 to .05. Some ρ's ranging from .15 to .50 were found among dentists examining boys' teeth. The dentists, though well trained, had not received special training that could induce greater uniformity of diagnosis. Not all medical items need necessarily be subject to large observer effects. There are wide differences among the effects for various illness items presented by Gray (1956); some are large and some negligible or absent. Horvitz (1952) reports a ρ of .02 for numbers of reported illnesses.

We end with two cautions regarding the use in Table 1 of ranges to summarize the ρ's for a number of items:

1. More of the items are near the lower than the upper end of the range.
2. A greater number of items in a set results in a larger range.
3. These sample values of ρ are generally subject to large coefficients of variation.

3. MAGNITUDES OF INTERVIEWER EFFECTS

Some investigators suggest that interviewer effects may vary markedly with differences in question form and content. In reviewing studies of response bias, Hyman et al. (1954, pp. 203–204) point out that "...the effects noted were not uniform. Significant differences were discernible on some questions and not on others...on most of the nonattitudinal questions, differences...are not significant, whereas on attitudinal questions most differences are highly significant..."

Hanson and Marks (1958) noted high effects for "Census questions containing one or more of the following factors—

(a) interviewer 'resistance' to the question, i.e., a tendency on the part of the interviewer to be hesitant about making the inquiry and possibly a tendency to omit or alter the question or assume the answer, or
(b) a relatively high degree of ambiguity, subjectivity, or complexity in the concept or wording of the inquiry, or
(c) the degree to which additional questioning tends to alter respondent replies."

The highest ρ's can be noted (in Table 2) for "measures of the number of improperly omitted Census entries, i.e., number of NAs (not answered entries) reported for a Census question"; their ρ's range from .01 to .07. For the relatively "difficult" items, including education, income, migration, farm residence and some occupation entries, the ρ's are mostly from .005 to .04; this range is similar to the range for attitudinal items in our two studies. For most of the "simple factual" items the ρ's were below .01 and for age and sex entries below .005.

Cannell (1954, p. 105 ff) offered the hypothesis that greater differences among interviewers would occur when the nature of the subject made extensive probing necessary—for example, when the subject had not given any prior thought to the question or when the response required making unaccustomed abstractions or generalizations. It seemed plausible that interviewer effect would tend to be greater when the interview situation lent greater salience to cues provided by the interviewer.

TABLE 2. Cumulative Percentages of ρ's on Two SRC Studies and on Census Study
Entries show for each study the percentage of items with value ρ less than the column heading. Thus the first column shows the percentage of negative ρ's.

.000	.001	.002	.005	.01	.02	.03	.04	.05	.06	.07	.08	.09	.10
					SRC Studies								
First study, 46 variables													
24	26	28	33	46	59	65	78	83	87	96	96	98	100
Second study, interviews, 25 variables													
16	28	28	32	48	60	80	88	100					
Second study, questionnaire, 23 variables													
57	65	65	65	70	87	91	96	100					
					Census Study								
All 96 variables													
2	18	34	55	68	83	89	94	95	97	100			
31 "Age and sex" items													
6	45	81	100										
19 "Simple"items													
11	16	68	84	100									
35 "Difficult"items													
	3	14	26	51	77	86	94	97	97	100			
11 "Not answered" entries													
					27	36	64	64	82	100			

In both of our studies, two categories of items appeared likely to enhance the effect of cues given by the interviewer. The first includes questions where the respondent might have doubts about the social acceptability, or the future personal consequences of his answer; when, for example, he had to decide whether or not to express criticism of the union or of the company. Answers to "*critical*" questions such as these might be expected to be influenced by the respondent's perception of the interviewer's expectations and intentions. A second category of items, suggested by Cannell's observation, included those which might be expected to involve a relatively unstructured cognitive field on the part of the respondent: items where he might be unsure of the facts, uncertain about the precise referent of the question or unclear about the relevant dimensions of evaluation. Wanting to give a "right" answer—or at least an appropriate one—the respondent might look to the interviewer for more guidance on these "*ambiguous*" items than on others. (Both the concepts and their application are the work of Dr. Carol W. Slater.) From our two studies, ninety-four items were chosen for analysis and all appear individually in Tables 3, 4 and 5. Three considerations governed their selection:

(1) to find all items common to the two studies, for the sake of comparability;
(2) to find "factual" items to contrast with the preponderance of attitudinal items;
(3) to select, among the attitudinal items, items whose content appeared particularly ambiguous, critical or emotionally "loaded."

TABLE 3. Interviewer Effects on 46 Items of the First Study

Var. No.	Description	Category	ρ
1	How much does R feel he should be earning? (II #38)	C	0.092
2	Does R mention more than one thing he does not like about working for ________________? [No]	C	0.081
3	Does R mention more than one thing he does not like about the kind of work he does? (II #7) [No]	C	0.068
4	How do R's wages compare with those of other people he knows? (I #13)	A	0.066
5	Total criticism score (II #25)	C	0.063
6	How interested are other men in the union?	A	0.060
7	Can the union and the company both achieve their goals?	A	0.059
8	Number of children (II #30)	O	0.053
9	If the company met hard times, how likely is it that the men would do what they could to help out?	A	0.042
10	Would the men who work here do as good a job for some other company? [R says he doesn't know]	DK	0.041
11	How satisfied is R with "bumping" procedures in the plant?	C	0.036
12	If the company met hard times, would R do what he could to help out?	C	0.035
13	How do the other men feel about a steward who becomes a foreman? [R says he doesn't know]	DK	0.033
14	Would R like to change jobs? [Yes]	C	0.032
15	Perceived chance of obtaining preferred kind of work.	A	0.030
16	Age (II #39)	O	0.030
17	Weekly take-home pay (II #36)	O	0.024
18	How does R feel about this kind of work? (II #36)	C	0.020
19	Does the union care how the company gets along?	A	0.021
20	How well do members of the union local work together? [R says he doesn't know]	DK	0.019
21	Does R mention intrinsic (task derived) satisfaction(s) as first thing he likes about his work? (II #14) [Yes]	?	0.017
22	Is any other member of R's family presently employed by ___________? (II #45) [Yes]	O	0.017
23	Does R foresee changes in union-management relations? [Yes]	A	0.016
24	Does R feel he is similar to or different from the men with whom he works?	A	0.011
25	How does R feel about the "special things" the union does for the men?	A, C	0.011
26	Is there any other job in the plant that R would rather have? [Yes]	C	0.009
27	Highest grade in school completed (II #41); graduated high school or more	O	0.009
28	How does R feel about his union steward?	C	0.008
29	Would R advise a young friend to come to work for ___________?	C	0.006
30	How does R feel about working for ___________?	C	0.006
31	How does R feel about the co-op store? [R says he does not know]	DK, C	0.005
32	How much does the union do for the men?	C	0.003

33	How does R feel about the foreman's attitude towards the union?	C	0.002
34	Would R like to be a foreman?	A	0.001
35	How do other men in the shop feel about the amount they are producing?	A	0.000
36	Years' seniority (II #22)	O	−0.001
37	How does R feel about his foreman? (II #34)	C	−0.003
38	How does R see the general economic outlook for the next few years? [R says he doesn't know]	DK	−0.003
39	Does R mention economic factor(s) as first thing he does not like about his work (II #6) [Yes]	?	−0.005
40	How does R feel about the union paper?	C	−0.006
41	Would R prefer an alternate method of computing his wages? [Yes]	C	−0.006
42	How well do the union and management get along together?	C	−0.006
43	Marital status (II #28) [married]	C	−0.012
44	Would R like to be a union steward?	A	−0.020
45	Does R mention intrinsic (task derived) satisfaction(s) as a second thing he likes about his work (II #23) [Yes]	?	−0.025
46	Does R mention economic factor(s) as first thing he does not like about working for ___________? (II #6) [Yes]	?	−0.031

Category Identification: A—High Ambiguity; O—Low Ambiguity; C—Criticism; ?—No Classification.

Figures in parentheses indicate the numbers of the corresponding items on second study. In the case of qualitative variables, ρ was computed on the basis of the proportion of respondents giving the answer indicated within square brackets. In no case was this an answer given by less than 15% of the total sample. For each respondent, the total criticism score was computed by counting the number of critical (unfavorable, negative) responses given in answer to 20 items; in Table 4, item 25 was similarly computed from seven critical answers.

The questions appear here in compressed form. R means respondent.

Items in these categories, however, proved to be so poorly distinguished that perhaps our array of ρ's does not differ remarkably from that which might have been yielded by a random selection of variables.

The items were categorized according to two basic hypotheses: potentially *critical* answers (C), and those given in response to relatively *ambiguous* items (A), were expected to show a greater degree of interviewer effect than those presumably involving less risk and greater cognitive clarity (O). The results do not support these predictions. As pre-coded, "critical" and "ambiguous" items appear to occur with roughly equal frequency at the high and low ends of the ranked orderings provided in the three tables. We hope that some readers will inspect these items, which have been rank ordered for their convenience with respect to the magnitudes of the observed interviewer effect, to find other modes of categorization and hypotheses for future testing.

The lack of very sensitive items may reflect in part lack of enough items with which to test the hypothesis. Several of the supposedly "critical" items produced almost nothing but "safe" answers. Thus, in the first study, only 2 per cent of those who expressed an opinion on item 34 said that union-management relations were any worse than average; on item 37, only 8 per cent gave negative evaluations of

TABLE 4. Interviewer Effects on 25 Items on Interviews of the Second Study

Var. No.	Description	Category	ρ
1	How many meetings a month does R attend?	O	0.044
2	Thinking of someone whose yearly earnings are different from his own, how satisfied is R with the comparison?	A	0.042
3	How likely is a general wage increase?	A	0.041
4	How often does time drag on the job?	C	0.037
5	Does R think he will have a chance to enter the training program at any future time? [Yes]	A	0.034
6	Does R mention economic factor(s) as first thing he does not like about his job? (I #39) (I #46)	?	0.023
7	Does R mention a second thing he does not like about his job? (I #3) [No]	C	0.022
8	Is R an officer or does he have any special job in any organization? [Yes]	O	0.020
9	Is R a member of any club or organization? [Yes]	O	0.020
10	How satisfied is R with his opportunity to do things when he wants to?	A	0.020
11	How many hours a week has R spent doing things around the house in the last two weeks?	O	0.014
12	Is there anything keeping R from doing as good a job as he would like? [Yes]	C	0.012
13	Thinking of two people whose earnings are different from his own, does R choose upwards, downwards or both? (I #4) [Both people mentioned earn less than R]	A	0.011
14	Does R mention intrinsic (task derived) satisfaction(s) as first thing he likes about his job? (I #21) [Yes]	?	0.009
15	To what extent does promotion depend upon how well a man can do the job?	A	0.008
16	Why does company have benefit plans? [Solely for the benefit of the company]	A	0.006
17	To what extent is a man himself responsible for how far he goes in life?	A	0.005
18	How satisfied is R with council representing workers?	C	0.002
19	If R inherited enough money so that it were no longer necessary, would he continue working? [Yes]	A	0.000
20	Thinking of someone whose weekly earnings are different from his own, how satisfied is R with the comparison? (Second comparison person)	A	0.000
21	How much variety does R experience on the job?	C	0.000
22	Years' departmental seniority (I #36)	O	−0.002
23	Does R mention intrinsic (task derived) satisfaction(s) as a second thing he likes about his job? (I #45) [Yes]	?	−0.003
24	Is R now in a training program? [Yes]	O	−0.005
25	Total criticism score (I #5)	C	−0.005

See notes at the bottom of Table 3.

TABLE 5. Interviewer Effects on 23 Items on Written *Questionnaire* of the Second Study

Var. No.	Description	Category	ρ
26	Weekly take-home pay (I #17)	O	0.040
27	Satisfaction with present chances of promotion	C	0.033
28	Marital status (I #43) [Married]	O	0.021
29	Number of children (I #8)	O	0.014
30	How well does management understand problems of the men?	C	0.014
31	Could R do better outside ___________ company?	A	0.011
32	Would R be earning more now if he had started in another department? [Yes]	A	0.011
33	How well does management *try* to understand problems of the men?	A,C	0.008
34	How does R feel about his supervisor? (I #37)	C	0.000
35	All in all, how does R feel about his job?	C	0.000
36	How does R feel about this kind of work? (I #18)	C	−0.001
37	Is there any way the company could make R's job safer?	C	−0.001
38	How fair is the pay for R's present job? (I #1)	C	−0.002
39	Age (I #16)	O	−0.003
40	Is the company more interested in efficiency or in looking out for the men?	A	−0.003
41	Highest grade in school completed (I #27) [High school or more]	O	−0.003
42	Total number of questionnaire items left blank—index of respondent motivation	?	−0.004
43	Total questionnaire criticism score	C	−0.004
44	Satisfaction with past promotion(s)	C	−0.005
45	Is any other member of R's family currently employed by ___________? [Yes] (I #22)	O	−0.006
46	Should the men have more to say about the benefit program?	C	−0.008
47	How does R feel about the amount of overtime he has worked in the past six months?	C	−0.010
48	Perceived chance for promotion two pay grades higher than present classification	A	−0.024

See notes at the bottom of Table 3.

the union paper, and on item 22, only 5 per cent expressed dissatisfaction with their foreman. The paucity of critical responses to items such as these—whatever its basis—creates obvious difficulties in testing the original hypothesis, because the scores reflect only different degrees of uncriticality. But chiefly these negative findings are important evidence that with a staff sensitive to the requirements of nondirective and nonjudgmental probing, even relatively unfamiliar or "touchy" material can be expected to show, on the average, no more marked interviewer effects than more "objective" subject matter.

We also computed ρ's for the proportions of "don't know" responses on those five items in the first study which showed a high proportion of them: ten per cent or more. These items are indicated in Table 3 by (DK) and it may be observed that these ρ's vary in rank, from the relatively high position of item 10 to the relatively low-ranking of item 38. This small sample of items suggests that whatever interviewer characteristics may be involved in raising or depressing the proportion of "don't know" responses, they do not operate with equal effect on our items.

The narrow limits we found for the interviewer variances—compared to their large sampling variability due to the small number of interviewers (20 and 9 in our two studies)—prevented us in part from finding differences among items and even classes of items. But this should not blind us to the importance of finding these narrow limits. The results summarized in Table 2 are important collective evidence in useful form for planning multi-purpose social surveys. Note first that for the written questionnaire items half (57 per cent) of the ρ's were negative; this bears out expectations of no interviewer effect here, although the three items (13 per cent) over .02 may indicate something beyond mere sampling variation. For the personal interviews of the two studies only 24 and 16 per cents respectively are below zero; thus interviewer effects are clearly present. The medians for both are near .01. But the first study, as expected, shows a greater proportion of items above ρ's of .02.

Generally, the distribution of items from the two studies resembles the distribution of the "difficult" Census items (line 7) that deal with items like education, income, migration, farm residence and some occupation entries. The existence of interviewer effects on these "factual" items is no surprise either to the Census statisticians or to us. But it is important to find similarly low effects for the subjective attitudinal items, including many ambiguous and critical questions, of our two studies. We expected to find greater effects and were pleasantly surprised. However, we should be cautious; the higher effects that experts—as Hyman et al. above—reasonably expect, may still hold widely if the training of interviewers is not thorough.

Finally, note in Table 2 the clearly smaller effects for the simplest age and sex items obtained by the Census and the much larger effects for "not answered" entries in the Census results.

4. THE EFFECTS ON RESEARCH DESIGN

We can first dispose of the question of the "statistical significance" of the interviewer variance as it affects individual variables. Do we have in these tables any evidence of effects sufficiently greater than mere chance variations to warrant explanation?

Of the 46 variables in Table 3, the usual F test of the analysis of variance shows 9, 15 and 17 items differing from zero significantly at P levels of .99, .95 and .90, respectively. These are all the first ranked items except No. 15, which misses merely because of its smaller n of 175. Of the 25 items in Table 4, the top 6, 11, and 12 items are significantly greater than zero at the P levels of .99, .95 and .90, respectively. Thus, in these two tables for personal interviews, there is clear evidence of variation beyond chance variability around zero effect.

In Table 5 the evidence for the presence of interviewer effect in the written questionnaire is weak or absent, as expected. Of 23 items the top 1, 3 and 5 items are significant at the .99, .95 and .90 levels. Incidentally, the last item in this table is the only one which shows a negative effect below the $P = .025$ level; we need not consider it other than a chance occurrence.

The values of ρ in the three Tables 3, 4 and 5 can be translated into approximate F values of the analysis of variance, then to the usual α levels of error of significance tests, by means of the relationships shown in Section 6. Then we get for Table 3, with $a = 20$ interviewers, $n = 440$ total sample size and an average workload of $n/a = k = 22$, the following approximate values of α, F and ρ:

α	.05	.10	.25	.975
F	1.65	1.45	1.2	.46
ρ	.03	.02	.01	−.025

For Tables 4 and 5, with $a = 9$, $n = 486$, $k = 52$, we get approximately:

α	.05	.10	.25	.975
F	2.0	1.7	1.3	.27
ρ	.02	.013	.009	−.013

What conclusions can we reach on the basis of these data? In considering individual variables, it would generally be wrong to use a low error level (such as $\alpha = .01$ or .05) for rejecting the null hypotheses of zero interviewer effect. That null hypothesis should not be accepted against reasonable evidence, when it is doubtful *a priori* and when the cost (or risk) due to acceptance is high. Perhaps one should operate at the high level of $\alpha = .25$ (or higher) for rejecting the null hypothesis; by that criterion, translated into ρ's of .01 for Table 3 and .009 for Table 4, about half of the items in each table can be suspected of showing some interviewer effect. A criterion of $\alpha = .5$, corresponding to ρ's of zero, would bring in about three-quarters of the items in Tables 3 and 4. (Incidentally, the last two statements are consistent with a hypothesis that half of the items are affected and half are not.) The mean of these ρ's is .020 for Table 3 and .014 for Table 4. In Table 5, the mean of the ρ's is only .003, and half of them are negative; these negative findings for the written questionnaire reassure our expectations. (But the somewhat high, though irregular values for three "factual" items, Nos. 26, 28 and 29, should arouse some curiosity.)

The individual ρ's are subject to wide variability, and this obscures the comparisons of individual items. This limitation, due partly to the small number of interviewers, characterizes most measurements of this kind. Another limiting factor is the small range of variation in interviewer effects when most values are under .05. Nevertheless, individual analysis can point out extreme items for which the average interviewer effects appear too large compared to others or to reasonable *a priori* expectations. Thus it can lead to corrective measures, either through interviewer training or through changes in the questionnaires. (For procedures to detect individual interviewers who contribute unduly to the variability, see Deming 1960).

Precise estimates of interviewer effects for specific items are difficult to obtain except from large-scale efforts, such as the Census data based on 705 interviewers (Hanson and Marks 1958). However, we believe that estimates for *many* interesting items, even if less precise (such as ours), are very much needed. Why? Because most social research is multi-purpose in character; hence considerations of interviewer effects in the design should typically relate to a wide variety of items. Furthermore, new research designs usually have new objectives and items; these can be more easily related to a spectrum of fairly similar items than to a single item.

How can our results be useful for planning other surveys? First, they show that responses without great interviewer effects on attitudinal items, some highly ambiguous and emotionally loaded, are feasible if the interviewers are carefully selected and well trained. The low values of ρ on these items compare favorably with those on some factual data, medical observations and others. Indeed, the effects for these attitudinal interview items are not generally higher than for supposedly "factual" items obtained in a good Census—except for the simplest items like age and sex. This speaks well for the prospects of obtaining attitudinal data with some assurance of reasonable and useful reliability.

Second, we can distinguish—albeit weakly—in the three tables concomitants of different interviewing situations. For newly hired and trained interviewers taking open-ended interviews, the ρ's in Table 3 range, in the main, from zero to .07, with an average of .02 or .03. In Table 4 we see more experienced interviewers taking a more structured interview; the ρ's vary mostly from zero to .05, with an average of .01 to .02. In Table 5 we find for the written questionnaire, that the *a priori* hypothesis of zero effect is generally acceptable.

Third, the results are valuable in disproving our beliefs (widely shared, we suspect) that the "ambiguous" and "critical" items comprise classes with distinctly higher susceptibility to interviewer effects than less sensitive items. Even informed intuition, it would seem, needs considerably more conceptual and empirical tools than are now available to evaluate the relative susceptibility of survey items to interviewer bias.

Fourth, our findings demonstrate again the existence of interviewer effects that, even if small, are nevertheless large enough to exert an important influence on the total variability of survey results. Even a small ρ, when multiplied by a moderate or large interviewer workload (k) can have great effects. This effect on the variance of the sample mean is about $[1 + \rho (k - 1)]$. Now consider an increase in the variance by a factor of 1.5 as "serious" and by 2 as "critical." With $k = 22$ in the first study, ρ becomes serious at .025 and critical at .045; and these values include the top 16 and the top 8 items respectively in Table 3. In the second study, with $k = 52$, $\rho = .01$ is serious and $\rho = .02$ is critical—thus including the top 13 and 10 items respectively in Table 4. The factors of 1.5 and 2.0 correspond to increases in the standard errors of $\sqrt{1.5} = 1.22$ and $\sqrt{2} = 1.41$. Neglecting them will necessarily result in underestimating "seriously" or "critically" the "true" standard error of the sample mean that would include the interviewers' contributions to the total variability.

Fifth, analysis of this type makes it possible to include interviewer effects in considering the economic aspects of survey designs. If the ratio of the cost of hiring and training an interviewer to the cost of a single interview is C, then the most economical plan (in analogy with optimum cluster size—Hansen, Hurwitz, and Madow 1953, Vol. I, p. 295) in terms of the least total variance $[(s_a^2/a) + (s_b^2/n)]$ of the sample mean, results in this optimum workload size per interviewer:

$$k = n/a = \sqrt{Cs_b^2/s_a^2} = \sqrt{C(1-\rho)/\rho}\,.$$

For example, if it costs \$180 to train an interviewer and \$10 to take an interview, then $C = 18$. For ρ's of .02 this gives an optimum workload per interviewer of $k = 30$. The actual workloads in our two studies were in this neighborhood; furthermore assigning smaller workloads for the first study than for the second study was fortunate in light of the larger ρ's found for the former. (This formulation assumes random choice from a pool of equally satisfactory interviewers; hence, it disregards the possibly greater over-all effect of hiring and training more interviewers.)

5. EFFECTS ON SUBCLASSES AND COMPARISONS AMONG THEM

Existing models of response errors deal mostly with the effects on the mean for the entire sample and most of our investigation was focused on that problem. (But Hansen, Hurwitz, and Bershad (1960) investigated the very large effects produced on small area data when these have been gathered by a few enumerators with large workloads.) The extension of the model and methods to the means of subclasses is straightforward, if they allow for the resulting unequal workloads per interviewer and insofar as the randomization of the entire sample among interviewers also randomizes its subclasses. Results for subclasses are generally important objectives of social research.

The difference of two means presents new problems. But the comparison of subclasses is frequently an important aim of research; indeed our colleagues often argue convincingly that they are more concerned with these comparisons than with the straight presentation of means. Even approximate answers are useful to this important question: What are the effects of variable interviewer errors on the difference of two subclasses? If the subclasses were covered by two different sets of interviewers, the variances (being variances of independent samples) would be simply additive. Complications arise because the same interviewers contribute their errors to both subclasses and this induces a correlation in the variable errors of the comparisons. If the individual interviewers' biases tend to be similar for members of both subclasses (which are randomly assigned) then the effects of these biases will tend to cancel out from the comparisons. In analysis of variance terms: if the interviewer effects are simply additive, lacking interaction with the subclasses, then their effects will tend to vanish from the comparisons. Our results are in good accord with this hypothesis.

We restricted the range of this investigation because of the amount of work it entailed and also because we were searching only for approximate answers with rough new methods. We chose the strategy of taking from Tables 3 and 4 several items with large ρ's for two reasons. First, for smaller ρ's the different effects on various statistics would be difficult to distinguish, because they would be hidden by sampling variability. Moreover, if the large ρ's tended to have vanishing effects in the comparisons, we expect *a priori* that items with small ρ's would likely not have larger effects; on the contrary, not much could be deduced from the small comparison effects of small ρ's. Thus the behavior in the comparisons of the larger ρ's provided tests which were more sensitive and more strategic. We chose four items from the first study and three from the second study. For each item we selected two ways of forming subclasses; two ways that were both relevant and contrasting—such as objective versus subjective criteria (column 2, Table 7). The subclasses split the sample into two groups which were not very different in size and which used most of the sample.

To illustrate the problem, consider some simple models, with two types of interviewers A and B, two subclasses 1 and 2; the two types and the two subclasses have equal size. Assume two kinds of possible biases which add either 0 or 1 to the value of the variable. (Any other two constants would make the same point, but more constants would permit more complex effects.) The different models possess different patterns of 0's and 1's and result in different "effects" on the standard errors. These effects are displayed in Table 6 for the separate subclasses; also for their sum which represents the entire sample and for their difference which represents the comparison.

Model I represents the simple additivity of response errors, where the interviewer bias is constant across the different subclasses. The effect on the sum of subclasses is the sum of the effects on the subclasses. On the other hand, in the comparison of subclass means the net effects, consisting of the difference of two equal effects, disappear. On the other hand, in both Models II and III, there is "interaction" between interviewers and subclasses, that is, the interviewer effects vary among the subclasses. Other models which add equal amounts for all interviewers (A and B) to a subclass result in net biases without interviewer variance; these cannot be detected with studies of interviewer differences.

Our results are consistent with the additive Model I. The evidence most readily appears in Chart 1, which presents graphically the data from columns 4, 8, 9, and 11 of Table 7. These provide grounds for approximate but useful conclusions on two important questions and we now add these to the five already listed in Section 4.

TABLE 6. Three Simple Models of Interviewer Variability

		Respondent subclass		Sum of subclasses	Comparison of subclasses
Model	Interviewer type	(1)	(2)	(1) + (2)	(1) – (2)
I	A	1	1	2	0
	B	0	0	0	0
	Effect	1	1	2	0
II	A	1	0	1	1
	B	0	0	0	0
	Effect	1	0	1	1
III	A	1	0	1	1
	B	0	1	1	–1
	Effect	1	1	0	2

Sixth, the effects of interviewer variance on the variances of subclass means tend to decrease proportionately to the decrease in the size of the subclass. These effects on the variances are shown separately for the two subclasses in columns 6 and 7 of Table 7 and their average in column 5; these are lower than the corresponding values for the sample totals in column 4. The effect on the variance can be measured approximately in terms of $[1 + \rho(n^*/a - 1)]$ where n^* is the sample size of the subclass. This effect decreases if ρ remains constant, where ρ expresses the interviewer *effect per element*. That ρ remains constant for the subclasses can be best seen on Chart 1 in the proximity of these values (marked by ×) to the 45 degree line, which denotes equality for the ρ's of the subclasses and of the entire sample. These values are the average of the entries in columns 8 and 9 of Table 7, plotted against the value, in column 4, for the entire sample.

Seventh, the effects of interviewer variance on comparisons between subclass means are reduced drastically to the neighborhood of zero. This important result seems to hold roughly, on the average and for subclasses randomized over the interviewers. Note that the marks (O) denoting interviewer effects on comparisons in Chart 1 fluctuate around zero effect. These come from plotting the effects per element of the comparisons, from column 11 of Table 7, against those of the entire sample in column 4. These values show a great deal of fluctuation, the causes of which should be sought in later investigations. Nevertheless, the tentative working hypothesis of zero average effects would lead to no serious underestimate of the variance.

TABLE 7. Effects of Interviewer Variability on the Variances of Subclasses and Their Comparisons

Var. and table no.	Bases of forming subclasses	Two measures of ρ for the entire sample		Measures of effects on the variances of means for entire sample and for subclasses 1 and 2			$^*\rho$ for subclasses		Effects on comparisons	$^*\rho$ of comparisons
		$\rho(\bar{y})$	$^*\rho(\bar{y})$	$e(\bar{y})$	$e(\bar{y}_1)$	$e(\bar{y}_2)$	$^*\rho(\bar{y}_1)$	$^*\rho(\bar{y}_2)$	$e(\bar{y}_1 - \bar{y}_2)$	$^*\rho(\bar{y}_1 - \bar{y}_2)$
(1)	(2)	(3)	(4)	(5)	(6)	(7)	(8)	(9)	(10)	(11)
1–3	Like working for ______	0.092	0.092	2.88	2.77	1.19	0.188	0.019	1.098	0.010
	Seniority		0.092	2.88	1.32	2.88	0.028	0.229	1.450	0.047
2–3	Should earn more	0.081	0.081	2.69	1.44	2.49	0.079	0.104	1.323	0.040
	Education		0.080	2.75	0.59	3.41	0.047	0.196	1.237	0.023
3–3	Present earnings	0.068	0.074	2.63	1.24	2.06	0.025	0.091	0.772	−0.022
	Rather have another job		0.065	2.33	2.18	0.77	0.114	−0.025	0.716	−0.029
5–3	Would like to be foreman	0.063	0.070	2.54	0.99	2.14	−0.001	0.080	0.572	−0.046
	Education		0.070	2.54	1.38	2.47	0.044	0.118	1.179	0.017
2–4	Would be earning more if had started earlier	0.042	0.026	1.95	0.96	1.14	−0.002	0.007	0.763	−0.013
	Present chances of promotion		0.059	2.49	1.84	1.93	0.057	0.035	1.221	0.012
3–4	Seniority	0.041	0.040	3.03	2.23	1.65	0.057	0.023	0.897	−0.004
	Shift		0.039	3.08	2.29	1.24	0.042	0.012	0.473	−0.021
4–4	Seniority	0.037	0.030	2.57	1.79	1.58	0.036	0.020	0.848	−0.006
	Absences		0.34	2.79	2.01	1.41	0.037	0.017	0.648	−0.014

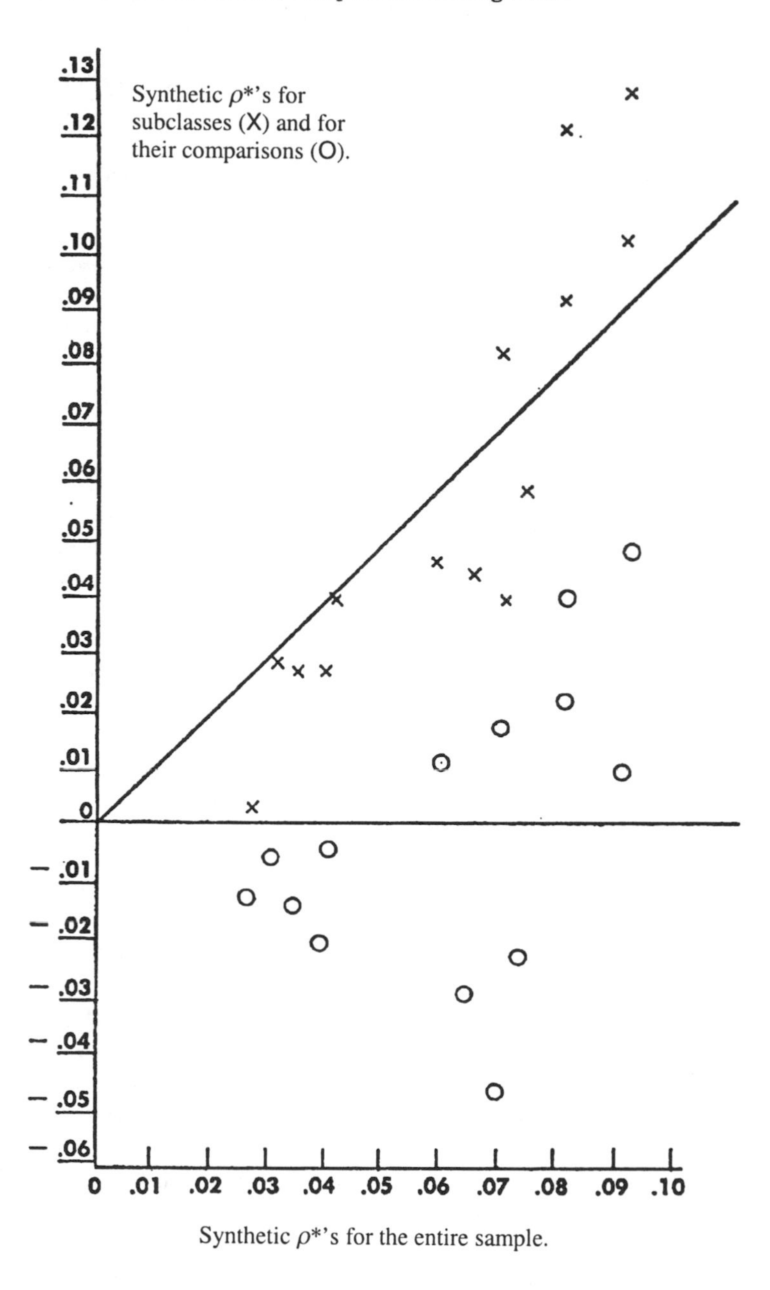

Chart 1. The effects of interviewer variability on subclasses (×) and on their comparisons (O) plotted against the effects on the entire sample. (The effects are measured as ratios to the total variance per interview—as synthetic equivalents of ρ's.)

This conclusion should apply also to comparisons of any two (or more) samples which have been randomized over the same set of interviewers. Important examples arise from the comparisons of periodic samples assigned to the *same* set of interviewers; these comparisons, too, will tend to be relatively free of the effects of the interviewer variance that affect a single sample. (Similar results were obtained for comparisons on the dental data, referred to earlier, which had very high initial ρ's.)

This result gains added significance in combination with the likelihood that *systematic* biases of comparisons are often also smaller than the biases of the means of either the subclasses or of the entire sample. That this is so for most kinds of comparisons seems *a priori* reasonable: if the interviewers' biases affect the subclasses equally (no interaction between interviewer bias and subclass) then both the systematic bias and the interviewer variance tend to disappear from the comparison of subclasses.

We consider these results as limited but welcome empirical evidence in favor of the hypothesis that the additivity of interviewer effects, without interaction with subclasses, exists in many practical situations. That *a priori* hypothesis is based on practical assumptions about the nature and distribution of subclasses in samples and the nature of interviewer effects. Clearly, it cannot be a theory about interviewer effects for all situations, variables and subclasses. The accumulation of empirical evidence can clarify the areas of its validity and its limitations. We disclaim sympathy with any naïve belief that comparisons are always free of both systematic and variable interviewer effects.

6. THE STATISTICAL MODEL

The model we used was a simple analysis of variance (*anova*) model for estimating components. Because each respondent is assigned to a single interviewer, it can be considered also as a two-stage "hierarchical" or "nested" model and with unequal frequencies (Anderson and Bancroft 1952). The response from the *j*th individual to the *i*th interviewer is expressed as $y_{ij} = y'_{ij} + A_i$. Here A_i is the average "effect" of the *i*th interviewer on any interview. Insofar as the interviewer has variable effects on different occasions and insofar as the respondent has "random" errors, these are confounded with the respondent's "true value" Y_{ij}—and inextricably for a single response per respondent. Hence, we cannot use a model like $y_{ij} = Y_{ij} + A_{ij} + e_{ij}$, because we cannot separate the last two components. Furthermore, the e_{ij} may also be considered as containing other sources of errors, such as coding errors, which have not been disentangled. In addition, any constant (or "systematic") biases common to all interviewers are not detected; we assume that the sum of the interviewer effects is zero for the population of A interviewers, from which the actual a interviewers are a random sample.

The total variance of each response is viewed as composed of two components, the sampling variance of the individual response and the component due to the variable interviewer bias, or interviewer variance: $s^2 = s_b^2 + s_a^2$. Assuming that of the n respondents n_i were assigned with simple random sampling to the ith interviewer, we have in terms of the usual *anova* table for computations (Anderson and Bancroft 1952):

Source of variation	Degrees of freedom	Sum of squares (SS)	Mean square	Components of the mean squares
Among interviewers	$a-1$	$\overset{a}{\Sigma} y_i^2 / n_i - y^2 / n$	$V_a = \dfrac{SS(a)}{a-1}$	$s_b^2 + ks_a^2$
Within interviewers	$n-a$	$\overset{a}{\Sigma}\overset{n_i}{\Sigma} y_{ij}^2 - \overset{a}{\Sigma} y_i^2 / n_i$	$V_b = \dfrac{SS(b)}{n-a}$	s_b^2

Here

$$n = \sum_i^a n_i, \quad y = \sum_i^a y_i \quad \text{and} \quad y_i = \sum_j^{n_i} y_{ij} \tag{6.1}$$

Then $s_a^2 = (V_a - V_b)/k$ where

$$k = \sum^a n_i^2 \frac{1/n_i - 1/n}{a-1} = \frac{n}{a-1} - \frac{\overset{a}{\Sigma} a_i^2}{n(a-1)}$$

$$= \frac{n}{a} - \frac{1}{n}\left[\frac{1}{a-1}\sum^a \left(n_i - \frac{n}{a}\right)^2\right] = \frac{n}{a}\left[1 - \frac{\text{var}(n/a)}{(n/a)^2}\right].$$

The last form shows that $k \doteq n/a$, the average of the n_i, except for a negative correction proportional to the *relvariance* of n/a. In our investigation the n_i did not vary much, hence the corrections were small. Computing a different correction factor for each characteristic would have been laborious and we suspected unnecessary for the very small differences in a small correction. We confirmed these suspicions with several computations: the ratio of k to n/a showed only negligible and irregular variations over a wide range of total n's for different characteristics :

n	486	484	483	480	475	400	383	377	318
$k \div (n/a)$	.993	.993	.993	.994	.996	.993	.987	.990	.987

In these nine computations the corrections average .008 and range from .004 to .013. This small variation came from workloads which varied as follows in one illustrative case: $n_i = 34, 44, 45, 49, 50, 53, 61, 63, 87$. Hence we used an average of the ratios of k to n/a to correct the value of n/a for the other variables.

How well does the simple model in the above table of *anova* represent the design actually used? In the second study the randomization was not carried out simply over the entire sample of n cases, but separately over each week's respondent workload. The week's available workload was not a random subset of the entire sample, but dictated by the work situation. This kind of practical problem exists frequently and it introduces complexities into the design. We could think of possible effects of these factors on our results: interaction between weeks and interviewers may exist for some reason; homogeneity within a week's workload could introduce the effects of stratification over the six weeks' workload of an interviewer. Our understanding of the situation tends to rule out both of these as important factors, but we wanted—and obtained—empirical reassurance. Therefore, we investigated the components computed within each week and compared it with the other computations, which assumed randomization over the entire sample of n cases. The comparisons of the two computations for individual items (using either the interviewer components s_a^2 or the ρ's) showed a great deal of variation, but all of it around the 45 degree line of no consistent trend; the overall distribution (our chief interest) of the two sets of results did not differ materially.

The effect of the interviewer variance on the variance of the sample mean $(\bar{y})$ may be seen as an increase of the individual sampling variance s_b^2 / n by a term that is inversely proportional to the number of interviewers a:

$$\text{var}(\bar{y}) = \frac{s_b^2}{n} + \frac{s_a^2}{a} = \frac{s_b^2}{n}[1 + \frac{s_a^2}{s_b^2}\frac{n}{a}]$$

$$= \frac{s_a^2 + s_b^2}{n}[1 + \frac{s_a^2}{s_a^2 + s_b^2}(\frac{n}{a} - 1)] \, . \tag{6.2}$$

The relative increase due to the interviewers' variance may be seen clearly in the second expression as proportional both to the ratio s_a^2 / s_b^2 and to the interviewers' workload n/a. The third expression gives a direct answer to the question: Accepting the total variance per respondent as $s^2 = s_a^2 + s_b^2$, what is the effect of increasing sizes of interviewer workloads in increasing the variance of the sample? It also gives the ratio by which the simple computation of the variance of the sample mean as s^2/n underestimates the actual total variability, including the increase due to interviewer variance. This increase is equal to $\rho\ (n/a - 1)$ where $\rho = s_a^2 /(s_a^2 + s_b^2)$, an expression of the interviewer variance as a ratio of the total variance. These relationships assume strictly an equal workload of n/a for each interviewer but, as we saw above, moderate departures permit the use of n/a as a useful approximation. This is also the effect of a two-stage selection with replacement, of a random clusters with n/a random elements from each. In all of our statements the s_a^2 and s_b^2 are sample estimates; they are the usual unbiased estimates of the corresponding parameters that could be written as S_a^2 and S_b^2. It is only a technical duty to point out that ρ is a consistent but not unbiased estimate of $S_a^2/(S_a^2+S_b^2)$.

Incidentally it may be noted that if the sample is simple random, the interviewer effects can be included in the variance by computing the interviewer's load as if it were a "cluster." In the case of actually clustered samples, where the interviewer is confined to a single primary selection (such as a county in a national sample) the proper computations based on primary selections automatically include this effect.

The usual test statistics, such as the F of *anova*, are not satisfactory measures for direct comparisons among studies of the magnitude of interviewer effects, because those statistics depend on the workload sizes and we want the comparisons free from this influence. The usual test statistic is

$$F=(s_b^2+ks_a^2)/s_b^2=1+k(s_a^2/s_b^2),\quad \text{hence}\quad s_a^2/s_b^2=(F-1)/k \tag{6.3}$$

and $\rho=(F-1)/(F-1+k)$. Using the values F and k we found in the articles, we computed approximate values of ρ's and used these in Table 1. We present the results in terms of $\rho=s_a^2/(s_a^2+s_b^2)$, but these can be converted easily to values of $\rho/(1-\rho)=s_a^2/s_b^2$; for our small ρ's these hardly differ from the ρ's.

For presenting the Census data in Tables 1 and 2 we used

$$\rho=\frac{V_a-V_b}{(n/1489)\,p(1-p)}. \tag{6.4}$$

Here V_a and V_b are the mean squares "between interviewers" and "between E.D.s, within interviewers." The E.D.s (enumeration district) were the sampling units randomized among the interviewers, but of no inherent interest to us; we wanted an approximation of ρ that a random sampling of individuals would have shown. We divided by $p(1-p)$, as a useful approximation for s^2 in this large sample. We also corrected by $n/1489$, the workload per E.D.; n is the sample size for the characteristic and 1489 E.D.s were assigned to 705 interviewers in the Census study (Hanson and Marks 1958). The authors kindly sent us values of V_a, V_b, n and p and full descriptions of their methods of computation. But we alone are responsible for the use we made of them in our approximation. We believe that more precise formulas would not change the first significant digits—which is all the precision we need—and probably not even the second digit.*

*For the Census data in Table 643 of Hanson and Marks (1958), here are the individual ρ's, multiplied by 10^4, for entries 1–97 (but No. 70 missing), in rows of 20 items each:

(1)	102, 40, 165, 43, 609,	330, 7, −1, 278, 9,	12, 2, −9, 30, 40,	26, 1, 1, 5, 14,
(21)	18, 17, 15, 16, 17,	7, 7, 1, 9, 14,	24, 26, 24, 19, 18,	17, 9, 9, 5, 107,
(41)	67, 30, 14, 378, 44,	249, 74, 145, 128, 37,	101, 300, 461, 151, 125,	64, 60, 27, 543, 40,
(61)	27, 37, 38, 13, 155,	16, 175, 27, 3, x,	49, 169, 19, 601, 98,	92, 31, 615, 298, 86,
(81)	50, 599, 313, 246, 9,	160, 59, 60, 87, 166,	110, 21, 345, 99, 16,	148, 267, x, x, x,

To measure the effects on the differences between the means of two subclasses we had to improvise approximate methods. To make these results comparable we began by computing the "effect" of interviewer variance as the ratio of the actual variance to the simple random variance:

$$e(\bar{y}) = \frac{\text{var}(\bar{y})}{s^2/n}, \text{ for which we compute } s^2 = \frac{1}{n-1}\sum^{a}\sum^{n_i}(y_{ij} - \bar{y})^2,$$

and

$$\text{var}(\bar{y}) = \frac{1}{n^2}\frac{a}{a-1}[\sum^{a} y_i^2 + \bar{y}^2 \sum^{a} n_i^2 - 2\bar{y}\sum^{a} y_i n_i]. \tag{6.5}$$

This last is the variance of the ratio estimator $\bar{y} = y/n$ of a randomly selected clusters. The computed effect on the variance is then expressed as $e(\bar{y}) = [1 + {}^*\rho(n/a - 1)]$ and from this we get

$${}^*\rho = [e(\bar{y}) - 1]/[(n/a) - 1]. \tag{6.6}$$

The values of $e(\bar{y})$ are presented in column 5 of Table 7 and those of ${}^*\rho(\bar{y})$ in column 4. These values of the synthetic ${}^*\rho(\bar{y})$ differ little from the values of ρ computed from the *anova* table and shown in column 3. Most of these differences between the values in columns 3 and 4 arise not from using different methods but from changes in the sample base: people who did not belong to either subclass were inadvertently excluded from all calculations in Table 7, except those of column 3, which come from Tables 3 and 4.

The values of $e(\bar{y}_1)$ and $e(\bar{y}_2)$ in columns 6 and 7 of Table 7 are the effects on the variances of the means for the two subclasses and are computed with (6.5) and (6.6) similarly to $e(\bar{y})$ in column 4. We computed actually $e(\bar{y}_1) = \text{var}(\bar{y}_1)/(s^2/n_1)$, and similarly for the other subclass; that is, we did not bother to compute separately s_1^2 and s_2^2, because they would not have differed enough from s^2 to justify that extra labor. Then we computed with (6.6) for each of the two subclasses the synthetic ${}^*\rho$. These values are shown in columns 8 and 9 of Table 7.

The variance of the comparison or difference $(\bar{y}_1 - \bar{y}_2)$ was computed, taking into account the covariances within the workloads of the a interviewers, as:

$$\text{var}(\bar{y}_1 - \bar{y}_2) = \text{var}(\bar{y}_1) + \text{var}(\bar{y}_2) - 2\,\text{cov}(\bar{y}_1, \bar{y}_2). \tag{6.7}$$

As with the variances, the covariance is computed for the ratio estimator of a randomly selected clusters:

$$\text{cov}(\bar{y}_1 \bar{y}_2) = \frac{1}{n_1 n_2}\frac{a}{a-1}[\sum^{a} y_{1i} y_{2i} + \bar{y}_1 \bar{y}_2 \sum^{a} n_{1i} n_{2i} - \bar{y}_1 \sum^{a} y_{2i} n_{1i} - \bar{y}_2 \sum^{a} \bar{y}_{1i} n_{2i}]. \tag{6.8}$$

From the two variances and the covariance we computed the variance of the difference $(\bar{y}_1 - \bar{y}_2)$. Then taking $s^2(1/n_1 + 1/n_2)$ as the variance of the difference without the interviewer effect, we computed the effect on the difference as

$$e(\bar{y}_1 - \bar{y}_2) = \frac{\text{var}(\bar{y}_1 - \bar{y}_2)}{s^2(1/n_1 + 1/n_2)}. \tag{6.9}$$

These values (entered in column 10) can be seen to be distributed rather symmetrically around and near 1.0. This consequence of the additivity of interviewer effects on the compared subclasses represents empirical results, not tautologies due to theoretical necessities. On the contrary, we possess many results of similar computations into the effects of clustered sampling on comparisons and these effects typically do not vanish, but result in ratios averaging more than 1.0.

Finally as a heuristic device, we computed a "synthetic $*\rho$" defined, in analogy with the preceding measures as:

$$\text{var}(\bar{y}_1 - \bar{y}_2) = \frac{s_1^2}{n_1}[1 + *\rho(\frac{n_1}{a} - 1)] + \frac{s_2^2}{n_2}[1 + *\rho(\frac{n_2}{a} - 1)] \tag{6.10}$$

and again using s^2 for both s_1^2 and s_2^2, solved this for

$$*\rho = [\frac{\text{var}(\bar{y}_1 - \bar{y}_2)}{s^2} - \frac{1}{n_1} - \frac{1}{n_2}] \div [\frac{2}{a} - \frac{1}{n_1} - \frac{1}{n_2}]. \tag{6.11}$$

These values appear in column 11 and in the (O) marks of Chart 1; they appear to vary around the *a priori* mean of zero—though rather widely and irregularly.

7. SOME PROBLEMS OF DESIGN

Research on interviewer variability can be designed to different degrees of symmetry and completeness. A very complete design might call for the simple random selection of equal numbers of interviews; the effects of other sources of errors, especially coding error, would be included in a symmetrical, clean (orthogonal) design; the questions could be chosen to test various hypotheses about them. The Gales and Kendall (1957) research lies near the upper end of the scale of completeness. Perhaps the lower end of that scale of completeness consists in simply contrasting the results of the interviewers without making further attempts at design and merely assuming randomization of respondents among interviewers without actually providing for—as in Veroff et al. (1960). This procedure leads to overestimating the interviewer effects insofar as convenience or preference led interviewers to select different types of respondents; a component due to the average differences among those types is confounded with the component for interviewer variability. This overestimation could be great in those sampling operations where the interviewer has wide latitude in choosing his workload; it might be small in surveys carried out at one limited site, where an approximation to randomization occurs automatically.

In our two studies randomization procedures were carefully carried out. To be practical, these were designed to minimize costs and interference with field operations. We sacrificed chiefly:

(a) equal size workloads which would have resulted in somewhat simpler computations and more efficient estimates;
(b) elimination of the complications arising because the randomized set (the workload for a day in the first study and for a week in the second study) is difficult to treat in the analysis; and
(c) the possibility of separating the components of the variance due to coding variability by randomizing coders in a neat design.

Perhaps we most regret not having achieved the last of these three improvements, which modest financial aid could have bought. But the procedure for assigning interviewers to coders did not depart enough from random to interfere seriously with our analysis; we checked the distribution of coders against interviewers and found it as even as a random assignment would have made it. In defense we plead that the choice was between little or nothing—as it often is.

With increased resources it is possible to climb higher toward a more symmetrical and complete design. Nevertheless, we are convinced of the desirability and economy of allocating near the lower end of the scale with the typically limited resources available for research in interviewer variability in surveys. This is not merely *post hoc* justification for our own work, but a belief based on expectations that a few—because expensive—"crucial experiments" will not yield definitive evidence about a small set of "basic parameters," because such don't exist. It is more likely that interviewer errors differ greatly for various characteristics, populations, designs and resources (this last including questionnaires, the nature and training of interviewers, etc.). Therefore knowledge about this source of variation, as about sampling variation, can be accumulated only from a great deal of empirical work spread over the length and breadth of survey work. This implies, together with the necessarily limited total means for this kind of research, that most research in this area must be done with marginal cost as appendages to the main aims and design of surveys. Hence our general strategy should aim at many investigations of modest scope. These should be widely communicated, at first in articles and later, when they become commonplace, in appendages to survey reports.

DOMAIN ESTIMATION

CHARLES ALEXANDER AND GRAHAM KALTON[1]

Most of the textbook and other literature on survey sampling deal with the estimation of parameters, particularly the mean and total, for the total population. Yet, in practice, survey analysis is much concerned with estimating parameters for various subclasses of the population. Thus all the tables that are found in survey reports present results for subclasses, as well as for the total population. Furthermore surveys are multipurpose, aiming to provide the data needed to produce estimates for many related aspects of the subject under study and their interrelationships.

One enduring theme in Kish's work concerns the design and analysis of multipurpose surveys that produce many estimates. Several of his papers draw attention to the different effects that stratification and clustering have on subclass estimates, and on differences between subclass estimates, as compared with their effects on total sample estimates, and the implications for the design of multipurpose surveys (Kish 1961a, 1969, 1976, 1980).

Many government surveys are required to provide not only national estimates but also estimates for separate population domains. Often these domains are defined geographically—such as regions or provinces—but this need not be the case. To produce domain estimates of adequate precision is likely to require the use of higher than average sampling fractions in small domains. However, this oversampling of small domains generally results in a loss of precision for national estimates and for estimates for subclasses that cut across these domains. When some domains are relatively very small, as is often the case, this loss of precision can be serious. Kish (1988a) discusses this issue in the paper entitled "Multipurpose Sample Design" that is reproduced as Chapter 13. In many aspects of sample design, it is often possible to find a compromise sample allocation that will satisfy competing objectives well, whereas the allocation that is optimal for one of the objectives may serve the other objective(s) poorly. In this paper Kish proposes an effective compromise allocation for the domain problem, and illustrates it with an application in which the domain estimates are estimates for Canadian provinces.

---

[1] Charles Alexander died in August 2002. Until his death, he worked at the U.S. Census Bureau. Graham Kalton, Westat, Rockville, MD 20850. grahamkalton@westat.com

Purcell and Kish (1979, 1980) developed a classification of domains by size: major domains comprise around 1/10th of the population or more, minor domains comprise between 1/10th and 1/100th of the population, mini-domains comprise between 1/100th and 1/1000th of the population, and rare items comprise less than 1/1000 of the population (see also Kish 1987 a, b)[2]. Standard survey estimation procedures can usually be employed to produce reliable estimates for major domains, but rarely for mini-domains and often not for minor domains. Rare items require either separate lists or censuses.

The growing demand for survey-based estimates for minor and mini-domains (e.g., many of the states and all of the counties in the United States) led Kish to pursue research in two directions. One direction took him into the field of small area estimation. The basis of small area estimation is to use data on related auxiliary variables that are available at the small area level in a statistical model to predict the small area estimands of interest. In the 1970s Kish directed three doctoral dissertations in this field (Erickson 1971, Kalsbeek 1973, and Purcell 1979; see also Ericksen 1973, 1974). He published two papers with Purcell (Purcell and Kish 1979, 1980), one of which, entitled "Postcensal Estimates for Local Areas (or Domains)" is reproduced as Chapter 14. This paper reviews and compares the methods for local area estimation extant at that time and presents a new approach termed structure preserving estimation (SPREE).

In recent years the demand for, and production of, small area estimates from statistical models have grown enormously, and considerable research has been conducted on the methods to be used. Schaible (1996) edited a report describing eight U.S. federal statistical programs that were publishing indirect small area estimates at the time that the report was prepared. Since that time two additional programs have emerged—the Census Bureau's Small Area Income and Poverty Estimates (SAIPE) program (see the Census Bureau's Web site http://www.census.gov/hhes/www/saipe/overview/html and National Research Council 2000a, b) and the Substance Abuse and Mental Health Services Administration's program for substance abuse estimates in states and metropolitan areas (Folsom and Judkins 1997). Rao (2003) describes in detail the wide range of small area estimation models, and the alternative approaches for estimating them, that have been developed in recent years.

The second direction for Kish's research was the idea of making direct estimates for small domains by cumulating sample cases from an ongoing survey over a period of time. He advocated this idea as a way to meet the demands for more up-to-date information about small domains than can be produced from a population census. In common with many other countries, the U.S. Census is conducted on a 10-year cycle and its results necessarily take some time to produce. Therefore its small area poverty estimates, for example, have been about two years out of date when they became available and 12 years out of date by the time that the estimates from the next census were released. Substantial changes can occur in many small area characteristics over relatively short periods of time. Thus, as the

---

[2] This classification is discussed in Chapter 14.

time since the last Census increases, obsolescence becomes a major limitation to the use of its outdated small area estimates.

Kish (1979a, b, 1981, 1990a) proposed a "rolling sample design" to address this limitation. A rolling sample or census would have nonoverlapping weekly or monthly panels, each with sufficient spread to yield good estimates for all small domains of interest when the panels are cumulated into annual, quinquennial, or decennial samples. His papers described how such a design could meet the needs for information about small domains in countries with a variety of official statistical systems. The rolling sample design was adopted for the American Community Survey now being tested in the United States as a replacement for the detailed information collected from a sample of households in the decennial census (Alexander 2000.) The idea influenced the proposal for a Census Continué in France (Isnard 1999.)

The paper "Cumulating/Combining Population Surveys," reproduced as Chapter 15, reviews the history of the rolling sample idea. Issues of how to cumulate information from rolling samples are presented in the context of the larger issue of how to combine data from multipopulation, multidomain, and periodic surveys for a variety of purposes.

13

MULTIPURPOSE SAMPLE DESIGNS

LESLIE KISH

Most surveys have many purposes and a hierarchy of six levels is proposed here. Yet most theory and textbooks are based on unipurpose theory, in order to avoid the complexity and conflicts of multipurpose designs. Ten areas of conflict between purposes are shown, then problems and solutions are advanced for each. Compromises and joint solutions fortunately are feasible, because most optima are very flat; also because most "requirements" for precision are actually very flexible. To state and to face the many purposes are preferable to the common practice of hiding behind some artificially picked single purpose; and they have also become more feasible with modern computers.

Key words: Allocations to domains; mean-square-errors; multipurpose allocation; multipurpose design; optimal allocation; periodic samples; sample size.

1. INTRODUCTION

Most studies involve several purposes during the planning stages and then typically many more purposes emerge later during the analyses of data and more during their interpretation and utilization. However, the real multipurpose nature of most studies tends to remain hidden under the surface of oversimplified, univariate discussions of study designs. This seems most clearly evident for sample surveys, which I shall discuss here; but I believe that this discrepancy also holds for other statistical designs, such as experimental and evaluation studies.

In practice, surveys are usually multipurpose. Why then are multipurpose designs neglected in sampling theory? Because multipurpose theory would be too complex and difficult, and sampling theory is rather complex already; specific exceptions will be noted later. Even the descriptions we read of actual sample designs tend to follow and to borrow the prestige of univariate and unipurpose sampling theory, rather than to portray faithfully the many compromises of complex reality. Many common designs (especially equal probability of selection

---

Reprinted with permission from *Survey Methodology* (1988), 14, 1, 19–32.

methods) probably serve robustly a variety of purposes; *explicit* planning of multipurpose designs seems to be rare, though much needed, I propose.

There are several aspects to the *multipurpose nature* of survey samples, and these are displayed in a hierarchy of *six levels* in Section 2. Then *ten areas of conflict* between purposes are specified in Section 3. Sections 4 to 9 deal with specific areas of conflict, presenting approaches to and solutions for them. Some of these solutions are attributed to widely dispersed articles of survey sampling; but others are more novel, hence less fully developed, derived, and referenced.

In this overall review I aim first and foremost to serve practitioners with handy references on approaches, methods and procedures for multipurpose designs; to alert them both to the importance and to the feasibility of such designs. Second, I also wish to provide a framework for integrated, theoretical future work on the many problems and conflicts of multipurpose designs. Imperfections of my methods can serve as stimuli to others for better derivations for them, as well as for developing new methods.

2. A HIERARCHY FOR LEVELS OF PURPOSES

To begin with, we need some clarification of the meaning of "multipurpose," because too many concepts are confused under this term in our literature. To reduce the confusion, I classified a score of purposes into six levels in Table 1. Most of the time either multiple variables or multi-subject surveys (levels 3 or 4 in Table 1) are discussed and "multi-subject" (4) has sometimes been distinguished from multipurpose (3) for the same or closely related variables (Murthy 1967). Each of these six levels is shown in several specific manifestations, which can be usefully augmented and discussed in more detail elsewhere (e.g., United Nations 1981; Lahiri 1963).

Integrated survey operations on level 5 are related to, but should be distinguished from multi-subject surveys, because they refer to organizations and institutions that conduct many surveys in diverse fields over longer periods of time (United Nations 1981; Foreman 1983). An earlier name was "continuing survey operations," when it was recognized that most large-scale, widespread sample surveys were conducted by continuing survey organizations like the U.S. Census Bureau, Statistics Canada, or our Survey Research Center. Such continuity has large advantages in costs and quality, with restraining effects on sample designs (Kish 1965a).

Master frames or master samples on level 6 refer to further extensions and specializations of multipurpose approaches. They may refer simply to using the same maps, or block listings, or area segments for several different surveys; or to the large-scale example of the "Master Sample of Agriculture" (King and Jessen 1945), where rural areas on the maps of all the counties of the USA were divided into segments of about four farms each; or to the firm that sells current listings of dwellings for most samples used in Western Germany. These very diverse examples have common bases in the savings from sharing the "startup" costs (of design, stratification, listing, etc.) for constructing sampling frames.

TABLE 1. Hierarchy of Purposes

1. Diverse statistics from the same variables
   - Totals or means or medians and quantiles, distributions
   - Analytical statistics: regressions, categorical analysis
   - Time aspects: static, macro-change, micro-change, cumulative
2. Diverse populations and domains (subclasses)
   - Proper classes and crossclasses
   - Comparisons of subclasses
3. Multiple variables on the same subject
   - Alternative measures of one variable; e.g., of income, or unemployment
   - Diverse periods—per day, week, month, year
   - Several aspects of one subject: income, savings, wealth
4. Multisubject surveys
   - Several subjects on same schedule, interview, operation
   - Health surveys of many diseases
   - Market research for several clients, many goods
   - Agricultural surveys of many crops
   - "Omnibus" social surveys
5. Continuing, integrated survey operations
   - NSS in India, CPS in USA, NHSCP of UN
   - Separate surveys from one office and field staff
   - Common source of surveys
   - Diverse methods, costs, operations, allocations, respondents
6. Master frames
   - Several samples from one frame or set of listings
   - Separate institutions, organizations
   - Separate field staffs? Same PSUs?

Diverse statistics based on a single variable and diverse domains (levels 1 and 2) have been typically neglected in the literature of multipurpose sampling, although they are the most common, but they can have the most drastic effects and cause the most dramatic conflicts, as we shall see later. The effect of designs can be very different for statistics like medians and quantiles or regression coefficients than the effects for means and for aggregates (Kish 1961a; Kish 1965a; Kish and Frankel 1974). Furthermore, designing for periodic samples brings on new considerations (Section 8). But most dramatic effects can be seen simply for the means of small "subclasses" (e.g., as small as 0.10 or 0.01) of the entire sample, representing similar "domains" in the population (Section 5).

Each of the six levels of purposes presents different aspects for designs and each level can be fruitfully explored for more specific meanings and examples, some of which are listed in Table 1.

The difficulties of multipurpose designs, which have caused them to be neglected and avoided, are of several kinds. First, the different purposes must be formulated *explicitly in statistical terms*, so that these may serve in formulas for their comparisons and for formulated compromises; but obtaining a (complete) list of such explicit, formal terms may be the principal obstacle. Second, estimates of *variance and cost factors* are needed for each purpose. Third, for some methods values must be assigned to the *"required" precisions* for all the purposes (Section 5). Fourth, the above values and estimates must be combined in a mathematical formulation in order to arrive at the *solution of a single "optimal" design* to be actually used. The computational tasks for such solutions have been eased by electronic computers, but the conceptual and theoretical tasks remain (Section 5).

The difficulties of these tasks help to explain why discussions of multipurpose designs have been largely neglected in textbooks. However, note later references and bibliography here and in Rodriquez-Vera (1982); also Cochran (1977) and Chatterjee (1967). Furthermore, also in descriptions of actual surveys, often a single statistic (e.g., the mean) of a single principal variable is presented as *the* only (principal) purpose for the study. In the framework of multipurpose design this is equivalent to assigning zero importance to all other purposes. The unreality of this pretense may be softened by assuming that other principal purposes would result in similar allocations; but this pretense should be buttressed with calculations of the four steps above.

3. AN OVERALL VIEW OF TEN AREAS OF CONFLICT

A brief overall view of ten areas of conflict, listed in Table 2, should be useful before we look at specific problems and possible solutions for each. The list will probably not prove exhaustive, and readers may well find other areas. Even more likely, they may find within these ten areas other problems and other solutions not explored here. It would be convenient if the ten areas of conflict should be linked rationally to the twenty purposes presented in six levels; we then could reduce this presentation to, say, twenty purpose/conflict nodes or to ten level/conflict nodes. Unfortunately the areas of conflict denote a perpendicular dimension to the purpose and all (or most) of the 10×6 cells have meaningful contents.

Of this long list of ten areas of conflict fortunately not all need to be formulated for every actual sample design. I believe that possible conflicts about (a) the sample sizes m_g and about (b) the relation of biases to sampling errors should always be considered, at least informally, because they are ubiquitous. Also (c) allocation among domains and (d) allocation among strata should receive at least a brief discussion, and often more. Computing sampling errors (j) should also be done on most surveys. However, in the common case of one-time surveys, conflicts (i) about design over time need not be considered. On the other hand, in a continuing operation with a continuing sampling frame, the decisions about (e), (f), (g), and (h) (stratification, cluster sizes and measures) may have been made a long time ago for a fixed design. However, the cluster sizes (f) used in intermediate stages (blocks and segments) may be open to flexible operational changes.

TABLE 2. Ten Areas of Conflicts (a)–(j)

(a)(4)* Sizes m_g or rates f_g are needed for purposes g

$$V_g^2 = S_g^2 D_g^2 / m_g \text{ and } m_g = S_g^2 D_g^2 / V_g^2 \text{ or } f_g = S_g^2 D_g^2 / V_g^2 P_g N$$

where m_g denote subclass sizes and $f_g = n_g / N = m_g / P_g N$ denote sampling rates

(b)(4) Relation of biases to sampling errors in RMSE = $\sqrt{(\sigma^2 + B^2)}$
- The bias ratio B/σ decreases as σ increases for subclasses
-For comparisons B/σ tends to be small as B decreases, σ increases

(c)(5) Allocation of m_g among domains

$m_t = \Sigma_g m_g$

(d)(6) Allocation of m_{gh} among strata h

$m_g = \Sigma_h mgh$

(e)(6) Choice of variables for stratification
Multivariate stratification

(f)(7) Optimal cluster sizes

$D_g^2 = [1 + \rho_g (\bar{b}_g - 1)]$ where $\bar{b}_g = P_g n_t / a$ for crossclasses

(g)(7) Measures for cluster sizes

(h)(7) Retaining sampling units (PSUs) for changed subjects, measures, and strata and for diverse subjects.

(i)(8) Design over time
How much overlap? Panels? Change versus cumulation.

(j)(9) Computing and presenting sampling errors

*The numbers (4) to (9) refer to sections with treatments.

It is also reassuring to know that compromises based on statistical methods can yield quite acceptable results, for several reasons (Sections 5–8). First, because moderate departures from optimal allocation result in only small or negligible increases of variance. Curves of efficiency tend to be flat within broad areas around the optimal points; thus great accuracy from separate designs, which would not be feasible, are not needed. Second, because wide departures from optimal allocations can, on the other hand, cause moderate to large increases in variances. Thus, ignoring important purposes can result in substantial losses of efficiency for them, and therefore those purposes should be included in compromise designs. Third, compromise designs, in accord with statistical methods, can reduce drastically the potentially large losses from allocations optimized for other purposes, and with only small increases over the separate optimal designs for each purpose (Section 5).

4. SAMPLE SIZES AND BIAS RATIOS (B/σ)

These two areas of conflict, (a) and (b) in Table 2, should perhaps be considered most important overall, because they can be most dramatic. We treat them together here only because they may be closely related through the effects of subclasses. Let us begin with the familiar (simple random sampling with replacement) sample size $m = S^2/V^2$ needed to yield a "required" precision = V^2 for a sample mean $\bar{y}$, with element variance = S^2. However, the S_g^2 depend greatly on the variables and on the domains, indexed jointly with g for the mean $\bar{y}_g$; and the "required" V_g^2 may vary even more. We also include design effects D_g^2 that also vary, and thus $m_g = S_g^2 D_g^2 / V_g^2$ expresses the sample size needed for the mean of the variable g. For the mean $\bar{y}_g$ of a domain g, comprising only the proportion P_g in the population, the *overall* sample size needed for the domain becomes $n_g = m_g / P_g$, and it is more practical to formulate the needed sampling fraction $f_g = n_g / N = S_g^2 D_g^2 / V_g^2 P_g N$. The factor $(1-f)$ may be neglected or included in D_g^2. The P_g become small and critical if high precisions are "required" for small subclasses.

For comparisons of subclasses the variances increase even more: $m_g = (m_a^{-1} + m_b^{-1})^{-1} = n(P_a^{-1} + P_b^{-1})^{-1}$, with the P_a and P_b denoting proportions in the sample n (assuming $S_a^2 = S_b^2$). For example, for the comparison of two subclass means of $0.01n$ and $0.10n$, we have the "effective size" $m_g = n(0.01^{-1} + 0.10^{-1})^{-1} = n/110$. For other statistics, such as medians and regression coefficients, formulating "required" sample sizes would become complex. It is more than we may discuss here, but some numbers may probably be specified.

Considerations for subclass statistics become greatly modified if, in addition to variances σ^2, we also include the biases B^2 in the Root-Mean-Square-Error = RMSE = $\sqrt{(\sigma^2 + B^2)}$ for measures of accuracy. Figure 1 is meant to portray a common tendency in the accuracy of survey data, although great differences in the relations of biases to sampling errors are possible; reading the legend is urged here. It occurs commonly that potential biases B_1 are greater than the measurable sampling and variable errors σ_1, for the entire sample. However, on the horizontal axis the standard error σ_1 is shown to increase by a factor of about 3 for σ_2 of a subclass of about 1/10 of the total sample. For comparisons (differences) of two such subclasses σ_3 increases by about 1.4 more.

However, the hypotenuses denoting the RMSE are shown to increase much less. In $RMSE_1$ the bias B_1 is shown to dominate, and this may happen for some variables in large total samples. However, the subclass $RMSE_2$, because the bias was kept constant at $B_2 = B_1$, increased only moderately and is dominated by σ_2. This is even more true for $RMSE_3$, where the σ_3 has increased, but the biases—assumed to have the same sign, because that is a common tendency—decrease B_3 in the difference of means.

Examples of these phenomena abound everywhere and for all purposes that are listed in Table 1. We choose the best known, critical statistics of unemployment where admitted measurement biases may completely swamp the low values (e.g.,

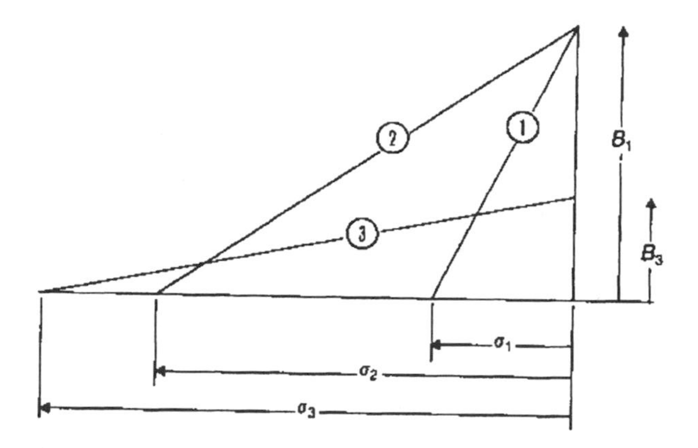

Figure 1. Variable errors (σ) and biases (B) in root mean square errors (RMSE)

The bases represent sampling errors and other variable errors (σ). For example σ_1 may be the ste($\bar{y}_t$) for the mean $\bar{y}$ of the entire sample, σ_2 may be a larger ste($\bar{y}_c$) for a subclass mean, and σ_3 may be the ste($\bar{y}_c - \bar{y}_b$) for the difference between two subclass means.

The heights represent biases (B) and the hypotenuse denotes the RMSE = $\sqrt{(\sigma^2 + B^2)}$.

(1) For the entire sample the bias B_1 may be large compared with the variable error σ_1, thus taking larger samples would not decrease the $RMSE_1$ by much.

(2) However with the same bias B_1, but with a smaller sample in the subclass, the ratio changes and the σ_2 dominates the $RMSE_2$; and this is not much larger than for (1) despite a much smaller sample.

(3) Furthermore, for the difference of means, the net bias B_3 may be much smaller; so that even with a larger σ_3, the $RMSE_3$ for the difference is but little greater than $RMSE_2$. This drastic change in the bias ratio B/σ tends to appear not only for differences between subclasses within the same sample, but also for differences between repeat surveys.

0.1 percent) of measurable fluctuations. However, for small subclasses (e.g., Black teenage boys) the sampling errors for small sample bases overtake the biases. For periodic comparisons the sampling variations become even more critical.

These relations among biases and variable errors assumed here are not logically necessary, but empirical and common. Neglect of these simple relations leads to a great deal of confusion concerning the need for sample surveys of adequate precision, i.e., with small sampling errors, σ. I propose Figure 1 as practical answers to some common questions, such as: Why do we spend money for large samples and on rigorous sampling methods in the face of large measurement biases? Why bother computing sampling errors when response biases dominate the total error? The implicit answers come from the domination of sampling errors in the subclasses, and even more in their comparisons. Let us make these implicit answers more explicit in future sample designs.

5. ALLOCATION AMONG DOMAINS

This most important and frequent area of conflict has several aspects. First, consider the allocation of total sample size (or effort or cost) among the domains that constitute a partition of the total population. A common example is allocation among the several (5, 10, 20 or 50) provinces or regions or states of a country; those domains typically have very unequal populations N_d, with ranges of 1 to 100 perhaps in relative sizes, though they may cover roughly equal surface areas. Often the question takes this form: Should the sample sizes n_d be roughly equal; or should the n_d be proportional to the N_d, with constant sampling rates $f_d = f$? Equal n_d tends to yield roughly equal errors, ste($\bar{y}_d$) for the means. On the other hand, constant $f_d = f$ tends to yield the lowest ste($\bar{y}_w$) for the overall mean $\bar{y}_w = \Sigma W_d \bar{y}_d$, because it yields lower errors for the larger domains. This error may be lower than "required" for $\bar{y}_w$, especially in view of potential biases (Figure 1), and may not justify large total sample sizes and costs. This is the contention of proponents of equal sizes n_d for provinces. However, increased sampling errors for $\bar{y}_w$ are also suffered by most other subclasses, especially "crossclasses" like age, sex, socioeconomic classes, etc., whose sizes tend to proportionality to the total. Those are common disadvantages of the highly unequal $f_d = n_d / N_d$ for provinces that result from the equal n_d values.

For example, in the Current Population Surveys of the USA, larger f_d are assigned to the smaller states. The resulting weighting increases the variances (for a fixed total cost) of the overall means and also of "crossclasses," such as young men and women, and especially of Black teenage boys and girls (with critically high unemployment rates). Similar conflicts between national and provincial needs occur in all countries, because provinces have widely different populations. The need for better provincial data, for fixed total cost, conflicts with greater precision for national and for "crossclass" statistics.

To reduce the usual confusion, I distinguish "domains" to denote partitions of the population, from "subclasses," the corresponding partitions of the sample. Then I distinguish "design domains" (and subclasses) to refer to partitions (like provinces and regions) that are contained in strata defined by the sample design, from "crossclasses" (like age, sex, occupation, income, etc.) that cut across the sample design, both clusters and strata, often almost randomly. The design effects differ for these two types of subclasses (Kish 1961a, 1980, 1987a).

In addition, other sources of conflict may arise from *domain* differences other than their sizes: in the distribution of variables, also in the variances $D_d^2 S_d^2$ precisions; but we need not enter into those complexities here. Beyond calling attention to the problems, we refer to two distinct technical methods for the joint solution of the conflicts in allocation (the fourth step noted at the end of Section 2). One approach uses iterative nonlinear programming in order to satisfy for *minimal cost* the "required" precisions jointly for all stated purposes. These elegant solutions to diverse problems exploit modern computers and have been published in many articles since 1963 (see reviews and references in Bean and Burmeister 1978, Rodriquez-Vera 1982, Cochran 1977). The "required minimal" cost often turns out much too high, because the "required" precisions were unrealistic. Then the solutions are drastically rescaled downwards. But such rescaling exposes the false pretensions (in my view) of this elegant approach that depends on unrealistic "required" precisions. Principally, I question the reality of "step functions" for "required" precisions that assign a constant value to any variance below the required V^2 and zero value to variances above it.

A very different approach calls for some form of *averaging* between all the "optimal" (preferred) allocations for various purposes, by *minimizing the combined* (weighted) *variance* either for fixed cost or fixed sample size. Of course, if the resulting combined variances turn out to be too high (or low), the solutions can be scaled up (or down) in total fixed cost or sample size. I prefer this solution, which compromises between different allocations, each of which would optimize for only one purpose (Yates 1981; Dalenius 1957). It involves assigning relative values of importance I_g to all the list statistics and this may seem difficult (but an "ignorant" decision-maker can assign equal I_g to all of them). But the other two alternatives are more extreme and they are bound to prove even more difficult: either to specify the "required" precisions of all statistics for the first approach, which then assigns arbitrarily equal weights of importance to all of them; or to specify one statistic for the total weight of one, and thus zero weights for all other statistics.

Furthermore, compromises for the average can be shown to be generally feasible and worthwhile, because the allocations are insensitive to moderate changes of weights of importance (as is often true in statistics). After all, changing the relative importance by ratios of e.g., 2 or 5 should be less drastic than assigning the total weight 1 to one variable and 0 to all others, a process that implies infinite ratios of importance.

First, denote with $\Sigma_i V_{gi}^2 / n_i$ the variance attainable for a statistic g with the allocations of sample sizes n_i for the ith component of variation. Then let $1 + L_g(n) = (\Sigma_i V_{gi}^2 / n_i) / V_g^2(\min) = \Sigma_i C_{gi}^2 / n_i$ denote the ratio of increase (with the allocation n_i) in the variance of the gth statistic over its own minimal variance, both for the same fixed Σn_i. Thus $L_g(n)$ is the *relative* loss over the minimal value of 1, and accepting the relative variances C_{gi}^2 / n_i as the functions to be minimized is a critical decision; those functions seem to me more reasonable than any others that I can imagine for the functions to be combined in (5.1) below. For example, I prefer them to the V_{gi}^2 which depend on arbitrary units of measurement, which are removed by the $V_{gi}^2(\min)$. But in rare cases we may be faced with $V_{gi}^2(\min) = 0$ or very small and this may make C_{gi}^2 wildly large and unstable; in these cases assign arbitrary values to the C_{gi}^2 or to the I_g below. These and the following including Table 3 are developed and discussed by Kish (1976).

Then with the weights I_g assigned for relative importance of the gth statistic for any set of allocations n_i of the sample sizes,

$$1 + L(n) = \Sigma_g I_g (1 + L_g(n)) = \Sigma_g I_g \Sigma_i C_{gi}^2 / n_i = \Sigma_i \Sigma_g I_g C_{gi}^2 / n_i = \Sigma_i Z_i^2 / n_i. \quad (5.1)$$

TABLE 3. Loss Functions (1 + L) for Two Populations (Kish 1976)

	(A) (1 + L) for $W_1/W_2 = 4$			(B) (1 + L) for 133 countries: 0.2 to 100 mn			
						Joint with weights	
Allocations m_i	$\Sigma W_i \bar{y}_i$	$\Sigma \bar{y}_i / 2$	Joint	$\Sigma W_i \bar{y}_i$	$\Sigma \bar{y}_i / 133$	1:1	I_c/I_d:1
mW_i	1	1.56	1.28	1	6.86	3.93	
m/H	1.36	1	1.18	3.34	1	2.17	
$\propto \sqrt{W_i}$	1.08	1.125	1.102	1.35	1.54	1.44	
$\propto \sqrt{(W_i^2 + H^{-2})}$	1.116	1.080	1.098	1.31	1.28	1.295	
$\propto \sqrt{(0.5W_i^2 + H^{-2})}$				1.47	1.17	(1.32)	1.27
$\propto \sqrt{(2W_i^2 + H^{-2})}$				1.20	1.44	(1.32)	1.28
$\propto \sqrt{(4W_i^2 + H^{-2})}$				1.12	1.66	(1.39)	1.23

In (A) there are two strata and domains ($W_1 = 0.8$ and $W_2 = 0.2$); note that the allocation $m_i \propto \sqrt{W_i}$ does almost as well for the joint loss as the optimal.
In (B) we have the populations of 133 countries, ranging in size from 0.2 to over 100 millions, a range of 500 in relative sizes. From this problem of allocation (for the World Fertility Survey) we omitted, for practical reasons, the four largest countries and a few under 0.2 millions. Their inclusion would raise the variance of relative sizes, W_i, from 2.5 to 12, and would make the results more dramatic. Note that the $\sqrt{W_i}$ allocation reduces losses quite well. Some compromise is better than none. But the optimal allocation, $\sqrt{(W_i^2 + H^{-2})}$, is considerably better. Different values of I_c/I_d(= 1/2, 2/1 and 4/1) increase slightly the variance of the joint loss function with (1:1) weights; but they remain steady for joint loss functions with their own weights I_c/I_d:1.
Two examples in Table 3 illustrate the surprisingly good compromise between conflicting allocations yielded by the method of weighted averaging: its results on the fourth row of Table 3 compare very favorably with the others. The reasons for the excellent results come from the very broad flat surfaces for the optimal allocations, as discussed in Section 2 and shown elsewhere (Kish 1976; Kish 1987a). For example, in Canada the 10 provinces vary seventy-fold from smallest to largest population sizes, and thus resemble (B) in Table 3; they serve as a graphical illustration in Figure 2. (See also Fellegi and Sunter 1974.)

After changing the order of summation, we created the new variables $Z_i^2 = \Sigma_g I_g C_{gi}^2$. This function may be minimized to give compromise solutions for fixed total cost $\Sigma c_i n_i$. For the conflict between $n_d = n/H$ of equal sample sizes for domains versus $n_d = nW_d$ proportional to domain sizes W_d, the optimal compromise allocations are found to be proportional to $\sqrt{(W_d^2 + H^{-2})}$, with equal values for I_g.

An important example was provided by the (otherwise excellent) World Fertility Surveys, which used roughly equal sample sizes for small and large countries: actual sample sizes varied only within the range of 3 to 10 thousand and

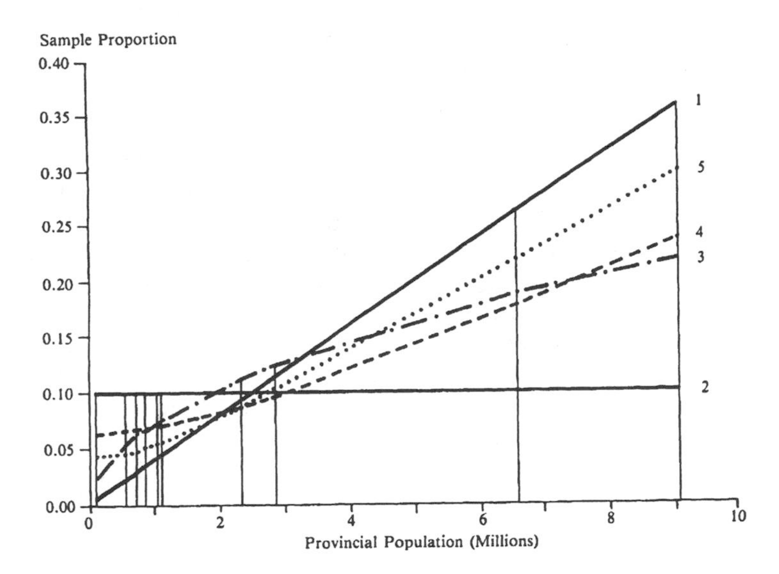

Figure 2. Five Alternative Allocations of Sample Sizes n_h of Fixed Total Σn_h
The ten provinces of Canada illustrate graphically the usual conflicts from major domains with unequal sizes, also the feasible successful compromises.

1. Allocation proportional to domain sizes $n_h \propto W_h$ is diagonal.
2. Equal allocation $n_h \propto 1/H$ is a horizontal.
   Divergences of the two allocations are large near the ends.
3. The square-root allocation, $n_h \propto \sqrt{W_h}$ yields compromises at both ends.
4. The "optimal" allocation $n_h \propto \sqrt{(W^{-2} + 1/H^{-2})}$ improves both ends, and especially with an appealing "floor" near the lower end.
5. A "weighted" optimal $n_h \propto \sqrt{(.8W^2 + .2/H^2)}$ improves the upper end considerably.

with no discernible correlation with population size. Consequently, there were two- or three-fold increases of variances in the continental averages of national surveys, their "main contributions to knowledge":

> "So far, the main contribution to knowledge has been to confirm the downward trend in fertility that characterized much of Asia and Latin America in the 1970s and to highlight the contrast with Africa where both fertility and the desire for large numbers of children remain high" (Macura and Cleland 1985).

6. ALLOCATIONS TO STRATA AND CHOICE OF STRATIFIERS

Domains and strata often get confused in discussions, but the two aspects should be kept distinct in practical work on designs. Domains refer to subpopulations for which separate estimates are sought, whereas strata are usually smaller partitions created for decreasing variances. For example, within provinces as domains more strata may be created to reduce province variances; but cross-domains like age, sex and economic status tend to straddle across the strata. Allocations of sample sizes to strata, though often not as crucial as allocations to domains, may be important in case of efficient disproportionate optimal allocations. The two methods of Section 5 for allocating sample sizes to domains can also be applied to allocations to strata, although the aims differ. Some of the references on nonlinear programming refer to domains and others to strata, and some confuse the two.

The presence of several variables and statistics among the purposes have clear implications for using more stratifying variables. Different survey variables will tend to have diverse optimal relations with the stratifiers; then it is best to use many stratifiers, even if each stratifier is used with only few stratum divisions (categories). Multipurpose design is the best reason for multivariate stratification (Kish and Anderson 1978). It may also best justify the need for "controlled selection" methods. The choice of stratum boundaries, called "optimal stratification," is a related topic, but of less importance in this condensed presentation.

7. CLUSTER SIZES; MEASURES OF SIZE; RETAINING UNITS

In descriptions of sample designs we find sometimes that the design effect has been approximated with $D_g^2 = [1 + \rho_g (\bar{b}_t - 1)]$, where ρ stands for a synthetic intraclass correlation of the "most important" variable g and $\bar{b}_t = n/a$, the average cluster size. This would yield the effective element variance $S_g^2 D_g^2$ and the variance $S_g^2 D_g^2 / n$ for the mean of the variable g. However, we must question the contents of n and of $\bar{b}_t$. If our population consists of married women of childbearing age, they may be only 10 percent of total persons and found in only 30 percent of dwellings; and much fewer than that for some rare populations. This situation has been treated in sampling for rare traits.

> Ordinarily we avoid large clusters, because of their adverse effects on the variance. But even large clusters of the entire population will yield only small clusters of a rare trait, if this is widely spread. For example, entire blocks may be sampled for persons over 65 years of age; entire villages may be searched for persons with an identifiable disease. If, on the contrary, the trait is concentrated in small areas, those areas often can be recognized and stratified accordingly (Kish 1965a).

In multipurpose designs, the crossclasses of the sample will be of variable sizes that are portions of the total sample size n_t, with $\bar{M}_g$ as their different proportions in the populations. Thus we want to estimate in the design not only $[1+\rho_g\ (\bar{b}_t - 1)]$ for diverse variables g for the total sample n_t, but also $[1 + \rho_g (\bar{b}_g - 1)]$ for many crossclasses. Here, as in Section 6, the index g is made to serve both variables and subclasses in order to simplify notation. Then we make use of some conjectures that have been shown to be good approximations in thousands of empirical computations for scores of samples:

$$[1+\rho_g(\bar{b}_g - 1)] \simeq [1+\rho_g(\bar{M}_g\bar{b}_t - 1)] \simeq [1+\rho_t(\bar{M}_g\bar{b}_t - 1)] \qquad (7.1)$$

That is, we use $\bar{b}_g = \bar{M}_g\bar{b}_t$ and $\rho_g \simeq \rho_t$ as rough approximations. True that this somewhat underestimates the average values of D_g^2 for crossclasses, because of variations in cluster sizes of crossclasses. But that is a small factor compared to the large variations of ρ_g between variables (Kish 1987a; Verma, Scott, and O'Muircheartaigh 1980; Kish, Groves, and Krotki 1976), and that underestimate has small effects on the efficiency of designs. It is important to consider efficiencies of estimates for subclasses as well as for the entire sample; these considerations point to considerably higher efficiencies for larger clusters than would be shown for $\bar{b}_t$ and n_t for the total sample only.

Measures of size are related to cluster sizes, but differ because of errors in the available measures, due especially to different population contents and to obsolescence. We must also note problems concerning measures of size for multisubject surveys and for "integrated survey operations" for different populations, which may especially need drastic compromises. Those two levels of purposes (Table 1) should be distinguished because multisubject surveys use single samples in one operation; but integrated survey operations may use different sizes of sampling units for different surveys (United Nations 1981). For example, consider integrated designs for total populations and for agriculture; also perhaps for ethnic subpopulations; also perhaps for industrial or business activities: the measures of size for each of these may differ greatly. Yet some compromise solution may be found to yield reasonable efficiencies for each.

Measures of size are also closely related to problems of "Retaining units after changing strata and probabilities" (Kish and Scott 1971). Those methods were designed to deal with changes over time of sampling units, both in measures of size and in stratifying variables; but the methods are also relevant for differences between survey variables:

> Unequal selection probabilities are often assigned to sampling units. Our methods, though more generally applicable, are especially needed for the selection of primary sampling units for surveys. Often these are selected separately from many strata, with one selection from each stratum.
>
> After the initial selection the units may be used for many surveys over several years. But as time passes, the needs of new surveys may be better served by new strata and new selection probabilities, based on new data, than by those used for the initial selection. The difference between initial and new data may be due to differential changes among the sampling units as revealed by the latest Census. Or the differences may be due to changes in survey objectives and populations; for example, a sample initially designed for households and persons may later be required to serve a survey of farmers, or college students. *Obviously our methods are also applicable to designing simultaneously a related group of samples with differing objectives.*

This method allows for using the best measures (for size and for strata) separately for each sample purpose, but maximizing the retention of the overlap of sampling units between the samples for separate purposes (especially PSUs). However, it would be possible to design a compromise that would average the measures in order to achieve a complete overlap of units, but sacrificing some efficiency for each of the purposes. A compromise between the two techniques may be even better than either: increase the overlap with small sacrifices of separate efficiencies by recognizing only differences of measures that surpass some arbitrary minimal criteria (Kish and Scott 1971).

8. PURPOSES AND DESIGNS FOR PERIODIC STUDIES

Periodic studies provide areas of conflict with great and growing importance as their numbers and sizes increase. It is wrong to assume that those expensive and influential surveys have only one of the five purposes listed in Table 4, because usually they are needed for several or all, if the design permits their use.

In Table 4 we note five purposes and six designs. The first four are paired with similar letters on the same four lines. These pairings call attention to designs that best serve, with reduced variances, each of the four purposes. Most periodic studies have several purposes and thus we should face, and perhaps solve, the difficult problems of multipurpose designs. Actually current levels (A) and net changes (C) can be served with any of the six listed designs, but with some increase in variances or in costs. However, individual (gross, micro) changes (D) need panels; and cumulations (B) need some changes of samples, and are fastest without any overlaps. For current levels (A) variances can be somewhat reduced with estimators using correlations from partial overlaps. Net changes (C) benefit from correlations from any overlap, and most from complete overlaps (Cochran 1977, Kish 1987a; Kish 1965a). Reasonable compromises often become possible, when purposes can be defined. However, extraneous considerations may rule out some designs (e.g., overlaps may be either prohibited or enforced) and thus force the use of less efficient—but still valid—designs.

TABLE 4. Purposes and Designs for Periodic Samples

Purposes	Designs	Rotation scheme
A. Current levels	A. Partial overlaps $0 < P < 1$	abc–cde–efg
B. Cumulations	B. Nonoverlaps $P = 0$	aaa–bbb–ccc
C. Net changes (means)	C. Complete overlaps $P = 1$	aaa–aaa–aaa
D. Gross changes (individual)	D. Panels	same elements
E. Multipurpose time series	E. Combinations, SPD	
	F. Master frames	

The chief variation in these six designs concerns the amount (and kind) of overlaps between periods. The rotation scheme of complete overlaps shows, with aaa–aaa, that the periods have all common parts; the nonoverlap with aaa–bbb shows none; and the partial overlap abc–cde–efg shows c and e as 1/3 overlaps between succeeding periods only. This section concentrates on the effects of varying proportions of overlaps P in diverse designs on different purposes; in complete overlaps $P = 1$, in nonoverlaps $P = 0$, and in partial overlaps $0 < P < 1$. The purposes are discussed in terms of variances for estimated means, because means (and percentages, rates, proportions) are both the most used and the simplest estimates. Effects on other estimates will not be entirely different but they are too many, diverse, and difficult to be explored here.

More discussions of panels is also available elsewhere, with its advantages, disadvantages, problems and solutions (Duncan and Kalton 1987; Kish 1987a). I call attention to SPD, or Split Panel Designs, that I am trying to promote for statistical designs. These would combine a panel sample P with new rotating or "rolling" samples, so that Pa–Pb–Pc–Pd would symbolize the periodic samples. The rolling samples a, b, c, d etc., could be cumulated into larger samples. The panel P serves primarily to provide micro (individual gross) changes. But it also serves as the partial overlap for better estimates of both current levels and macro (mean, net) changes *for any pair of periods.*

9. COMPUTING AND PRESENTING SAMPLING ERRORS

It seems questionable to include this topic under design, but I have no doubt that this is a multipurpose problem. The strategies for computing and presenting sampling errors deserve separate listing as an area of conflict among the many statistics given generally for the results of surveys. It is not enough to present standard errors for only one or a few of the most important statistics: they are too many and too diverse. Because of that diversity, the practice has grown up to compute from the variances other expressions of sampling variability, especially estimates of the "design effects" d_g^2; also sometimes from the $d_g^2 = 1 + \rho_g(\bar{b}_g - 1)$, estimates of the synthetic intraclass correlation ρ_g.

Briefly, I advise:

(a) Compute sampling errors for many variables, because the variances, the design effects (d_g^2), and the intraclass coefficients (ρ_g) can and do differ greatly between variables.
(b) You may have to do some averaging of sampling errors, because it may be inconvenient or confusing to present them all.
(c) It may be neither feasible nor necessary to compute sampling errors for all subclasses, because they can often be approximated with reasonable models.
(d) It is necessary to present sampling errors for subclasses and for other statistics to guide the readers of the reports (Kish 1965a; Kish 1987a;

Verma, Scott and O'Muircheartaigh 1980). I hope that this topic will receive in the future from theorists and methodologists some of the attention it needs.

10. CONCLUSIONS

For the ten areas of conflict of Section 3 approaches and solutions are proposed in Sections 4 to 9 that are very diverse. Averaging allocations among domains in Section 5 seems to give surprisingly good compromise solutions. The advice in Section 6 to use more stratifiers can also yield worthwhile gains. In Sections 4 and 7 considerations for subclass estimates lead to drastically different decisions for sample designs. In Section 8 we note how periodic designs can be best suited to purposes, and the best compromise for multipurpose aims. We looked at the different levels of purposes and at the various areas of conflicts jointly. Asking the right question is the core of most problems. I propose multipurpose design as a new paradigm, to replace "optimal" solutions to artificially partial questions such as: What is the optimal allocation for the mean $\bar{y}$ or the total $\hat{Y}$ of "the most important" variable?

14

POSTCENSAL ESTIMATES FOR LOCAL AREAS (OR DOMAINS)

LESLIE KISH AND NOEL J. PURCELL

Summary: Demands are growing for more timely and varied statistics for administrative units and for other small domains. These demands exist in several countries now, and they are bound to spread to all others as useful methods become known. Possibilities depend on the availability of data and of other resources, but they are also being broadened with the introduction of newer methods. We present a review and classification of those methods together with an extensive bibliography. Then we describe a new method; iterative proportional fitting (IPF), which fits sample data with iterative techniques to a flexible nonlinear model that preserves specified relationships found in an earlier census. The model and method are justified within a broad theoretical framework. Great improvements over current methods are shown in empirical results, and are promised for many situations.

1. STATEMENT OF THE PROBLEM

Timely and complete social and economic data can be obtained from national sample surveys, but usually only for *major domains* of study. Domains can be local areas, often administrative units, such as geographic areas, for which separate estimates are planned, and which also tend to be partitions in the collection process, whether of censuses or of sample surveys. Or, on the contrary, they can be "crossclasses" of the population and of the sample, which cut across the partitions of the collection and the sample design; for example, age and sex categories. Less commonly used than either of the above are domains that have not been distinguished in the sample selection, but tend to concentrate unevenly in the primary sample units. Estimates for *small domains*—any "small" subclass or subdivision of the original domains of study—are generally unavailable from typical samples, and are obtained from population censuses or from administrative registers, and sometimes from special-purpose surveys. However, effective planning

First published in the *International Statistical Review* (1980), 48, 3–18. Reprinted here with permission.

of social services and other activities cannot depend on these traditional data sources; the data must be more current, more complete, and more relevant than these sources provide. Estimates for local areas as administrative units appear as the most salient and common concern for detailed data, but crossclasses and other small domains can also be important. Several of these methods are equally pertinent to both.

Since the sizes of the small domains influence the choice and applicability of methods, a classification of domains, based on their sizes, is presented here (Purcell and Kish 1979a). It can remind us of the practical differences between the types of domains and help us avoid the common mistake of considering "statistics for domains" as one homogeneous problem. The boundaries of this classification are stated very roughly to orders of magnitude and should not be taken too seriously; they depend on the variables and the statistics estimated, on the sizes of samples and of populations, on the precisions and decisions involved, etc.

1. *Major domains*, comprising perhaps 1/10 of the population or more. Examples: major regions, 10-year age groups, or major categorical classes, like occupations.
2. *Minor domains*, comprising between 1/10 and 1/100 of the population. Examples: state populations, single years of age, two-fold classifications like occupation by education, or a single small classification like the unemployed or the disabled.
3. *Mini domains*, comprising from 1/100 to 1/10,000 of the population. Examples: populations of counties (more than 3000 of them in the U.S.A.), a two-fold classification like state by work force status, or a three-fold classification like age by occupation by education.
4. *Rare domains*, comprising less than 1/10,000 of the population. Examples: populations of small local areas, perhaps all classified by various ethnic groups.

For major domains, standard estimates, basically without bias, are generally available from probability methods of survey sampling. However, frequently for minor domains and usually for mini domains, the standard methods of survey estimation break down, because the sample bases are ordinarily too small for any useable reliability, and new methods are needed. For rare domains, sample surveys are usually useless; separate and distinct methods are required (see, for example, Kish 1965a, Section 11.4). Thus this discussion is directed principally to classes 2 and 3, and we use the general term small domains to refer to these two classes throughout this paper.

2. EXISTING LOCAL AREA ESTIMATION TECHNIQUES

2.1 Symptomatic Accounting Techniques (SAT)

These techniques utilize current data from administrative registers in combination with their statistical relationships established with earlier census data. A series of techniques using diverse registers have been developed and tested by the U.S. Bureau of the Census, and several good reviews of them are available (U.S. Bureau of the Census 1975a,b; Ericksen 1971; Kalsbeek 1973). Most of these techniques deal with small area estimates of the total population, though other estimates could similarly be designed with appropriate data and techniques.

The diverse registration data used in the U.S.A. have included the numbers and rates of births and deaths, of existing and new housing units, of school enrollments, and of income tax returns. These "symptomatic" variables need not concern births and deaths directly, nor would these account for the large effects of migration. But fluctuations in the symptomatic variables should be highly related to changes in population totals or in its components. Thus school enrollments are highly related to numbers of children and of young parents, and death rates to older persons. Diverse techniques have been described: Census Component Methods I and II (U.S. Bureau of the Census 1949, 1966); the vital rates technique and the composite method (Bogue 1950; Bogue and Duncan 1959); the housing unit method (U.S. Bureau of the Census 1969); and the administrative data records method (Starsinic 1974).

Clearly the relative advantages of diverse techniques will depend greatly on the nature of the desired estimates, and especially on the kinds of registration data which are available in each country.

2.2 Synthetic Estimation

This name has been used for a form of ratio estimation which combines recent sample means, $\bar{x}'_{.g}$, for subclasses (g) at the large domain level with Census proportions $Y_{hg}/Y_{h.}$ of the subclasses in the small areas h to obtain estimated means $\bar{x}'_{h.}$ for small areas (domains):

$$\bar{x}_{h.} = \Sigma_g (Y_{hg} / Y_{h.})\bar{x}_{.g}$$

The sample size is presumed to yield useful estimates of the $\bar{x}'_{.g}$ because these are reasonably large and widespread, but not of the $\bar{x}'_{hg}$ nor of the $\bar{x}'_{h.}$. The estimates $\bar{x}_{h.}$ can refer to the general population or to some subset, related (preferably highly and linearly) to the auxiliary statistics $Y_{hg} / Y_{h.}$. These statistics can refer to the general population in the Census, or to some other source and to other populations.

For estimates of total persons and of unemployment see Gonzalez (1973); Gonzalez and Waksberg (1973); Schaible, Brock, and Schnack (1977); Gonzalez and Hoza (1978); also work from Norway by Laake and Langva (1976) and by Laake (1978).

For estimates in the health field see U.S. National Center for Health Statistics (1968, 1977a,b); Levy (1971); Namekata (1974); Namekata, Levy, and O'Rourke (1975).

A different form of the synthetic estimator is $\hat{x}_{h\bullet} = \Sigma_g \hat{x}_{hg} = \Sigma_g (Y_{hg} / Y_{\bullet g}) x'_{\bullet g}$, where $x'_{\cdot g}$ are sample estimates for subclasses g. This has been proposed and investigated in Australia by Purcell and Linacre (1976), and in Canada by Ghangurde and Singh (1977).

2.3 Regression—Symptomatic Techniques

These methods use multiple linear regressions computed on ratios of change between two Census base periods ($t = 1, 2$) and between the predicted ratio of change, R_{hi}, and the predictors r_{hi} ($i = 1, 2, \ldots, p$). The p predictor ratios are based on "symptomatic" variables, which are available from administrative registers for local areas, not only for Census years but also currently for postcensal periods.

For each symptomatic variable (i) the *proportion* for area h, $P_{h1i} = Y_{h1i} / Y_{\bullet 1i} = Y_{h1i} / \Sigma_h Y_{h1i}$ is computed for Census period 1. A similar proportion is computed as $P_{h2i} = Y_{h2i} / Y_{\bullet 2i}$ for Census period 2. The ratio of change $r_{hi} = P_{h2i} / P_{h1i}$ for the proportions in area h also equals $(Y_{h2i} / Y_{h1i}) / (Y_{.2i} / Y_{.1i})$, the proportion of the change ratio which belongs to the hth component area. A similar ratio R_h is computed for the predicted variable. Then the multiple regression

$$R_h = B_0 + B_1 r_{h1} + B_2 r_{h2} + \ldots + B_p r_{hp}$$

is computed to obtain the regression coefficient $B_0, B_1, B_2, \ldots, B_p$, linking changes between the two Census periods in the predicted population ratio of change R_h to those in the p predictor changes r_{hi}.

Then the ratio of change R'_h in the local area h occurring in the postcensal period (between periods 2 and 3) is predicted from the computed ratios of change $r'_{hi} = P_{h3i} / P_{h2i}$ in the symptomatic variables. For this equation we must use the coefficients B_i obtained from the relations observed between the earlier intercensal periods 1 and 2:

$$R'_h = B_0 + B_1 r'_{h1} + B_2 r'_{h2} + \ldots + B_p r'_{hp}.$$

The ratio-correlation techniques follow a paper by Schmitt and Crosetti (1954). References and modification (averaging of univariate estimates) are given by Namboodiri (1972). Stratification was investigated by Rosenberg (1968) and dummy variables by Pursell (1970). Martin and Serow (1978) applied the method to estimate age and sex compositions in local areas. O'Hare (1976) investigated the advantages of differences, $d_{hi} = P_{h2i} - P_{h1i}$, in the place of ratios, r_{hi}, and was supported by Swanson (1978), whereas Morrison and Relles (1975) used a logarithmic form.

2.4. Sample-Regression Methods

The regression method above uses the coefficients B_i from structural relations established from earlier Census periods, but changes in those relations bring errors into the current postcensal estimates. These biases of obsolescence in the B_i can be overcome by computing b_i from large current samples, but at the cost of sampling errors in those b_i. The ratios r'_{hi} are computed from postcensal changes in symptomatic variables for all local areas, and the R_h are computed only for each local area in the sample. Then from

$$R_h = b_0 + b_1 r'_{h1} + b_2 r'_{h2} + \ldots + b_p r'_{hp}$$

the regression coefficients b_i are estimated from the local areas in the sample. These values of b_i are then used with all values r'_{hi} to estimate R'_h for all local areas in the population, not only in the sample.

This method was investigated by Ericksen (1971, 1973, 1974); in his studies the gains from avoiding obsolescence slightly overcame losses from sampling. This balance depends on the size and quality of the samples, on the dynamics of the structural model, and on the value of the variables. The computed values of the b_i are subject to sampling variances which have two components: a between-area component when only a sample of areas is sampled, and a within-area component due to sampling within areas (Ericksen 1974).

Combinations of the method with other regressions and with synthetic estimators are possible. Nicholls (1977) added the synthetic estimate as an additional independent variable, and Gonzalez and Hoza (1978) also reported on improvements over synthetic estimates with this combination.

In the above, the same units (counties in the USA) were supposedly used both for primary selections (PSUs) and for local area estimation. But Kalsbeek (1973) proposed that greater precision may be had with a *base unit method*, which utilizes the secondary (or later) selection units. Furthermore, he avoided linear regression for greater flexibility with a suitable clustering algorithm to sort the base units into a set of homogeneous groups. Later Cohen (1978) used minimum variance stratification for homogeneous grouping.

2.5 Stein-James, Bayesian and "Shrinkage" Estimates

Here we note jointly several kinds of estimates which all involve weighted means of the sample estimates x'_h for H domains combined with an auxiliary estimator p_h in the form $\bar{x}_h = Cx'_h + (1-C)p_h$, where $0 \leq C \leq 1$.

The auxiliary p_h may be the overall mean $\bar{x}$ of the sample, and then the method is based on data from the sample alone, and needs no auxiliary data. This can be an advantage for some situations (some countries and some kinds of statistics) where auxiliary data are judged to be lacking. Valid and useful choice of C presents difficult problems of balance between increased variances and basically unknown biases. Three lines of approach are "shrinkage" methods (Thompson

1968), "empirical Bayes" methods (Efron and Morris 1973), and "Bayesian" methods (Box and Tiao 1973, chap. 7).

The auxiliary estimators p_h may also come from outside the sample. They may be synthetic or sample regression estimates, as in a study by the U.S. Census Bureau (Fay and Herriot 1977; Fay 1979). In an earlier study a method of double sampling with regression was tried (Hansen, Hurwitz, and Madow 1953, Vol. I, Section 11.5; Woodruff 1966).

Other possibilities may be advanced for future investigations with this method of potentially great flexibility. Instead of the overall mean we can use the means of large domains for estimates of the small domains within them; a kind of stratified estimator. Furthermore, multifactor classification may be used in

$$\bar{x}_h = Cx'_h + \Sigma_i C_i p_{hi},$$

where each C_i identifies and weights a different "predictor" of the variable $\bar{x}_h$, and

$$C + \Sigma_i C_i = 1.$$

3. CHOICE OF METHODS IN DIVERSE SITUATIONS

From the variety of available methods several lessons may be learned. First, that one may find among them a better method than the one he is arbitrarily using at present for small domains; this is often the passive "null" method of continued reliance on the last decennial census which may be 12 years out of date.

Second, there is no single method that is best for all situations. Great differences between countries exist in the sources and quality of data available: the scope and quality of its census; the extents, contents and sizes of its sample surveys; and especially the scope and quality of its administrative registers. However, passive and negative attitudes are generally unjustified since every country has some resources, and ingenuity and effort can find unused resources of data. These may be of apparently different origins, but potentially useful because of high correlation with population sizes. It may be possible to find diverse data for different portions of the population; in Section 2.1 we note an example from the U.S.A. with school enrollments more useful for young age groups, and death registration for old ages. Perhaps in some countries one source is highly correlated with the rural population and another with the urban, or different states or provinces may have diverse sources.

Furthermore, the choice of sources and methods should vary with and depend on the nature of the statistics, on the desired estimates, and also on the domains to which they pertain. Note also the effects of the lapse of time since the last census; this may be seen in Figure 2 in divergences between estimates. More generally, the balance between the biases of a census and the variance of a large sample survey will move in favor of the latter during the 10 years between decennial censuses. The balance will also move in favor of less accurate but more current registers. The balance of sample surveys versus censuses or registers should depend on the sample size, but a fixed sample size has relative advantages in smaller populations. A general treatment of the relative advantages of censuses, of sample surveys and of administrative registers would be too complex and too long here but obviously the choice should depend on many factors and criteria (Kish 1979a).

Finally, the choice between methods is more difficult because the "best" is often not clear even after the event. Errors of the estimates arise from biases chiefly, and "true values" are usually not available for measuring the biases directly. Tests must be combinations of empirical and model bases, often depending eventually (and uncertainly) on results from decennial censuses. Better methods and criteria must be pursued with several methods and over the long range with an evolutionary approach and with patience.

The models of Section 4 should clarify the nature of the choice, and the new methods shown in Sections 5 and 6 should prove the "best" in many situations, we believe.

4. A NEW APPROACH: STRUCTURE PRESERVING ESTIMATION (SPREE)

Most small domain estimates concern variables represented by frequencies (count data), and these can readily be structured within a framework of categorical data analysis. Thus our framework is designed for and has been tested on frequency data. But it can also be adapted to estimate non-frequency characteristics (such as per capita income), as we demonstrate later in this section.

Categorical data analysis has been used for other purposes, but it is a new method for small domain estimation. A form of it appears in a study conducted at the Australian Bureau of Statistics, as initially reported on by Chambers and Feeney (1977); and Bousfield (1977) in a related context has proposed a similar approach. In a more unified treatment, Purcell (1979) fully documents this methodology and clearly develops and formulates the properties of the resulting estimates and their link to the ordinary, synthetic estimates.

This approach can utilize wide varieties and forms of information on variables that are associated with the estimand, i.e., the variable of interest. For example, age, sex and race are highly related to rates of employment of local areas. These *associated variables*, which effectively divide the population into crossclasses (cutting across local areas), will be seen to play an important role in the proposed estimation procedure.

Two explicit assumptions concerning data availabilities are made. First, we need an *association structure*: to establish at some previous time point (usually a census) the relationships between the estimand and associated variables, at the required small domain level. Second, we need an *allocation structure*: to establish current relationships between the variable of interest and the associated variables at the large domain level, accumulated over the small domains of interest. In addition, we may also utilize other current information on the variables, such as the small domain population sizes. The current information is typically obtained from large scale sample surveys, but other sources can be utilized. In general, the data constituting both the association and allocation structures are obtained directly from hard data sources (e.g., the census), but they may also be model based.

The information constituting the association structure is completely specified by the three-dimensional contingency table $\boldsymbol{N}$ with cell counts N_{hig} ($h = 1, \ldots, H$, $i = 1, \ldots, I$, $g = 1, \ldots, G$), where the subscripts refer respectively to the hth small domain, ith category of the estimand variable, and gth category of the associated variable dimension. The allocation structure (current information) actually represents updated margins for the association structure, $\boldsymbol{N}$, and we denote the set of cell counts constituting these margins by $\boldsymbol{m}$. For example, if the allocation structure is defined by the margin of the estimand variable i and the associated variables g, then $\boldsymbol{m}$ is represented by the cell counts $m_{\cdot ig}$, where the dot subscript is used to denote summation over the subscript h for local areas.

Given the two data structures, $\boldsymbol{N}$ and $\boldsymbol{m}$, we first need updated estimates of the association structure, $\boldsymbol{N}$, and we denote these estimates by $\boldsymbol{x}$, with cell counts x_{hig}. Our ultimate interest, however, is in the margin of $\boldsymbol{x}$ for small domains by the estimand variable; that is, in the estimated cell counts $x_{hi\cdot}$, which are obtained by simply summing over the associated variable dimension, g.

Example: For a graphical illustration only, consider the simple situation of estimating the number of employed and unemployed persons for each of two counties (local areas). Race is assumed to be highly associated with employment status and is used as a single associated dichotomous variable. Diagrammatically then, our available information is represented as in Figure 1. The association structure, $\boldsymbol{N}$, in $2^3 = 8$ cells, represents the data obtained from the previous census, while the allocation structure, $\boldsymbol{m}$, given by the margin of employment status by race (summed over the counties in four cells) is derived from the current survey. In other words, we obtain current information on the back margin $m_{\cdot ig}$, in Figure 1, but such data are not available, or are of unacceptable reliability at the county level.

The procedure then is to adjust the association structure (original table) to conform to the current data in the allocation structure (updated margin), while preserving (as much as possible in some way) the relationships between the variables in the association structure, N_{hig} (as established at the previous census). The resulting adjusted table x_{hig} is not stable and it is summed across the associated variable, g, to obtain the required estimates $x_{hi\cdot}$. In terms of Figure 1, our required estimates are represented by the four cells on the right hand margin (the margin for employment status by county).

Extensions of this illustration to large numbers of small domains, also of categories of the estimand, and of categories of the associated variable dimension, are straightforward. We note that where several associated variables are used, the combination of categories of these variables are strung out in a single dimension. The important point of this approach is that the estimation process is completely specified by the two data structures; the association and allocation structures. In summary, we require our estimation procedure:

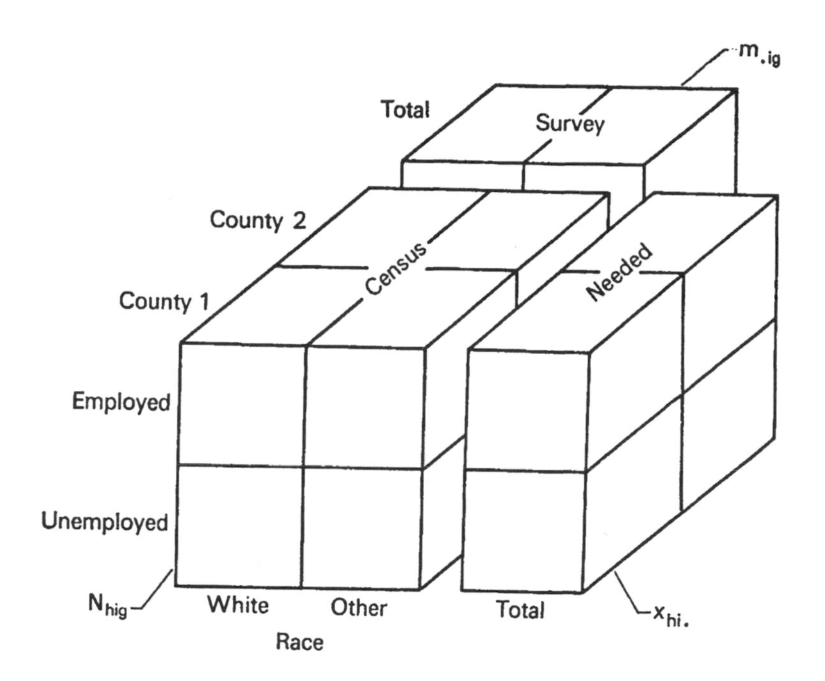

Figure 1. A simple example of the data structure for small domain estimation.

(i) to conform to the current information in the allocation structure; and
(ii) to preserve somehow the earlier relationships present in the association structure without interfering with aim (i) above.

Then depending on what information is incorporated into the association and the allocation structures, the form of the resulting estimates that satisfy (i) and (ii) will differ. For this general method we now examine the principal specific techniques distinguished by the nature of the information incorporated into the association and into the allocation structures.

(a) $\boldsymbol{N} = \{N_{hig}\}$ and $\boldsymbol{m} = \{m_{.ig}\}$.

We present this first as the most important situation: information is full (h, i, g) in the association structure, but the small areas (domains) dimension h is unavailable from the current sample data in the allocation structure. We shall show the superiority of this full structure over the ordinary synthetic estimator, the BASE estimator of model (d), in Section 5.

Several criteria can be used to estimate $\boldsymbol{x}$ in this important case and they all lead to the same estimate. Probably the simplest one is a weighted least squares approach to minimize a weighted sum of squared differences between the original cell counts, N_{hig}, and the resulting modified cell counts, x_{hig}, subject to the marginal constraints specified by the allocation structure. That is, by standard Lagrangian analysis we minimize the function

$$F(x_{hig}) = \Sigma_{hig} w_{hig}^{-1} (x_{hig} - N_{hig})^2 - \lambda(\Sigma_h x_{hig} - m_{.ig}),$$

where w_{hig} are some prespecified weights. If we choose $w_{hig} = N_{hig}$, then our estimator becomes

$$x_{hig} = \frac{N_{hig}}{N_{.ig}} m_{.ig}.$$

The resulting estimates for our variable of interest, i, in each small domain, h, are therefore

$$x_{hi} = \Sigma_g \frac{N_{hig}}{N_{\bullet ig}} m_{\bullet ig} \tag{4.1}$$

The same estimator is obtained if we use maximum likelihood, also if we minimize a discriminant information criterion (see Purcell 1979). These estimates are seen to be members of the general class of *segmented ratio estimators*, where the associated variable dimension defines the segments.

(b) $\boldsymbol{N} = \{N_{hig}\}$ and $\boldsymbol{m} = (\{m_{.ig}\},\{m_{h..}\})$.

In addition to the information in (a), there are current estimates of totals for the local areas h. This addition can greatly improve the estimates (as we will see in Section 6). Current information on the sizes of the small domains may come from registers or from sample surveys, and may be on all areas or only on a sample of them. In this case, we could proceed by using any of the criteria mentioned in case (a), but the resulting Lagrangian expressions are not very tractable; obtaining an analytic solution is difficult. An alternative approach follows from an iterative procedure of Deming and Stephan (1940) for a similar adjustment problem. Their procedure is generally referred to as *iterative proportional fitting* (IPF), although other names are also used: iterative scaling, marginal raking and marginal scaling. The IPF solution only approximates the weighted least squares solution, but it simultaneously maximizes the likelihood equation of the multinomial distribution and minimizes the discriminant information (in the context of simple random sampling assumptions).

As we have already seen in (a), when there is only one set of margins defining the allocation structure, a simple solution exists which is just a segmented ratio estimator. The approach proposed here utilizes this fact by successively adjusting the cell frequencies, N_{hig}, to agree with the two sets of marginal constraints, $\{m_{.ig}\}$ and $\{m_{h..}\}$, through the use of (4.1) in an iterative cyclical fashion. Thus the procedure is as follows:

At the first step the starting values are set equal to the cell counts specifying the association structure. That is,

$$x_{hig}^{(0)} = N_{hig}.$$

These cell counts are then adjusted to the first set of marginal constraints, specified by the allocation structure, $\Sigma_h x_{hig} = m_{.ig}$, then to the second set of constraints, $\Sigma_{ig} x_{hig} = m_{h..}$, in an iterative fashion. Thus an iterative cycle consists of two steps, and at the kth iteration we have

$$_1x_{hig}^{(k)} = \frac{x_{hig}^{(k-1)}}{x_{.ig}^{(k-1)}} m_{.ig} \quad \text{and} \quad x_{hig}^{(k)} = \frac{_1x_{hig}^{(k)}}{_1x_{h..}^{(k)}} m_{h..},$$

where $_1x_{hig}^{(k)}$ are the estimates resulting from adjusting to the marginal constraints, $\{m_{.ig}\}$ at the kth iteration. The resulting estimates, $x_{hig}^{(k)}$, are then used as inputs into the next cycle. The iteration is continued until some convergence criterion is satisfied following an iteration cycle. These estimators can be said to be members of the general class of *two-step raking ratio estimators*. Once again we sum the resulting estimates, x_{hig}, across the associated variable dimension to give our final *two-step segmented raking ratio estimates*.

(c) $\boldsymbol{N} = \{N_{hig}\}$ and $\boldsymbol{m} = (\{m_{.ig}\}, \{m_{h \cdot g}\})$.

In this case there is more additional current information, and the solution follows directly from the developments in (b). The only difference is that at the kth iteration the second step adjustment is given by

$$x_{hig}^{(k)} = \frac{_1x_{hig}^{(k)}}{_1x_{h \cdot g}^{(k)}} m_{h \cdot g}.$$

Again by summing the solution over the associated variable dimension we obtain two-step segmented raking ratio estimates.

(d) $\boldsymbol{N} = \{N_{h.g}\}$ and $\boldsymbol{m} = \{m_{.ig}\}$.

This represents a common situation when the full association structure $\boldsymbol{N} = \{N_{hig}\}$ is not available for the estimation process because the estimand data have not been collected at the earlier (census) time; for example, the use or non-use of contraceptives. Rather, the census data define an *incomplete association structure*, which is represented by the cross-tabulation of the associated variables by small domains at the previous census. Thus the information we have on the association structure is defined by $\boldsymbol{N} = \{N_{h.g}\}$. Because this structure lacks (or ignores) information for categories i of the estimand variable, a model must be used to define a *dummy association structure*. The simple and common procedure is to assume proportionality across the i categories and use

$$x_{hi} = \Sigma_g (N_{h\cdot g} / N_{\cdot\cdot g}) m_{.ig}. \tag{4.2}$$

This is a special case of the segmented ratio estimator, which we term the basic synthetic estimator (BASE). We discussed this in Section 2.2, as well as another "synthetic" estimator,

$$\bar{x}_{hi} = \Sigma_g (N_{h\cdot g} / N_{h..}) \bar{m}_{.ig}. \tag{4.3}$$

Both these estimators use only $N_{h.g}$ instead of N_{hig}, which is useful when the N_{hig} are not available. But when these are available, the estimators (4.2) and (4.3) impose a rigid model needlessly and wastefully, as we show in the increased biases in Section 6; the alternative model (a) yields considerably reduced biases in our empirical study.

(e) $\boldsymbol{N} = \{N_{h\cdot g}\}$ and $\boldsymbol{m} = (\{m_{.ig}\}, \{m_{h..}\})$.

(f) $\boldsymbol{N} = \{N_{h\cdot g}\}$ and $\boldsymbol{m} = (\{m_{.ig}\}, \{m_{h.g}\})$.

The estimators in these cases follow on directly from cases (b) and (c), except that we utilize the dummy association structure, as in (d). Thus, the only difference is that our starting values are now defined by

$$x_{hig}^{(0)} = N_{h\cdot g}, \quad i = 1, \ldots, I.$$

This approach, as we mentioned before, can be easily modified to address the problem of estimating small domain non-frequency characteristics such as total income or expenditure. Essentially, this involves redefining the association and allocation structures in terms of the total value for the variable of interest in each cell, rather than cell counts (frequencies). For example, the association structure would be defined in case (a) as $\boldsymbol{N} = \{Y_{hig}\}$, where the components Y_{hig} represent the total value for the ith category of the estimand in the hth small domain and gth category of the associated variable dimension. Usually, for non-frequency characteristics the estimand variable would only have one category. Per capita estimates can then be obtained by dividing the estimates of total for each small domain by the respective estimated (or known) population sizes, in the appropriate cells.

Each of the estimators derived above can be expressed in log-linear form as shown by Purcell (1979). This fact can then be utilized to show that all the proposed estimators preserve (carry over) all the interactions specified by the association structure except those that are respecified by the allocation structure. In other words, all the relationships between the variables that are incorporated into the association structure (past data) will hold in the current estimates, except those that are restated by the allocation structure (current data).

Consequently, as a function of this structuring of the small domain problem within a framework of categorical data analysis the resulting estimators all carry over certain interactions from the association structure. Thus all the estimators developed here can be said to belong to a class of *structure preserving estimators* (SPREE). The effects in the different cases can be compared in Table 1. In case (a) the one-way effects *i* and *g* and their two-way interaction *ig* come from the current (new) data $m_{.ig}$ but the others (*h*, *hi*, *hg* and *hig*) come from the earlier (census) data. Case (d), the BASE synthetic estimator, differs in that the *hi* and *hig* interaction are both forced to be zero in the association structure, and in the estimates.

The implementation of the SPREE estimation procedure is a straightforward process, and Purcell (1979) documents a Fortran program that carries out this estimation. Basically, it requires only the input of the association and allocation structures.

5. COMPARISON OF DIFFERENT SPREE ESTIMATORS

The variances of the SPREE estimates depend mostly on the variances of the marginal constraints specified by the allocation structure, because the association structure is derived from past census data. Thus, the variances of these estimates will, in most practical situations, be small compared to the biases which are less controllable, and will have a greater influence on the efficiency of the different SPREE estimators.

The SPREE estimates tend to be biased estimates to the degree that the underlying association structure, which gets imposed on the estimates, does not correctly represent the true structure present in the current population. The bias

Table 1. Structure Preserving Properties of Different SPREE Estimates

	Effects						
	One-way			Two-way			Three-way
Estimator	*h*	*i*	*g*	*hi*	*hg*	*ig*	*hig*
(a)	c	n	n	c	c	n	c
(b)	n	n	n	c	c	n	c
(c)	n	n	n	c	n	n	c
(d)	c	n	n	o	c	n	o
(e)	n	n	n	o	c	n	o
(f)	n	n	n	o	n	n	o

Note: *n* = new; *c* = census; *o* = zero.

question is obviously closely related both to the choice of marginal constraints imposed through the allocation structure, and to the form of the association structure used. Thus, it can be expected that the different SPREE estimates developed above will have considerably different magnitudes of bias.

Purcell (1979) has carried out an empirical investigation into the bias of the different SPREE estimates developed in the previous section. We present a brief summary of its major findings and of their implications for efficient small domain estimation. In the study, vital statistics and census data were used to obtain "synthetic" estimates of mortality due to each of four different causes (the estimands i) and for each state (domains h) of the United States. The associated variable dimension, g, was defined as 36 convenient age-sex-race groups. Estimates and biases were calculated for five individual years ranging over a full 10-year postcensal period, 1960 through 1970, so that, as the association structure becomes increasingly out of date, the performance of the different estimators could be evaluated.

Due to the nature of the data sources used, the resulting SPREE estimates have zero sampling variability and no conceptual bias. The only source of error in the estimates is the *SPREE estimate bias*, which can be measured in terms of the *percentage absolute relative difference* (%ARD) defined in this case as

$$\%\text{ARD} = \frac{|x_{hi.} - X_{hi.}|}{X_{hi.}} \times 100,$$

where $x_{hi.}$ and $X_{hi.}$ are the SPREE estimate and the true number, respectively, of total deaths by specific cause i, for the hth state. Several different SPREE estimators were evaluated in this study and they are specified in Table 2.

We use medians of the state %ARDs here (Table 3) because of the skewness of the distributions, but the means yield very similar results (Purcell 1979). It is clear that the accuracy (in terms of the method bias) of the different SPREE estimates varies considerably. Differences are also found within each estimator, both between the four causes of death studied, and with the length of the postcensal period. However, only the level of the bias is different, and the pattern of the differences is basically the same for each of the four causes of death. Figure 2, which

TABLE 2. Classification of the Different SPREE Estimates Evaluated

	Allocation structure	
Association structure	$(\{m_{.ig}\},\{m_{h..}\})$	$\{m_{.ig}\}$
N_{hig}	b	a
$N_{h.g}$	e	d

illustrates these differences for deaths due to malignant neoplasms, is typical of the pattern for the other causes of death.

The most striking feature of the analysis of the different SPREE estimates is the strong performance of the estimates based on the full association structure ((a) and (b)). While the (a) estimates (and to a lesser degree the (b) estimates) did show a tendency to deteriorate over time, they were clearly superior to the corresponding estimates based on the incomplete association ((d) and (e)). The performance of (b) compared to (a) points to the importance of incorporating through the allocation structure the maximum available current information into the estimation process. The association structure, however, must be accurate as the SPREE estimation procedure is very sensitive to approximations (and bias) in the allocation structure (Purcell 1979).

The overall performance of the estimates based on the incomplete association structure ((d) and (e)) is inferior. The performance of the commonly used BASE estimator (d), in what can be regarded as a well-behaved data set, raises some concern over its wide use. Its performance, however, is seen to be somewhat

TABLE 3. Medians of the Percentage Absolute Relative Differences of Different SPREE Estimates

		Cause of Death			
Estimator	Year	Malignant neoplasms	Major CVR diseases	Suicides	Total other
(b)	1960	—	—	—	—
	1961	1.853	0.732	6.488	1.386
	1964	2.221	1.028	8.642	2.199
	1967	3.222	1.197	6.317	3.355
	1970	2.752	2.218	8.518	3.854
(a)	1960	—	—	—	—
	1961	1.975	1.470	5.556	1.923
	1964	3.502	1.981	8.983	3.275
	1967	5.581	3.471	7.761	4.892
	1970	8.183	4.716	13.415	6.650
(e)	1960	6.310	3.362	12.162	7.357
	1961	5.620	2.963	12.941	7.163
	1964	5.783	2.513	14.444	6.152
	1967	5.588	2.215	15.238	5.033
	1970	5.606	2.739	11.943	4.867
(d)	1960	9.255	8.417	11.862	6.700
	1961	9.171	7.506	18.519	5.717
	1964	10.053	7.116	15.306	7.191
	1967	10.070	7.916	17.333	6.457
	1970	8.503	11.111	19.417	7.736

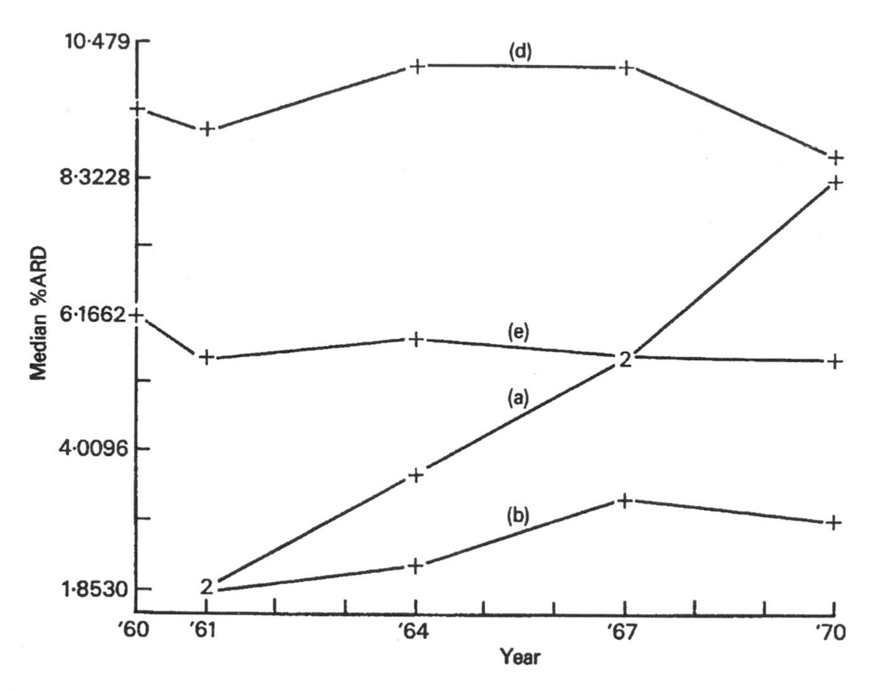

Figure 2. Median %ARD of different SPREE state estimates of the deaths due to malignant neoplasms by year.

strengthened by the use of additional accurate current information as in the (e) estimates.

Clearly, estimates with low bias can be obtained if accurate association and allocation structures are available on which to base the estimation. But when we seek to input less information and approximate these structures, either for simplicity or due to data availabilities, the performance falls away.

6. SOME MODIFICATIONS OF THE SPREE APPROACH

The use of the dummy association structure, when the variable of interest has not been collected in the previous census, is in effect equivalent to using an unsaturated log-linear model to approximate the full association structure; various interaction terms are forced to be zero. Such reduced association structures may, however, be utilized for several other reasons:

1. The full association structure may not be available because the required data were not collected in the census (e.g., in models (d) and (e) in Section 4). Or, even if collected, those data from the census were simply not

published at the required levels. For example, the three-way cross-classification, $h \times i \times g$, may not be available from census tabulations, but the three two-way cross-classifications, $h \times i$, $h \times g$, and $i \times g$ may be published. And it may be too expensive or difficult to go beyond published tabulations. Special tabulations from census tapes can be extremely costly. But from the available partially overlapping lower order cross-classifications a "dummy" association structure can be built by raking methods. In this, however, those higher order interactions that are not specified by the lower order cross-classifications are set to zero. The appropriateness of this procedure depends on whether these interactions are actually negligible. Otherwise bias results.

2. When the number of cells in the association structure is very large, it can frequently occur that many of the cell frequencies are zero or very small. This problem of "thin" data is often handled by collapsing across adjacent cells. Alternatively it could be handled by using some reduced association structure; this again usually involves setting certain interactions to zero.
3. The SPREE procedure, discussed in Section 4, preserves all the interactions specified in the association structure except those redefined by the allocation structure. However, by assuming that all "preserved" interactions are stable over the postcensal period we are introducing a significant bias, whenever unstable subsets are present. One possible approach, where we suspect "unstable" interactions, is to preserve only the interactions regarded as stable. This can be done by fitting an unsaturated log-linear model to the observed association structure, from which adjusted starting values are obtained for SPREE estimation, which reflect the required reduced association structure. However, a difficulty with this is that the "unstable" interactions are set equal to zero, and it is unclear whether this leads to more harm than leaving in the "wrong" interactions, as we do when basing the association structure on the complete saturated model.
4. It should be possible, instead of either accepting obsolete interactions or setting them to zero, to modify them according to some theoretical/empirical model. But this is for future research.

Case 1 is of prime interest concerning the question of simplified data input requirements for efficient small domain estimation, and the mortality data were used to test it (Purcell 1979). The full association structure was ignored and we assumed the past data only provided us with the three two-way cross-classifications of the variables. By raking of these margins we are effectively fitting the reduced association structure that specifies that the three-way interaction is zero. On this reduced association structure, we can then impose the two different allocation structures used in the evaluation in Section 5. The estimates based on the reduced association structure are denoted by

Estimator	*Allocation structure*
(g)	$\{m_{.ig}\}$ and $\{m_{h..}\}$
(h)	$\{m_{.ig}\}$

To investigate the equivalence of the corresponding estimates based on the full and reduced structures, the estimates were compared directly to each other. The main results of this comparison are presented in Table 4, where the median state %ARDs between the full and reduced association structure-based estimates are summarized. The %ARDs calculated are given by

$$\%\text{ARD}(1) = \frac{(\text{g}) - (\text{b})}{(\text{b})} \times 100$$

and

$$\%\text{ARD}(2) = \frac{(\text{h}) - (\text{a})}{(\text{a})} \times 100.$$

From Table 4, we see that the full and reduced association structure-based SPREE estimates are essentially equivalent, and it was only the suicide estimates that show any real differences (the maximum median %ARD was still only 2.82 per cent). However, the differences that did occur showed a slight tendency to increase uniformly over time, for all causes of death and for all comparisons. While the reasons for this are not entirely obvious, we suspect it is mainly a reflection of the increase in the variability of the estimates themselves, over time.

While these results are specific to these data, they do lend support to the contention that, in other small domain data, higher order interactions can be ignored and set to zero. Clearly, a wider scope of data needs to be examined,

TABLE 4. Medians of the Percentage Absolute Relative Differences Between Full and Reduced Structured SPREE Estimates

		Cause of Death			
%ARD	Year	Malignant neoplasms	Major CVR diseases	Suicides	Total other
%ARD(1)	1961	0.0406	0.0142	0.5450	0.0398
	1964	0.0637	0.0350	1.1594	0.0806
	1967	0.1081	0.0623	1.9084	0.1503
	1970	0.1145	0.0897	2.7027	0.1807
%ARD(2)	1961	0.0456	0.0204	0.5063	0.0381
	1964	0.0737	0.0334	1.1765	0.0517
	1967	0.1546	0.0810	1.9553	0.1282
	1970	0.1945	0.1344	2.8210	0.1514

as such a result is of importance from the point of view of reduced data input requirements for efficient SPREE estimation.

7. SOME CONCLUDING REMARKS

It should be stressed that we view the categorical data analysis approach as a valuable though not exclusive method for efficient small domain estimation. We recognize the need for composite procedures combining the SPREE estimation approach with other methods, in order to fully utilize the different strengths of the various estimation procedures. This can best be facilitated by formulating a unified framework, incorporating considerations of cost effectiveness for evaluating the different estimation strategies.

The importance of such a unified framework for future small domain estimation developments cannot be stressed enough. Among other things, this framework should incorporate resource/methods considerations relating to the use—either singly or in combination—of both direct and indirect estimation for small domains. Also important is an assessment of the cost-benefits of straightforward versus complex computational algorithms for both the small domain estimates and their subsequent evaluation.

Direct estimation can be strengthened with more current censuses (quinquennial?) or microcensuses (annual?), or with larger sample surveys, or with more complete and better registers. But various and legitimate constraints of resources are bound to intervene. Demands for timely, detailed and complete data are bound to run ahead of the supply of data. Thus there will be considerable scope and need for strengthening indirect procedures for small domain estimation; such as regression and especially SPREE methods. The framework provided here should be useful not only for better analyses, but also for pointing out the kind of data that would be most useful for better estimates. In general, the central issue involved in using indirect methods is the appropriateness of the underlying models, arising either implicitly through the basic assumptions on which the techniques incorporating this information are based, or directly through our assumptions about the data structures. Therefore, the quality of indirect small domain estimates depends largely on being able to identify associated variables that are highly related to the variable of interest, and on understanding the data structures (relationships) existing between the variables.

Such a total design approach—incorporating considerations of survey design, data analysis and data management considerations simultaneously—will enable us to better meet the increasing small domain data requirements of our societies. Within this framework, we anticipate further developments that will expand theory and applications to new problems and to new variables. The ultimate aim is more efficient procedures for combining the strengths of the diverse sources of data—samples, censuses and registers—to construct needed *synthetic small domain data bases*. Estimates derived from these data bases should be simultaneously more detailed, timely, and accurate than those currently available.

Finally *forecasting* for local areas and other small domains poses a clear and present challenge to our methods. It appears as an extension of postcensal methods which also go beyond the last census. But our postcensal methods are based on current marginal data which are missing for forecasts. However, it should be possible to base forecasts on models which forecast a continuation of the movement of current and past interactions (relationships) into national projections of total populations.

15

CUMULATING/COMBINING POPULATION SURVEYS

LESLIE KISH

Designs for and operations both of multipopulation surveys and of periodic surveys have become more common and important. The needed large resources, both financial and technical, have been organized only in recent decades, and the great values of both became recognized. For both types of designs the developments have concentrated on comparisons between surveys. Yet the coordination and harmonization needed for comparisons also makes the combinations of the survey statistics possible, desirable, and practiced. But the combinations of surveys have been achieved and presented largely without a theoretical/methodological framework, and often poorly. Here such a framework is attempted. Some closely related designs are also discussed: multidomain designs, rolling samples, combining experiments, and combining several distinct survey sites.

Key words: Multipopulation design; multidomain design; periodic surveys; rolling samples; combining experiments.

1. Multipopulation Models

A paraphrase of the standard model in all books on survey sampling goes roughly thus: "The aim of survey samples is to produce an estimate of the total Y (or the mean $\bar{Y}$) for a variable Y_i in a population of N elements." Such statements are misleading because they fail to describe the actual purposes and practices of survey sampling. First, most surveys treat many variables, and second, survey results use diverse kinds of statistics; thus sample surveys are "multipurpose" on several dimensions (Kish 1988). But instead of discussing all the omissions of the standard model, I want in this paper to concentrate only on its insufficiency and inadequacy because of its restriction to a single, finite population. Among the several examples below of multipopulation expansions that are possible with a new and different model, I begin with two important examples of survey samples

First published in *Survey Methodology* (1999), 25, 129–138. Reprinted with permission.

that achieve a variety of treatments and results on different dimensions. First is the emergence of multinational designs since 1965, best illustrated by the World Fertility Surveys (Section 3), which involve combinations across national spatial boundaries. Second are combinations of periodic samples, best illustrated by "rolling samples" (Sections 4 and 5), which concern combinations across temporal dimensions.

The designs and operations for periodic surveys require large resources and new methods. Those for multinational surveys are even more demanding. Both of these types of complex surveys are rather late arrivals among sample surveys and both types are growing in numbers and in importance. Furthermore, both types have been designed and used mostly for comparisons: temporal and spatial comparisons, respectively. The concept of using them additionally for combinations and cumulations is new, and is often encountered initially with doubt and disbelief (Sections 3, 4, 5). For both types, the variations between the populations are commonly affirmed as obstacles to combinations or cumulations, and thus are then used for restricting the sample estimates to single populations, because typically methods for combining them are unknown or unavailable. Or even when they are combined, only *ad hoc* methods are used, without justifying them. References to several papers indicate my concern for designs of multinational surveys and of rolling samples. In this paper the emphasis will be on combinations for multinational samples and on cumulations of periodic and rolling samples.

You may notice that I use the terms "cumulating" and "combining" interchangeably and perhaps confusedly. "Combining" seems to fit the multipopulation and multidomain situations better, whereas "cumulating" seems better for periodic and rolling samples. It would be better to have one word to cover both spatial and temporal combinations/cumulations, but neither seems to be exactly right. Also "combinations" serves uses other than joining populations—the usage I wish to emphasize here.

I am also not clear if it is better to consider the enlargement of the scope of samples from one population to several as a new model or as a paradigm shift. Discussions with a few philosophers here left me confused about this choice. And my fellow statisticians probably do not care whether we write the word model or paradigm. In any case, a new model instead of the standard model of sampling from a fixed frame of a stable, finite population is the radical proposal I am pursuing in this paper.

2. MULTIDOMAIN DESIGNS

Statistics for national samples are commonly based on combinations of domains, and these are often quite diverse. But because these combinations are simple and familiar, they can also serve as heuristic examples for the less familiar combinations I want to discuss, such as multinational and multiperiodic statistics. The diversity of domains may be recognized within national sample designs; e.g., provinces, which may number from 5 to 20 in most countries. In samples of smaller populations (cities, institutions, firms, etc.) similar partitions into major domains also are typical. But for smaller and more numerous domains (e.g., the 3000 counties of the USA) deliberate sample designs are not feasible for most samples of limited size. For these small domains, methods of "small area estimation" have been developed (Kish 1987a, 2.3; Platek, Rao, Särndal, and Singh 1987, pp. 267–271). There are great practical differences in both design and estimation between large and small domains, and it is careless to use the adjective "subnational" to cover both. Furthermore, these distinctions between large and small domains exist not only for national designs, but also for samples of smaller populations. It seems that the structured (nonrandom, grainy) natures of populations persist also on smaller scales. This conforms to the proposed new model of populations, and is supported with empirical analysis of multistage components (Kish 1961b).

Although practical for provinces, deliberate designs are not feasible for most domains, whether few and large or many and small, of the kind we call "crossclasses," such as sex and age or occupation, social class, education, etc. These "crossclasses" are often important both for their relations (correlations) with the survey variables and for their great diversity. Thus samples of national (or other) populations are mosaics of domains that are diverse and often highly variable; and we must depend on the properties of large probability samples to yield reliable representations of them. In this sense we perceive that all population samples consist of combinations of subpopulations.

Subclasses designate the representations in the entire sample of the domains that compose the whole population. Crossclasses are commonly the most common types of subclasses in survey analysis: partitions of the sample, for which deliberate selection designs are not feasible. For example, occupation and education classes, behavioral and attitudinal categories, and so on. These can be strong explanatory variables for survey analysis; yet we lack the data and resources not only for pre-stratification, but also even for post-stratification methods. From that extreme of lack of controls at one end, we can move to the other extreme of strong controls by separate samples, which can be designed for major provinces.

For example, different methods of sampling can be used in the different provinces. But more common are designs that use different sampling rates; for example, higher sampling rates for small, or for especially important provinces. Sometimes equal sample sizes $n_h = n/H$ are designed for all H provinces in order to obtain (approximately) equal precisions for all provinces, regardless of their sizes. This equal allocation results in sampling fractions n_h/N_h that are inversely proportional to province sizes. But for fixed total sample size n the consequences are higher variances for the entire sample, as well as for crossclasses; see Section 8 (Kish 1988). We assume here, that the statistics $\bar{y}_h$ of the provinces (domains) get weighted with population weights $W_h = N_h / \Sigma N_h$ for the overall statistics $\bar{y}_w = \Sigma W_h \bar{y}_h$, as is commonly practiced for national statistics. This serves as a useful introduction to the multinational statistics coming next.

3. MULTINATIONAL SAMPLE DESIGNS

National "representative" samples were started by Kiaer (1895a, b) only in 1895 and, after much opposition, they became widespread only after 1945 (Kish 1995b). Since then the efforts of the samplers were encouraged and supported by statistical agencies of the United Nations, especially the UN Statistical Office and the FAO. Their spread then led naturally to multinational comparisons of surveys; yet the deliberate design of multinational samples that could provide valid comparisons is recent, starting only around 1965 (Szalai 1972; World Fertility Surveys (WFS) 1984; Kish 1994). The new demands for survey designs for multinational comparisons create many new difficulties: in resources—financial, institutional, cultural; and also in methods. Those difficulties encountered with comparisons reappear also in similar form for multinational combinations, our main concern in this paper.

It is interesting to compare these difficulties with ones with which we are familiar in multidomain designs. From a theoretical perspective, combining the provinces of a country is similar to combining the nations of a continent. Indeed we should profit from those similarities by using metaphorical arguments from the familiar multidomain designs to the proposed multinational combinations. However, from a practical view we find great differences between the two efforts because of five fundamental practical obstacles that make multinational designs much more difficult to achieve, discussed below.

1. The centers of decisions reside in separate national offices, both for setting policy targets and for obtaining funds. Further, within any nation the agencies for policy setting and for resource allocation may be distinct and separate; e.g., the Education Ministry may share participation in a school survey, but the Parliament or the Finance Ministry may fail to allocate funds.
2. The needed technical resources reside in and are staffed and developed by separate national offices. These separate offices may have very different levels and types of technical development, as well as distinct organizational structures and different social connections.
3. The survey variables can vary immensely across national boundaries, due to different cultures, religions, economic and educational levels, legal and social relations, etc. Achieving comparable results demands immense efforts—but the task is not impossible, as multinational surveys have shown.
4. The crossnational translations of concepts and of questionnaires, also of codes and analysis, are daunting challenges that need ingenuity, knowledge, and devoted effort.
5. Separate samples must be designed and operated to meet distinct national conditions, with local resources, sampling frames, and field operations. This subject needs volumes; more discussion and study than is possible here.

Multinational comparisons probably go back many years, based on diverse kinds of observations—by travel, wars, conquests, etc. But probability sample surveys of entire nations have become common over all continents only during the past half century. As the second phase of development, those national surveys soon led to multinational comparisons. The third phase of deliberate multinational designs dates only from 1965: the Time Use Surveys of 1965 (Szalai 1972); the World Fertility Surveys of 1972–82 (World Fertility Surveys 1984; Cleland and Scott 1987); the Demographic and Health Surveys since 1985 (Macro International Inc. 1994); the Labour Force Surveys of the European Community (Verma 1992, 1999); see Kish (1994). Other multinational survey designs are also emerging, with the funding and technical resources increasingly meeting the growing effective demands. I am heartened and amazed at the emergence of the International Surveys of Psychiatric Epidemiology, a field that I had feared was beyond the reach of probability surveys in my lifetime! (Heeringa and Liu 1999).

Now, for the new fourth phase, I propose deliberate designs for combinations of multinational surveys. Multinational combinations of surveys are now being produced and published; e.g., European unemployment rates or birth rates; African or Sub-Saharan birth rates or death rates; world growth rates; and many other rates, means, and totals. The data for each nation may be based on probability samples (phase two), or even designed for multinational comparisons (phase three). But the methods used for combining them seem to be completely *ad hoc*; and current usage for the relative weights for combining national statistics seem to be in order of A, B, C, D, E, F from most common to the least. I made no actual counts, nor an empirical study, but glaring examples appear weekly. Very often the methods and weights for combining the national samples are not even mentioned in the media, even in respectable and scientific journals. To the contrary of the above order, our preferences may be almost in the reverse order of E, F, D, C, B, A—and very much depending on the situation, sample sizes, etc.

Allow me, with due modesty, to propose that phase five should be the development of solid theory for choosing among those preferences, and also others. But the need for methods for combinations cannot wait for the future better theory; and it is usual in statistics (and in the sciences) for practice and methods to develop before, and thus both to precede and to stimulate theory. Meanwhile, the discussions below may lead to some improvements in methods, even if they are not quite "optimal."

Here then follow six possible alternative ways and weights for combining national statistics.

A. Do not combine: publish only separate national statistics. This is the most common treatment for several reasons. 1. The authors have not thought of the possibility or need for combination, or rejected them. 2. Perhaps they could not decide on the "best" method, and wanted to leave that to the reader, user, customer. This may be defended by "caveat emptor," or "Bayesian" arguments. However, I reject them. The authors should do no worse in choosing than the average users—who in any case can reject the authors' combination if the national statistics are also published. I believe that when the reader's eye roves over the usual horizontal (or vertical) bars in graphs or over data in tables, it tends to yield a simple mean, hence this roving reduces Method A to Method C in effect. This tendency can perhaps be improved if the width of the bars is made proportional to population weights.

B. Even in the absence of combining populations, designs for multinational comparisons should be "harmonized" in survey measurement methods, to allow for proper comparisons (Kish 1994).

C. Use equal weights ($1/H$) for every country. This method is also common and also avoids (like A) the difficult questions of how to choose population weights W_h, with h denoting country. Probably its use is seldom based on deep reflection, but is widespread mostly because it appears to be a "common sense" approach. Perhaps it would be justified with models, where the between-country variation is paramount, and the population sizes are not relevant. However, I have no faith in such models.

D. Weight with sample sizes n_h. Thus $\bar{y}_w = \Sigma n_h \bar{y}_h / \Sigma n_h$, which results automatically from simply cumulating sample cases from separate countries, or sites, or surveys. This is also done frequently, and can be justified when elements are drawn essentially from the "same population" or when per-element variance is the only (or prime) component of variation. It denotes "cumulating cases," as distinguished from combining statistics (Kish 1987a, 6.6). This approach can be extended to situations where there are serious differences of element variance due to "design effects"; and then "effective sample sizes" $n_h/deff_h$ may be substituted for the n_h. The "effective sample size" may also be applied if the σ_h^2 differ between populations in order to use weights with precisions $n_h / \sigma_h^2 = 1/(\sigma_h^2 / n_h)$. In most situations, however, the variations in sample sizes n_h depend on arbitrary, haphazard factors; and C may be a worse choice than using equal weights $1/H$ for all countries (surveys, sites).

E. Use population weights W_h. Thus $\bar{y}_w = \Sigma N_h \bar{y}_h / \Sigma N_h$ and $W_h = N_h / \Sigma N$. This method has the most commonly understood meaning when the N_h represents total numbers of persons in population h. However, sometimes the population content may be quite different. For example, for grain (or wine) production it may be total number of farmers, or wheat (or grape) farmers, if those numbers are available, or can be estimated; these populations may yield potentially interesting meanings either for comparisons or for combinations. The population extent also needs to be determined; for example, all persons, or only adults, or only women, or only married women; only urban or rural, or both? Also the timing (date) of the surveys needs standardization, e.g., censuses are conducted in '0 (or '9, or '1) years. Often the population weights are not persons, but acres of land, or tons of steel, or barrels of oil, and so on.

F. Use post-stratification weights. Often in multipopulation situations we encounter the same problem as described later in Section 7 for multiple sites. And we may consider the same hierarchy of alternative treatments, the last of which (F) is using "post-stratification" weights. We may well have comparable surveys from several diverse countries of a continent (or the world), but neither all the countries, nor a probability sample of them. (For example, the African or South American countries in the World Fertility Surveys or the Demographic Health Surveys.) One may think of constructing "pseudo-strata" from which the available countries would be posed as "representative selections." Some one stratum could have only a single, available, large country. Another stratum could have 2 (or 3) countries, but with only one available representative that would get the weight of all 2 (or 3) countries. This artificial "pseudo-stratification" procedure may be preferable to simply adding up the available countries into an artificial combination with W_h (E) or with $1/H$ (C). The rationale for this preference is not very different from methods of adjustments for nonresponses.

Several questions and decisions remain concerning the choice among alternative weights. First: the choice should be made chiefly on substantive grounds. What must the combination represent mainly? My own preferences tend strongly toward D and E, and I deplore the prevalence of A and B that we encounter daily. However, I cannot support my preferences on technical grounds. Also I have faced grave problems with the extremes posed by the giants China and India, each more like a continent, and neither solutions E or C seem adequate. I advise defying the geographer's classification of Asia and leave both of them out of Asia, considering them as separate entities. For example, I have omitted all four countries greater than 200 millions in total population (including the USA and USSR) in my computations in 1970 (Kish 1976, Table 4; Kish 1987a, 7.3D).

Second: Is the bias due to using incorrect weights important? This would be difficult to prove, as the bias is a function of correlations between the weights and specific survey variables. However, the proof should belong to the denial, as it does with the biases of nonresponses or of poor sampling methods. Ignorance of sources of bias does not imply their absence. I believe that using equal weights instead of population weights can often lead to important biases.

Third: When samples are (roughly) equal-sized, weighting up to population sizes can greatly increase variances. These increases in variances due to unequal weights can be measured quite well (see Section 8). They should be balanced against probable biases in models for reducing mean-square errors. In small samples the large variances may dominate the MSE.

Fourth: It seems clear that the combination of population surveys into multipopulation statistics needs a good deal of research, both empirical and theoretical—and especially together.

4. CUMULATING PERIODIC SURVEYS

Periodic surveys have been designed and used mainly for measuring periodic changes, and also for "current" estimates, exploiting the advantages of partial overlaps. But here we shall explore their design and use for cumulated estimates. Furthermore, I include periodic surveys here in order to emphasize their basic similarities to surveys combined over space, such as multidomain and multipopulation surveys. We cannot enter here into the philosophical issues involved in repeated studies of the "same" population, except to note that the "stability" of any population differs greatly for diverse variables (Kish 1987a, Chapter 6); and the stability for any one variable will also differ greatly, depending on the length of the periods, which may be weekly, monthly, or quarterly. These are common and useful man-made periods. But there exist only two global "natural" cycles of variations: the diurnal and annual cycles, based on the earth's rotation around its own tilted axis, and around the sun.

I must note four practical, rather than theoretical, differences between cumulating periodic surveys and combining multinational or multidomain surveys.

1. Periodic surveys are designed for the "same" population, which tends to retain some stability between periods. The "sameness" and "stability" are only relative, and with many exceptions; e.g., epidemics in health data or fluctuations in stock prices. They differ greatly between variables and decrease for longer periods.
2. These stabilities imply positive correlations between periods, encouraging designs with "overlapping" sampling units in order to reduce both unit costs and variances for estimates of change and of current values. These overlaps are not desirable for cumulations, so this conflict between the two designs must be resolved.
3. Because similar methods and designs are feasible and generally preferred, they are used over all the periods; on the contrary, harmonization of methods is difficult to achieve between national samples. I emphasize here cumulating periodic surveys, but these aspects also apply to comparisons.
4. Methods for periodic surveys for comparisons have been widely published, in contrast to the novelty both of multinational designs and of periodic cumulations.

There now exist several cumulated representative samples (CRS) of national populations: samples designed for cumulations over large populations. These remain restricted within selections of primary sampling units in order to reduce field costs, whereas "rolling samples" (Section 5) are spread deliberately over all sampling units in the population. The Health Household Interview Surveys (HHIS) of the USA are separate weekly samples of about 1000 households, cumulated yearly to 52,000 households (U.S. National Center for Health Statistics 1958, pp. 15–18). These samples are selected by the U.S. Census Bureau within their large sample of PSUs. The yearly samples of over 150,000 persons constitute a remarkable example of multipurpose surveys, representing even rare diseases. The Australian Population Monitors have quarterly nonoverlapping samples that are cumulated to yearly samples, and these are also confined into primary sampling units (Australian Bureau of Statistics 1993). The new Labour Force Surveys of the United Kingdom publishes each month the cumulation of three separate, nonoverlapping monthly samples (Caplan, Haworth, and Steele 1999). There are other examples as well, and the applications of cumulative representative samples (CRS) are increasing in scope and diversity, although until now they have lacked a common name and literature. Nevertheless, I propose to differentiate the CRS from rolling samples for practical reasons (Section 5).

Two problems and methods associated with cumulated samples deserve brief mentions, but with references to more adequate treatments. Asymmetrical Cumulations refer to proposals and some actual practices of reporting large aggregates frequently, but reporting on small domains only after cumulating over longer intervals. For example, the HHIS above may report some national averages each week or monthly, but smaller regions, or specific diseases, only for annual aggregates (Kish 1990a).

A serious conflict can arise if periodic samples are to be used (as they should) both for measuring periodic changes and current levels and for measuring cumulations over the periods. This double use has been proposed and practiced, although I do not yet know of any deliberate double designs. Most periodic surveys use partially overlapping samples with some kind of rotation design. One reason often given for these overlaps is the reductions in variances per sample element both for measuring changes between periods and for making current estimates. These reductions depend on positive correlations between the overlapping sampling units. Such reductions are well documented in sampling textbooks and articles since the original papers on this topic (Jessen 1942; Patterson 1954). But even greater reductions are possible in element costs, when the later interviews are much cheaper than the first contacts; for example, if the later contacts are by telephone. On the other hand, separate new samples will be much preferred for cumulations in order to avoid the positive correlations. One may imagine different compromises that may be efficient, when:

(a) most of the positive correlations are not high;
(b) reinterview costs are not much cheaper; and
(c) reinterview response rates are discouraging.

However, consider also a new design that I call a Split Panel Design (SPD) that adds a panel *p* to a parallel series of nonoverlapping samples *a–b–c–d, etc.*; with the combination then denoted as *pa–pb–pc–pd, etc.* The panel replaces the overlaps of rotating designs and provides the useful correlations for measuring net (macro) changes. Further, it also serves to measure individual (gross) changes, which are lacking in the usual designs of overlapping sampling units, because of the mobility of persons and households. Including panels of individuals (persons, elements) would bring considerable advantages for SPD over all current overlapping samples, which usually use merely the same sampling units (Kish 1987a, 6.5; Kish 1990a).

Another considerable advantage of SPD is that these overlaps would be based on the correlations from all periods, rather than only for the arbitrarily chosen periods for the rotation designs. How arbitrary are these? Some decisions use one-month groups, some three months, others 12 months, etc., etc. It is most unlikely that these disparate overlaps are actually "optimal" for those countries. It seems most likely that the "optimal" overlap cannot be predetermined for any single variable, and a single optimal period is even less likely for multipurpose designs.

5. ROLLING SAMPLES AND CENSUSES

These should be considered as special types of the related cumulative representative samples (CRS); but rolling samples (RS) should be distinguished, because they are designed for different and specific functions. CRS have been confined to designs of PSUs. They are spatially restricted for cost reasons and for fitting the designs of labor force surveys, and other surveys associated with them. However, RS designs must aim at a much greater spread in order to facilitate maximal spatial range for cumulations over time. Rolling samples must be designed specifically to readily yield good estimates for all small spatial units, when the periodic samples are cumulated into annual or decennial larger samples or censuses.

First let us define a rolling census: it consists of a combined (joint) design of F separate (nonoverlapping) periodic samples, each a probability sample with fraction $f = 1/F$ of the entire population, and so designed that the cumulation of the F periods yields a detailed census of the whole population with $f' = F/F = 1$. Intermediate cumulations of $k < F$ periods should yield rolling samples with $f' = k/F$ and with details intermediate between 1 and F periods.

Imagine weekly national samples, each designed with *epsem* selection rates of $f = 1/520$. The cumulations of 52 such weekly samples would yield an annual sample of 52/520 = 10 percent. Then ten of these annual samples would yield a census of 520/520. I have proposed in several papers to have these rolling samples replace both kinds of the most important forms of official statistics that are either used or planned in many countries: the monthly surveys of population and labor force and the decennial censuses. Even more important, these surveys could also provide annual detailed data, perhaps with 10 percent samples, which are badly lacking, and needed in many countries (Kish 1990a, 1997, 1998). Providing spatially detailed annual statistics for a variety of economic and social variables, not a mere population count of persons, would be the chief aim of rolling samples in many countries. These are needed even in countries that can provide fairly good estimates of population counts and a few simple statistics either from registers or with estimation methods. In countries without good frequent (monthly or quarterly) surveys of labor force and population, rolling samples could also serve them as efficient vehicles.

I must admit that the above basic ideas provide merely the skeleton for any actual national design for rolling samples. But such actual national samples have been recently designed—the largest and best of which is the American Community Survey (ACS)—now undergoing a 37-area pilot study by the U.S. Census Bureau (Alexander 1999). This aims to provide monthly surveys of 250,000 households and detailed annual statistics based on 3,000,000 households, after year 2003; and also to provide quinquennial and decennial census samples later. The National

Statistical Office of France is working on plans for a Census Continué (Isnard 1999). The Labour Force Surveys of the United Kingdom are now based on cumulated monthly surveys. Some other countries are examining different but generally similar possibilities.

It is also proper to add references to two early publications describing "rolling samples" of large sizes, although not national in scope (Mooney 1956; Kish, Lovejoy, and Rackow 1961). Others probably exist that I have not seen.

How to cumulate periodic surveys? This topic must receive serious technical consideration in the future, because so far they have been done only with *ad hoc* procedures. Perhaps for cumulating over a single year, *epsem* samples with the same sampling fraction *f*, and simple cumulation of cases may serve as a simple model: averaging over seasonal and random variations may outweigh secular trends. However, averaging annual statistics over 10 years may have to consider secular trends in population size.

Consider several alternative sets of weights W_i to be assigned to yearly means $\bar{y}_i$ for a decennial mean $\bar{y}_t = \Sigma W_i \bar{y}_i$ $(i = 0, 1, 2, \ldots, 9)$ and $\Sigma W_i = 1$.

A. $\bar{y}_{ta} = \bar{y}_9$, with $W_9 = 1$ and the other nine $W_i = 0$, utilizing only the final year. This could be used for national and large domain estimates, and for highly fluctuating variables (unemployment, epidemics, stock prices), where the need for timeliness dominates sampling precision.

B. $\bar{y}_{tb} = \Sigma W_i \bar{y}_i$, with all ten $W_i = 0.1$. For variables without time trends, and for small domains, obtaining a stable average over time may be good strategy.

C. $\bar{y}_{tc} = \Sigma W_i \bar{y}_i$, with $W_0 \le W_1 \le W_2 \le \ldots \le W_9$, monotonically increasing (or nondecreasing) W_i. The curve of increase may be determined with a model or with empirical data. Thus $\bar{y}_{ta}$ and $\bar{y}_{tb}$ may be viewed as two extremes of $\bar{y}_{tc}$. They all seem better than the present practice of giving full weight $W_0 = 1$ to a decennial census that may be from 1 to 10 years old and obsolete.

Furthermore, with rolling censuses, the statistical office need not wait to publish only decennially. It can publish annually the results of the latest rolling samples, with several available alternatives from those above: either the latest year $\bar{y}_{ta}$; or $\bar{y}_{tc}$ an average that favors the latest years. Or "asymmetrical cumulations" favoring $\bar{y}_{tb}$ for smaller domains, but $\bar{y}_{ta}$ for larger domains and totals. It could conceivably publish both $\bar{y}_{ta}$ and $\bar{y}_{tb}$ and let the reader choose (perhaps publish electronically). Clearly technical research will be needed to search for "optimal" solutions to support the applications already appearing.

6. COMBINING EXPERIMENTS

(A) This topic has been the subject of three early and good papers by Cochran and has also received attention from both Yates and Fisher at Rothamsted (Cochran 1937 and 1954; Yates and Cochran 1938). These dealt with experiments relating crop yields (predictands) to fertilizers (one or more predictors), conducted over different populations, fields, and years. They used ANOVA methods for statistical analyses and for combining the several independent experiments.

(B) Fisher's test for combined probabilities, from 2×2 Chi-square tests of the "same" null hypothesis is even older. It can use entirely different populations, and even diverse variables, for testing the "same" null hypothesis. This well-known test can be found in most statistics textbooks.

(C) Methods of meta analysis are newer, and increasingly used. They combine experimental results from different samples and populations for the same predictand (outcome) variable from one or more predictors (inputs) (Glass 1976; Hedges and Olkin 1985).

Methods for combining sample surveys are just emerging, much later than methods for combining experiments. The two fields, however, have many similar aims, which should be noticed, in order to see useful relations between the two distinct topics. Perhaps these relations can be best perceived by looking at the differences between the aims and the problems that have been the subjects of the two methods. There seem to be three main differences between the two methods, as they have been applied.

1. Combining surveys (CS) needs a great deal of advance preparation, planning, and coordination. This is true of multinational surveys for both the comparisons, which have been already achieved, and for their combinations, which are new. For national multidomain surveys the coordination comes naturally, but for multinational surveys the coordination of the separate national designs is difficult, but necessary (Kish 1994). On the contrary, a great virtue of combined experiments (CX) is that they can be performed on the reports of experiments already performed, as the name meta analysis signifies. That analysis is based on the relations of the predictand/predictor pair of experimental variables. The Fisher test needs only the probabilities P_i achieved by the tests of significance.
2. The second difference between the two methods is related to the first. The CX are based on experiments, whereas CS concentrates on surveys. Thus CX emphasizes experimental control through randomization of variables over subjects. However, CS are based on probability sampling with randomized selections of subjects—not variables—from defined populations. Usually these two kinds of randomizations are difficult to achieve in any research study and one must be sacrificed (Kish 1987a, 1.1). The population base of CS is specified, whereas those for CX usually are not and cannot be.
3. Third, CS involves a full statistical analysis, and even a full survey method, designed for similarity and comparability in order to facilitate the joint

analysis. On the contrary, the methods of CX can use the very end of the statistical analyses, often even from published statistics. The extreme of this kind of abstraction is shown by the combined Fisher test, based only on the terminal P_i values of the separate statistical tests.

Because of the large, consistent and interrelated differences between Combined Experiments and Combined Surveys, it may be best to keep the two methods separate. Some may propose that the gap between the two subjects is only an historical accident and that the gap can be closed sometimes. But I believe that it is more useful to maintain the separation of the two methods, even if sometimes a compromise may be usefully adopted.

That still leaves open the question whether the three methods of combined experiments (A), (B), and (C) above should be called "Combined Experiments," as Cochran, Yates and Fisher called them since the 1930s, or if it is better to distinguish them all as "Meta-Analysis," now a widely known and accepted joint designation. Happily we need not decide here, but perhaps meta-analysis is the best, provided we also recognize the earlier successes.

7. COMBINING SEPARATE SITES

Suppose that similar data have been collected in several sites of a combined population, but not in all of the sites, nor in a probability selection of them. The sites may be cities, provinces, or districts of one country. Or they may be institutions, such as schools, or hospitals, or factories. Or the sites may even be entire countries of a continent. I have seen a variety of such situations when the sites are either chosen arbitrarily, or are simply "volunteers." Often the sample sizes per site are similar, though the population sizes of the sites vary greatly. Here follows a list of possible alternative treatments of the data.

A. Separate survey estimates $\bar{y}_i$ may be presented only. Usually this is all that is done, especially if the data have not been coordinated, or "harmonized." Any comparisons and any combinations of the separate statistics are left to the readers, to use their own methods or resources.

B. Comparisons between the separate sites require harmonization (of variables, measurements, timing, populations) to render the differences $(\bar{y}_i - \bar{y}_j)$ meaningful.

C. Simple cumulations $\bar{y}_t = \Sigma y_i / \Sigma n_i$ of all sample cases amount to assuming that the populations N_i of the sites can be considered parts of the same population of ΣN_i elements. Note that the sample means $\bar{y}_i$ are weighted by the sample sizes n_i. Often these are nearly equal and then C approaches D.

D. Equal combination $\Sigma \bar{y}_i / k$ of k sites weights each of the sites equally, disregarding both the sample sizes n_i and the population sizes N_i.

E. Weighted combinations $\bar{y}_w = \Sigma W_i \bar{y}_i / \Sigma W_i$ weight the sites with some measure of their relative importance. Population sizes N_i seem reasonable, but others may be used. However, we may object to the combination of an arbitrarily selected set of sites.

F. Post-stratification weights $W_i \propto \Sigma_j N_{ij}$ can save attempts to overcome the above objections by constructing pseudostrata $\Sigma_j N_{ij}$, composed of "similar" sites, from which the unit N_i may be considered a valid selection. Thus the total sample then is considered a sample from the larger population of total size $\Sigma_i \Sigma_j N_{ij}$. Such model building resembles the attempts to reduce nonresponse bias with nonresponse classes.

Three sets of decisions must be made, and this order is chronological in activity, but not necessarily in planning.

1. The allocation of sample sizes, especially whether equal sizes for the sites, or proportional to relative population sizes (W_i).
2. Whether the samples should be combined, or to merely accept alternative A.
3. What weighting to use among alternatives B to F.

The above alternatives resemble those in Section 3 and multinational combinations may be viewed as special cases of multi-site combinations, but a very special case, for the reasons given there. Furthermore, the alternatives listed above deal not with academic or idle speculation, but with many practical, actual problems. I have advised and argued on problems of every kind, and felt the need for and lack of dependable references on combinations and cumulations, whether technical and published or oral and authoritative. Some examples I have encountered:

(a) The World Fertility Surveys had national sample sizes without much (any) relation to population sizes. Should they be combined and how? I thought yes and with N_i (method E).

(b) Samples of several hundred households were selected in each of 12 large cities of the USA (which had "racial riots" in 1968). Should they be combined and how? I thought yes and with N_i (E).

(c) In each of 13 counties of the USA samples of a few hundred 4-year-old children were selected for a study of preprimary learning situations. They were combined with method F.

(d) In 11 of China's 30 provinces probability samples averaging 1,000 4-year-old children were selected for studies of preprimary learning situations. They were combined with method F.

(e) In 5 of Nigeria's 30 states small urban and rural samples were selected for studies of preprimary learning situations of 4-year-olds. After examining the 5 × 2 small samples the sample cases were merged with method C into urban and rural samples.

(f) Coordinated survey designs and university resources are being planned for 5 to 8 large cities of China. The designs are planned both for comparisons and for combination, with either method E or F.

8. ERRORS, LOSSES, COMPROMISES

The Mean Square Error of a weighted combination of means may be written as

$$\begin{aligned}\text{MSE}(\Sigma W_i \bar{y}_i) &= \text{Bias}^2(\Sigma W_i \bar{y}_i) + \text{Var}(\Sigma W_i \bar{y}_i)\\ &= \{\Sigma W_i [E(\bar{y}_i) - \bar{Y}_i]\}^2 + (\Sigma W_i^2 D_i^2 S_i^2 / n_i)\,.\end{aligned}$$

This holds for distinct countries (i) and distinct domains like provinces. But for some domains there may also exist covariances (S_{ij}), positive or negative. The relative weights are W_i, and S_i^2 and n_i are element variances and sizes, with design effects D_i^2 to compensate for the effects of complex designs. On any study all these values can differ greatly between variables. Note that the bias of the combined mean is the weighted average of the individual biases. For periodic samples these may be fairly constant. For multipopulation and multidomain samples this emphasizes the need for reducing biases for the larger units, with large W_i. The variances of means decrease in proportion to the number of units being averaged, and thus they decrease in importance relative to the biases.

The situation is different for comparisons, where

$$\begin{aligned}\text{MSE}(\bar{x} - \bar{y}) &= \text{Bias}^2(\bar{x} - \bar{y}) + \text{Var}(\bar{x} - \bar{y})\\ &= [E(\bar{x} - \bar{y}) - (\bar{X} - \bar{Y})]^2 + \text{Var}(\bar{x}) + \text{Var}(\bar{y})\\ &= [\{E(\bar{x}) - \bar{X}\} - \{E(\bar{y}) - \bar{Y}\}]^2 + D_x^2 S_x^2 / n_x + D_y^2 S_y^2 / n_y\,.\end{aligned}$$

Note that the biases of differences tend to vanish if the biases are similar, even when not small. The variance is the sum of two variances (and a small n_x or n_y can increase it), hence may dominate the bias term. When there are overlaps (in periodic surveys) the covariance term $-2\text{Cov}(\bar{x}, \bar{y}) = -2D_{xy} S_{xy} n_c / n_x n_y$ tends to decrease the variance.

I have emphasized in some detail elsewhere the need for the utmost "harmonization," for the coordination of survey methods: in variables, measurements, and in populations. On the other hand, there is great freedom to choose different sampling methods for the different populations, provided they are all based on good probability samples (Kish 1994).

In multipopulation combinations, frequent and serious conflicts arise, because the relative sizes W_i of the populations (of countries or of provinces) often vary greatly; ranges of 1 to 50 or more are common. But the sample sizes may be (roughly) equal for all H populations. Then weights k_i may be introduced to adjust the combinations to the W_i. These inequalities of sampling rates increase the variances of combinations by a relative factor $1 + L = 1 + C_k^2$; where L denotes relative loss (increase in variances) and C_k the coefficient of variation among the weights k_i. Both are zero when all k_i are the same, i.e., for proportional allocation of the n_i to the W_i. But then the average variance of the populations and their comparisons suffer even greater losses than the sum. This conflict can be resolved with compromises, especially an "optimal" compromise with $n_i \propto \sqrt{(W_i^2 + 1/H^2)}$ (Kish 1976; Kish 1994).

A good numerical example below comes from the 10 provinces of Canada, whose total population (in 1991) of 27, 211,000 with an *epsem* selection of f = 1:2721 would yield roughly these 10 values of n_i in row 1 for a total of n = 10,000. You see that the largest province of 3,706 cases is about 75 times greater than the smallest with 49. This range seems common for provinces within most countries. Also for multipopulation cases; e.g., in the European Union, Germany is about 200 times the size of Luxembourg. The proportions are $W_i = n_i /10{,}000$; and a proportional sample would yield a minimum variance for $\Sigma W_i \bar{y}_i$, hence a relative loss function $1 + L = 1$, with loss $L = 0$. For simplicity and to concentrate on weights, we can assume that element variances $D_i^2 S_i^2$ and costs c_i are similar, or can be averaged out. However, these proportional n_i values would result for average provincial means $\Sigma \bar{y}_i / 10$ or for average comparisons of provincial values $(\bar{y}_i - \bar{y}_j)$ of $1 + L = H^{-2}\ \Sigma(1/W_i) = 3.9785$ for a relative loss of 2.9785, a 300% increase in average variances. These losses come mostly from the 6 small provinces (derivations in Kish 1976).

Row 1	3,706	2,534	1,206	935	401	363	331	266	209	49
Row 2	2,437	1,730	995	869	684	676	669	657	648	636

Thus, some people (in Canada and in other countries too) ask for equal size samples, n_i = 1,000, so that each province can provide the same precision. Then the means $\Sigma \bar{y}_i / 1000$ will all have variances $\Sigma(1/1{,}000)$ and relative efficiency of $1 + L = 1$, with loss $L = 0$. However, the national mean will have a variance of $\Sigma W_i^2 / 1{,}000$, with a relative loss of $1 + C_k^2 = H \Sigma W_i^2 = 2.3003$, or a 130% increase in variance. We must also remember that all crossclasses, such as those by age, education, occupation, etc., will also tend to suffer similar losses.

However, some remarkably good compromises can be had, and the best is a least-square solution with the $n_i \propto \sqrt{(W_i^2 + H^{-2})}$ These give the $1 + L$ values of $1 + L = 1.2424$ and $1 + L = 1.2630$, for $\Sigma W_i \bar{y}_i$ and $\Sigma \bar{y}_i / H$, respectively, only a 25% loss for each! The n_i values in Row 2 show a "floor" between 600 and 700 for the n_i for the 6 small provinces, and a roughly proportionate increase (but below 10,000 W_i) for the largest 4 provinces. This optimal allocation has in fact been used for some of the surveys of Statistics Canada (Tambay and Catlin 1995). It is interesting that the mathematical solution also makes good common sense (Kish 1976, 7.6; Kish 1987a, 7.3; Kish 1988). However, the mere common sense solution of allocations proportional to $\sqrt{W_i}$ is less efficient than the optimal allocation.

PROFESSIONAL LEADERSHIP AND TRAINING

STEVEN G. HEERINGA AND GRAHAM KALTON[1]

The subject matter of statistics is wide-ranging, extending from abstract mathematical statistics at one extreme to the procedures for collecting and analyzing data at the other. In his Presidential Address to the American Statistical Association in 1977, reproduced as Chapter 16, Kish described his statistical philosophy. He emphasized the importance of applications and bravely chastised the university statistics departments of the time for their concentration on abstract theory. Although the situation today is much better, with theory and applications combined in many departments, his observations remain pertinent. He also argued for improvements in statistical theory and methods in government statistical offices, and outlined training and other means to achieve these improvements.

Throughout his career, Kish devoted a great deal of his energies to education and training. References to the work of his Ph.D. students have been given in earlier chapters and commentaries. Here we note his long-standing dedication to training survey statisticians in developing countries and later also transition countries. In 1961 he established the highly successful Sampling Program for Foreign Statisticians (SPFS) at the University of Michigan. This two-month intensive summer program has now trained over 500 survey statisticians from more than 105 countries, and has contributed greatly to the use of sound probability sampling methods around the world. The SPFS is ongoing and continues to thrive. The contents and structure of the program that Kish established were much influenced by his extensive experience as a sampling consultant in many countries. His insights on "Developing Samplers for Developing Countries" are given in his paper of that title, which is reproduced as Chapter 17. This paper was the 1994 Morris H. Hansen Memorial lecture, a lecture named in honor of another of

---

[1] Steven Heeringa, Survey Research Center, University of Michigan, Ann Arbor, MI 48106-1248. sheering@isr.umich.edu.
Graham Kalton, Westat, Rockville, MD 20850. grahamkalton@westat.com.

the world-renowned pioneers in survey sampling. Kish's choice of subject matter for this lecture is an indication of his commitment to promoting the use of sound probability sampling methods around the world. He had a conviction that the use of these methods would help to yield the reliable data that policy makers need in order to make informed decisions to improve the conditions of their populations. That was always his ultimate aim.

16

CHANCE, STATISTICS, AND STATISTICIANS

LESLIE KISH

Statisticians and statistics deal with the effects of chance events on the treatment of empirical data. I am not attempting to add still another definition to the many already in print; Kendall (1948, p. 1) notes that "Willcox listed well over a hundred definitions" in 1935. I merely hope my description is neither too narrow nor too broad to describe your own activities as statisticians. Some of your other activities may not be well covered because the statistician's work so often penetrates other fields, a feature of statistics I shall emphasize. My aim today is to illuminate the consequences of that description for the practice, for the writing, and for the teaching of statistics. Also I shall relate its consequences to the structure and activities of our Association.

I perceive it as both my task and privilege to present here a broad and philosophical view, rather than the kind of narrow, technical exposition we statisticians usually give in our sessions. I want to entertain not only my fellow statisticians but also their husbands—or their wives.

All of us think perhaps daily—or at least once a year—about what we statisticians do—and why. Indeed you have often been stimulated to do so from this platform. Even if my contribution adds little to what you already know, it may suggest a new way of putting it all together and stimulate you to do the same. Here then is my apologia, plea, justification for statistics and statisticians. I am not deliberately seeking controversies, but neither am I avoiding them.

I know that emphasizing the central role of chance in all aspects of statistics is controversial. Does that emphasis exclude the collection and tabulation of censuses, of health, and of other similar data? Not at all. Those activities gave birth and name to our field and they still occupy central and vital areas. But in those areas also it is the recognition of the effects of chance, of variability, of fluctuations, that distinguishes the work of statisticians from the neighboring fields of accounting, bookkeeping, and inventory taking.

Reprinted with permission from the *Journal of the American Statistical Association* (1978), 73, 1–6 (Presidential Address).

Controversy also exists about the sources of chance effects. We may agree about the prevalence of errors of sampling and of measurements, and also about errors due to our current ignorance of disturbing factors. However, many of us, but not all, also postulate a fundamental source of indeterminacy that underlies all empirical work, and a basic role for such indeterminacy in our outlook, models, and designs.

Furthermore, the remedies I suggest for closing the gaping wound between academic mathematical statisticians and practicing empirical statisticians are bound to be especially controversial.

STATISTICS CONTRASTED WITH OTHER SCIENCES

Statistics is a peculiar kind of enterprise of contradictory character because it is at the same time so special and so general. Statistics exists *only* at the interfaces of chance and empirical data. But it exists at *every* such interface, which I propose to be both necessary and sufficient for an activity to be properly called statistics. It has a special and proscribed function whenever and wherever empirical data are treated; in scientific research of any kind; in government, commerce, industry, and agriculture; in medicine, education, sports, and insurance; and so on for every human activity and every discipline. This widely spread yet specialized character differs from other disciplines, which tend to cover in depth all aspects of their special domains.

Thus statistics differs fundamentally from other sciences. The data of other scientists come chiefly from their own disciplines—though they may also take side trips into other fields. In stark contrast, statisticians have no field of their own from which to harvest their data. Statisticians get all their data from other fields, and from *all* other fields, wherever data are gathered. Because we have no field of data of our own we cannot work without others, but they also cannot do without us—or not very well, or for very long.

Ours is a symbiotic way of life, a marginal and hyphenated existence. We resemble the professional harvesters of wheat and grains on our Great Plains, who own no fields of their own, but harvest field after field, in state after state, and lead a useful, rewarding, and interesting life—as we do.

Instead of the word chance we could use probability, except that today I would rather avoid the theological controversies surrounding that word. Yet we can agree that statistics does not claim to deal with all kinds of uncertainties. Not, for example, with the uncertain aspects of unique historical events such as brought on the American, French, Russian, Mexican, or Chinese Revolutions. (Nor, for a more humble example, with the uncertain events that brought me to this platform tonight.) Statistics deals with chance effects in empirical data concerned with classes of events, not single historical events; it treats numerical data from classes, from populations (Kendall 1948). I also believe that we should deal chiefly with large classes of many events, and not with classes of size 2, 3, or 6.

My aims are entirely pragmatic: to describe what we statisticians do in our practice, in our teaching, in our writing—and what we should be doing; also what our Association tries to encompass, and in what directions we should plan to move. The descriptions and plans are not simple because statistics is such a diffuse enterprise. At one end it has vital roots in abstract mathematical theories of probability and stochastic processes; it shades without definable boundaries into several profound areas within mathematics proper. At the other, it branches into procedures for collecting and analyzing empirical data of all kinds; thus it gets embedded in the methods and interpretations of all substantive fields of measurement.

Statistics and statisticians must remain in touch with both ends, but they can neither encompass nor reside exclusively in either end. Without data we are within the deductive arena of mathematical probability. Without chance effects we are in one of the many areas where mankind has learned to observe, measure, and count. All types of counting, all processes of observation, collection, tabulation, and analysis of data are subject to variability, to errors, and to mistakes. But statistics does not necessarily include all kinds of counting, such as those found in accounting, banking, voting, or scoring in sports. When dealing with measurements, experts in the specific substantive fields concern themselves with refining instruments, and with eliminating mistakes and cheating. Statisticians also get involved in those efforts, but the statistician brings a different and special view to the study of errors of observation: an explicit recognition of chance effects, a probabilistic view and treatment of them, and especially the incorporation of that view into the research design and the interpretation of observations. Note then that to err is human, to forgive divine—but to include errors in your design is statistical.

In any scientific enterprise one can distinguish a hierarchy of four problems and decisions. First comes the choice of variables to consider, second the design of the model for relations between the variables. These two decisions belong to the scientist or expert in the substantive field. Third comes the estimation of parameters, and fourth the assessment and analysis of the variability of errors of those estimates. For these latter two stages statisticians are needed, and they must be either bought or borrowed. (Otherwise, the role is taken by the scientist; and he may perform it well, but he may not.) Consider, for example, the model $d = (½)\ gt^2$ for the distance covered in t seconds by an object falling freely to the earth's surface. The constants ½ and 2 come from the physicist's model, but the statistician gets involved in designs for measuring d, t, and especially g, the gravity parameter. Furthermore, to measure the distribution of g may pose a formidable challenge for the sample design, requiring both a background in the relevant variables and foresight about future uses of the model.

The situation is similar in more complex models, whether physical, biological, or social. Consider, for example, the highly controversial separation of an IQ test into components for genetic heritability, environment, situational factors like age and culture, interactions, and error terms. I maintain that the basic difference between the social sciences and the "exact sciences" lies not in the magnitudes of the error terms, but in the multivariate nature inherent in the models. In the social sciences, most models must be multivariate because disturbing variables cannot be excluded as the physicist excludes air resistance and wind from his *freely* falling body.

STATISTICS AND PHILOSOPOHY

We may agree then that statistics, in both its theory and its practice, plays important roles in many or in most human activities. We generally also agree that it requires as much learning, experience, and intelligence as any other profession. Still, is it an interesting and noble occupation? Is it an exciting activity for a young person and is it dignified for a mature person? Would you want your daughter to become a statistician? Or your son to marry one?

It may seem like a humble task to hunt for and measure errors while the search for the main patterns and variables belongs to other scientists. I believe and shall argue that statistics is the *most mature* of the sciences. Learning to live with the inevitable uncertainties of chance effects and to include them in our patterns, plans, designs, assessments, and inferences is among the latest and most mature products of the human mind. We shall note that the late arrival of statistics among the sciences has important consequences for its position among university departments and for its problems today.

In all her (his) endeavors the statistician cannot avoid the basic philosophical problem of empirical science: to make inferences from limited sets of empirical data to large finite populations and to infinite superpopulations of random variables induced by causal systems. Survey samplers, in particular, cannot avoid the basic questions posed by David Hume, then by Karl Popper (1959) and by Wesley Salmon (1967). The fundamental ideas of statisticians like Fisher, Neyman, and Savage are not only mathematical but profoundly philosophical. The converse is also true: in the current philosophy of science you find fundamental ideas coming from statisticians and from the theory of statistics.

A closer link between the disciplines of statistics and the philosophy of science, at least in joint seminars, would be stimulating to both. The diverse schools of inference—frequentist, fiducial, Bayesian, likelihood, and others still to come—may be viewed as valiant efforts to capture within a mathematical framework the "Hume-an" problem of inference from variable empirical sample data to universal statements.

The powerful emergence of the new ideas may be seen in Darwin's theory of evolution through natural selection, and in the later developments of the theory. Consider genetics from Mendel to Morgan and to Watson. Scardovi (1976, p. 35) writes:

> Mendel's was the first explicit indeterministic paradigm in the history of science,... [and in 1866] ...was thirty years ahead of the philosophy and science of his time. The complete lack of recognition accorded him by his contemporaries can thus be explained. Mendel also surprises us with his statistical insight: he was capable of seeing in his numerical data random oscillations about a limiting value, and thus was able to extract the ideal proportion toward which the outcomes in the combined heredity of several traits tend...

All biological diversity of species and of individuals may now be viewed as the results of many throws of the DNA dice interacting with the environment. Monod (1971, p. 118) writes:

> Even today a good many distinguished minds seem unable to accept or even understand that from a source of noise natural selection alone and unaided could have drawn all the music of the biosphere. In effect natural selection operates upon the products of chance and can feed nowhere else; but it operates in a domain of very demanding conditions, and from this domain chance is barred. It is not to chance but to these conditions that evolution owes its generally progressive course, its successive conquests, and the impression it gives of a smooth and steady unfolding.

SYNTHESIS OF CHANCE AND NECESSITY

Statistics is ubiquitous because events everywhere involve chance factors. All nature and all human activities resemble games like bridge, baseball, football, or tennis in combining chance with skill or causal factors. Björn Borg said on winning at Wimbledon in 1977, "I think I am number 1 for the moment."

The structure and language of sports show people's widespread intuitive knowledge of combining chance with skill. In football the top team must win about 8 of 10 games, and we speak of "upsets." I liked the sophistication of a simple headline one Friday in our Michigan Daily: "Upset Unlikely." But in baseball games we see no "upsets," and winners average perhaps 60 percent of about 160 games. Examples abound in sports and games of sophisticated designs for combining chance with skill.

Yet statistical views are still emerging in new fields, when mature minds discover the effects of chance. Those views come late, because (as Monod said) "minds are unable to accept or even understand" the *joint roles*, instead of mutually exclusive roles, for pure chance and for natural laws. Einstein, the wisest mind of our century, is quoted (Hoffman 1972, p. 193) as saying, "God does not play dice with the universe." Yet the Einstein-Bose statistics define a basic game of chance for elementary particles.

The delays in the synthesis of chance effects with causal systems present a paradox whose examination is fruitful for us. Primitive man's fascination with chance, fate, and fortune was early and widespread. He was first filled with wonder and awe of hidden powers, magic, miracles, and ad hoc explanations. Only after discovering rules for connecting cause and effect, and formulating natural laws, did he learn to *counterpose* chance effects to explain deviations from the rules. Much later came the combined view of causal laws with chance effects. "At the end of the seventeenth century the philosophical studies of cause and chance...began to move close together," and were developed in sciences like astronomy and demography in the eighteenth and nineteenth centuries.

> This development had an important impact on the theory of chance itself. Previously chance was a nuisance, at least for those who wished to foresee and

> control the future. Man now began to use it for the other purposes,...to bring it under control, to measure its effect, and to make due allowance for it. (Kendall 1968.)

A similar development of the idea of chance in children was investigated in 1951 by Piaget (1975). At first

> the child does not distinguish the possible from the necessary. During the first period of life there is no differentiation between what is deducible and what is not because the intuitive anticipation remains halfway between operation and chance itself. During the second period [after seven years] there is differentiation and hence an antithesis between chance and operations,...chance defines...the unpredictable. In the course of stage III [after eleven years], on the contrary, there is a synthesis between chance and operations,...

Thus the complementary roles of chance and necessity appear basic and ubiquitous. That the discoveries of the diverse syntheses occur late to mature minds has several consequences. First, I believe that this vital outlook should belong to the philosophical, humanistic, and moral heritage of all scientists, and further of all citizens; also I believe that statisticians have natural responsibilities for teaching it. Consider the broad implications of several recent discoveries about the role of chance: the genetics of individual health, intelligence, and behavior; the statistics of population genetics (for which Fisher has been called the foremost biologist of our century); the occurrence of cancer; the lengths of our lives.

Appreciating the pervasiveness of chance should help us cope better with diverse aspects of life. About success in affairs, in business, even in science, we may say: "Fortune favors the prepared mind." The winners of millions of dollars or of Nobel prizes tend to accent the prepared mind, the losers emphasize fortune. This tendency for biased judgment, between winners and losers, is also found in national surveys of attitudes (and these self-protective biases may be psychologically healthy).

The tragic biblical Job might have been happier and wiser if he knew that his plagues were due to chance. The triumphs or the problems of your children may be due to chance, not only to your behavior—despite what Freud may say; a statistical view may protect parents against false pride or against guilt and despair. But we are not mere helpless puppets of chance, and we can improve our chances—for example, by quitting smoking, with regular exercise, and by losing weight. Recognition of the interplay of chance with discernible causes may yet lead us to a better way of life and to a better moral philosophy. Somebody may even start a new religion of Statisticology!

RECENT DEVELOPMENTS OF NEW FIELDS

A second consequence of the late recognition of chance effects may be found in recent discoveries of new fields for statistics. Within the past two generations, field after field has been laid open with the statistical recognition or introduction of chance effects. For example: Fisher with randomized experimental designs; Neyman with randomized designs for complex samples; Shewhart with statistical control for quality of products; stochastic analysis for many fields; time series and spectral analysis; Monte Carlo simulation; response surface analysis; and so on. There will be still more as our era of statistical discoveries continues.

Within the major fields many new techniques were also developed. In survey sampling, for example: multiphase sampling, interpenetrating samples, jackknifes, balanced repeated replications, controlled selection, measurements of interviewer variance, randomized response techniques, randomization for confidentiality, etc. Each of these techniques uses chance and randomization to tackle some problem in application. In addition, new procedures are designed as survey sampling penetrates new fields such as anthropology, archeology, and geology.

I view the entire field of survey sampling as another synthesis of chance and structural effects. It is a set of strategies for representing actual populations of variables, which are ill-defined mixtures of the two components. Instead of assuming a well-mixed urn of random variables, the sampler designs a complex randomized design over the frame population, and controls the chance effects through stages of randomized operations of selection.

Many other problems and fields still remain to be conquered with new statistical techniques and theories to capture and to tame the wild errors of chance. For example, modern epidemiology has created many new ideas; McNeill (1976) used those ideas to write a new interpretation of world history!

STATISTICS IN UNIVERSITIES TODAY

But even if we grant broad importance to it, does the discipline of statistics need autonomy? Does statistics need separate departments in our universities? The question is being asked by financially pressed universities, which notice that students elect most of their statistics courses outside the statistics departments. With one foot in mathematics, a second in computing, a third in philosophy, and seventeen more in other departments, do we have a leg to stand on?

I say YES—BUT. Yes enthusiastically, because a statistician trained in the theory and in the applications of, for example, survey sampling and experimental design, can design better surveys and experiments in just about any field of application than either a mathematical statistician or an expert within his (her) own field. The knowledge and the approach of the statistician are reasonably portable from field to field.

But I also believe that departments of statistics must change drastically from their present narcissistic concentration on abstract theory. Research and teaching, in many departments, have become divorced from applications, from empirical data, from the real world. This centripetal academic disease is chronic and contagious, and is also common in other disciplines and in other countries (although British statistical tradition seems less affected even today). Typically the teaching of statistics in universities is based on the assumption that graduates with good theoretical backgrounds can readily learn applications afterwards, while "on the job." But "on the job training" for unprepared statisticians can be very expensive both to graduates and to their clients and employers. Teaching applications within an academic setting may be more difficult than teaching theory, but with care it can be provided, and ultimately it should be much cheaper than the total costs of "sink or swim on the job" for unprepared graduates. Would you accept a physician or surgeon with only theoretical training?

At the same time, I also believe universities should offer more theoretical statistics as continued education for statisticians—and for others who practice statistics without statistical degrees. The practitioners working on applied problems will bring their own experiences, views, and needs to their studies of theory. To satisfy their needs, to answer their questions, and to help solve their problems should also provide the stimulation that academic departments of statistics need. I believe those efforts hold great promise for universities.

The history of our discipline exhibits significant paradoxes in autonomy. Our Association was founded in 1839, only the second national scientific body after the Philosophical Society (Wallis 1966). Yet more than 100 years passed before the first department of statistics was established in a university. In the 30 years since then more than 100 departments have mushroomed. The late arrival of this mature science was not accidental; after Galton, Pearson, and Fisher it was bound to come. But the portentous accident was the coincidence of its arrival with the historically unique explosion of American universities, which just now ended. Consequently our academic statisticians have been nurtured in a fantasyland that they mistook for normalcy. Departments concentrated on turning out academic Ph.D.s and imported others to fill the hungry pipelines.

> [O]ne had the curious situation where the highest objective of the teacher of statistics was to produce a student who would be another teacher of statistics. It was thus possible for successive generations of teachers to be produced with no practical knowledge of the subject whatever. (Box 1976.)

But this geometric explosion had to end, and statistics departments must now reform their teaching strategies accordingly. I doubt that the future lies in concentrating exclusively on mathematics and theory, and having students find their own way to applied statistics in other departments. Some biostatistics departments have done remarkably well in teaching applied statistics to students in all fields, but this service function interferes with their primary responsibilities to the biomedical sciences. Also the mislabeling is a hindrance to students and graduates in other fields.

Reformation must come and it will probably take diverse forms. One solution consists of one large department having many joint appointments with departments of application, of mathematics, and of computing. In one university I visited, a good and happy department consists entirely of joint appointments.

Many statistics departments should change, I propose, from their primary concentration on self-contained and self-perpetuating theoretical Ph.D. degree programs. Instead they should develop strong Master's level programs in applied statistics, designed first for statisticians in government, industry, and other applications, and second as vital components of Ph.D. programs in other fields.

If a statistics department insists on confining itself to mathematical statistics, then a decent regard for truth to its clients, the students, demands that it relabel itself accordingly.

BETTER THEORY AND METHODS IN STATISTICAL OFFICES

In asking for changes both in universities and in work situations I agree with Bjerve's (1975) presidential address to the ISI, and his "worries about the wide gap existing between official statisticians and academic statisticians. Professionally, these two groups seem to be living in two different worlds without communication in between."

The time has come for better integration of theory with application. Theory should improve the quality and efficiency of applied work. Simultaneously, applications must be the testing ground for statistical theory and methods, and must provide the motivation for further advances. Several paths toward integration are possible.

1. Programs of continued education should be encouraged. They can transmit new techniques and also satisfy needs discovered on the job by practitioners. Beyond the traditional devices such as books, journals, and sessions at annual meetings, we need more special and short courses both at universities and elsewhere (note the very successful short courses at our ASA meetings). More short courses, seminars, workshops, and meetings should also be organized within the offices themselves. ("Offices" refers to places—in government, industry, hospitals, etc.—where applied statistics are being performed.)
2. Consultations of practitioners with academic experts on current problems can be beneficial and stimulating to both parties. It seems to me that the offices could profitably increase those contacts and the sizes of their consulting budgets. Incidentally, such consultations may help to redress the imbalance between giant organizations that have their own specialists and small ones with none.
3. More statisticians strong in theory and methods are needed in offices. These positions, when created, should be filled better and more easily in the coming job market than in the past—especially if universities prepare their graduates for practical work. In many cases part-time positions for academics, or visiting or rotating positions for one or two years, may be preferable for both parties. All these situations exist, but they are underutilized because they run against the rigid structural habits both of offices and of academia.
4. More internships for students are badly needed and can greatly benefit all three parties (students, universities, and offices) and society. Because of conflicting drives of the three parties, the programs need care, skill, and empathy—and finances. But the outlook for the coming years is more propitious than in the past.

5. Better recognition is also needed for the status and roles of statistical methodologists. In too many offices the ladder of promotion takes good statisticians out of technical positions into largely administrative roles.

OUR ASSOCIATION AND PUBLICATIONS

What should our Association do to close the great gap, to better serve our discipline and all statisticians? I throw this question back to you. Write to our Board, try reforms in our sections, chapters, and committees.

I note only that the coming years are propitious for bringing our mathematical colleagues back into the Association. But our main efforts should be directed to bringing in all hyphenated statisticians from other fields who have strayed from our fold. Their presence and activities are needed for vigorous development in our field and in our Association. Our sections, chapters, and committees must help. A central role in this unifying task must be played by our publications.

We are proud of our publications, but here again I appeal to you to help restore a better balance toward applications. Our editors and the Publications Committee are working toward those aims, but we need your help. We face several major obstacles.

1. We all respect mathematics because we each have had some, but we also fear mathematicians. I find that the more mathematics one has, the more one craves (an insatiable desire in accord with psychological reward systems). Statistical publications have been flooded with mathematical virtuosity, and that flood has been permitted to impede statistical creativity.
2. We must work against a common but false double equality whereby methods are equated with theory and then theory with mathematics. Both equalities are only partly true—and misleading and harmful. Statistical theory needs strong philosophical foundations and orientation toward empirical data. And methods are valuable only insofar as they are applicable. I would strongly prefer that methodological contributions be actually applied to empirical data to prove (probe, test) them. That, I believe, was Karl Pearson's demand as editor of *Biometrika*.
3. Academicians have more time, experience, and motivation to write, edit, and referee than most practitioners. They deserve our gratitude for performing—very well on the whole—a difficult task and a necessary evil. But papers on applications and methods too often suffer from lack of understanding by the referees. The imbalance can be redressed with more help for, and more pressure on, our publications from our practitioners.

What can we do about the gaping chasm between the statistics departments and statistics in the applied world? The very words statistics and statisticians seem to have entirely different meanings in the two worlds. Some friends say that it is too late and hopeless to try to reunite the two. Some favor surrendering the very name statistics in universities, and starting out fresh in new departments with names like "Data Analysis" or "Statistical Computing." Others even dare to deprive the practicing, applied statistician of the name of statistics and to appropriate it exclusively for their own academic mathematical pursuits.

I disagree with these pessimists. Iago may be talking of statisticians when he says to Othello:

> Who steals my purse steals trash;...
> But he that filches from me my good name
> Robs me of that which not enriches him,
> And makes me poor indeed.

We need the good name of statistics for our field and our Association, and we need it to unite the efforts of theoretical and applied statisticians. There is a common chain that links mathematical probability with statistical theory, then with our methods, and these with our applications. To weaken or break that chain anywhere is harmful to all.

But I shall not end on a pessimistic note. As a statistician I am an optimist, with a statistical definition of the species: "An optimist is a person who thinks the future is uncertain."

17

DEVELOPING SAMPLERS FOR DEVELOPING COUNTRIES

LESLIE KISH

Summary: Samplers for less developed countries have been educated and trained in the last 50 years in various settings: in universities, in national statistical offices, on research projects, etc. Also both in their own countries and in "host" countries, such as the USA, India, France, etc. Methods of survey sampling are eminently "transferable" between countries (and disciplines), but are not readily accessible (portable) in printed (etc.) form, because the procedures must be modified to fit local resources. Therefore, travels by "learner" statisticians and/or by the "experts" are needed. The teaching/training process should combine academic subjects, individually selected, with practical training. These are difficult to arrange, and survey sampling also suffers neglect in universities, which teach statistical analysis but not data collection. Recruiting suitable candidates and career placement also pose challenges. But the rewards are great, because survey sampling is spreading all over the world.

Key words: Survey sampling; statistical education; developing countries; LDCs.

1. MORRIS HANSEN

Between 1940 and his death in 1990 I met Morris Hansen often, and every meeting was a stimulating and pleasurable learning experience for me. At the Census Bureau in 1940, when still a young man, he was already recognized as a natural leader by the "best and the brightest" group of statisticians. In the forties I took a course or two from him, at the US Department of Agriculture Graduate School. Later I often sought his advice, and he was always free and warm with it. The last few years I met him twice

---

First published in the *International Statistical Review* (1996), 64, 2, 143–162. Reprinted with permission.
This paper was the Fourth Hansen Memorial Lecture, 1994.
This lecture was based on my experiences and those of many others. But it does not represent comprehensive scholarly research, for which I had neither requests nor inclination.

a year on the Methodology Advisory Committee to Statistics Canada. We had many arguments and he was right, usually if not always, but he always won them.

He often showed lively interest in my Sampling Program for Statisticians for developing countries. The group he organized at the Census Bureau "became a center of research and training, with hundreds of its trainees performing statistical sampling throughout the world" (Kish 1965a, p. 378). Among his earliest ardent students from abroad were Zarkovich, Dalenius, Foreman, and many others and they went on to train still others.

2. COVERAGE OF PAPER

For educating survey samplers, my emphasis will be on survey sampling, not on other statistical methods, though these must also be related to the training. I emphasize survey *sampling*, not other survey methods; and *survey* sampling and sample *design*, rather than the foundations of sampling theory; also survey *applications*, rather than the publication and writing of academic articles; also, I shall discuss developing samplers for the LDCs (less developed countries), not for the USA or other developed countries. Finally I focus chiefly on aid from the USA, rather than from sources in other host countries, or within the LDCs themselves, or through home study and individual initiative alone.

3. VARIETIES OF EDUCATION/TRAINING

I would not attempt to list all possible varieties. Good examples exist for each variety and I regret the many omissions. The varieties discussed in this section concern prolonged contacts between LDC statisticians and USA experts, but omit briefer contacts and occasional lectures. The list also ignores the role of universities within the LDCs, also the considerable benefits from self-study from books and articles, as forms of continuing education.

Let us begin with shorter programs of a few weeks' duration that do not confer university degrees, but may confer certificates or diplomas, which may be meaningful to and valued by the participants. The program may be at a university. A good example is the Summer Institute for Survey Research Techniques, organized at the University of Michigan in 1948, which included one or two courses in survey sampling in eight weeks. That first year the sampling courses were given by Dr. Benjamin Tepping and Roe Goodman. These I expanded in 1961 into a formal Sampling Program for Foreign Statisticians which specialized in samplers for LDCs and has had over 400 graduates in 95 countries over 30 years.

Among programs beyond universities I single out the International Statistical Programs Center at the US Census Bureau. There are other examples of brief courses for survey sampling in the USA, in other countries, and at statistical meetings, and I had taught in several. These are more important for survey sampling than for other statistical subjects, because that subject is neglected and absent in most universities worldwide. Yet sampling needs statistics courses for its base, hence its neglect by the departments of statistics is harmful. I am keenly interested in teaching sampling, because it is the most cost-effective method for improving data quality with limited resources anywhere, and especially in the LDCs.

Graduate degree programs at universities in the USA (and elsewhere) are a vast subject and I have strong opinions on it, but here concentrate only on a few points most relevant to sampling for LDCs. First, only a few universities offer adequate training in survey sampling. Second, the MA degree is more cost-effective than the PhD, because you can get three MAs for the price of one PhD. Third, for many statisticians in LDCs the doctorate may be even counterproductive with negative values for practical work back home, because they were "overtrained" for the conditions both in their offices and in their universities.

We face conflicts here because our universities are overwhelmingly oriented toward academic PhDs and undervalue the MA programs and candidates. This may be one reason to favor departments of biostatistics, which are often more oriented toward both MAs and applications than statistics departments are. These are usually only prepared (in both senses of able and willing) to offer theory, and they neglect methods of data collection. Nevertheless, we can and should help a few exceptional students to PhDs and to careers in research and teaching in their own countries; and we at Michigan have so helped. Finding the right USA university and financial support for each of them are difficult and individual tasks.

We have neither space nor capacity to discuss sampling training within the LDCs, the statistics programs at their universities, and sometimes at their statistical offices. These must be sporadic, because both teachers and students of sampling are few, except in India and China.

Finally I call attention to the teaching that occurs in LDCs between the visiting experts and the local statisticians during consultations and collaborations. This one-on-one or one-on-ten teaching is a fringe benefit during the weeks or months of the expert's visit to the LDCs. I propose that such teaching should be brought from the fringes of projects closer to their center in the form of classes, formal or informal, say for five or even fifty students. I also propose that this training function, when feasible, be made part of the visiting expert's budget.

It is different when teaching statistics at an LDC university is the main assignment for a visitor. Then he/she may be consulted on some specific practical problem. The chief caution is for the visitor to avoid appearing as an "expert" in a field where he/she is actually a "novice."

4. TRANSFERABILITY AND PORTABILITY OR ACCESS

With these two terms I propose to distinguish two criteria that can be easily confused, and both of which are involved in the transfer of various methodologies. "Transferability" should denote the ease with which a method or technique can be adapted from a "donor" country to a "receiving" (LDC) country or culture; also for the transfer between disciplines. My colleagues like the separation between the two criteria but some would prefer other terms, and *applicability* and *generality* have been proposed as substitutes for transferability. Also the verbs *translate* and *transmit* come to mind. However these have other meanings also, and I propose to continue with *transferability*.

On the other hand, I propose now to use *access* or *portability* to refer to the low cost of the transportation (e.g., by mail) or access to the method. Rather than abstract and difficult definitions, let me illustrate with a few pertinent examples of why we need this distinction and how we can use it for rational, efficient use of the limited funds available for technology transfer. You don't need to agree with my placements in Figure 1, so long as you find playing the game worthwhile.

Mathematics represents a favorable extreme, because it is highly transferable, nearly culture-free, and also easily portable or accessible. Therefore, you don't need to send the mathematicians, just sending their books and papers should suffice. If the trainees come to mathematics classes in the USA, the professors can deliver their usual lectures. Even good knowledge of English is often not crucial, because symbols speak clearly. Computing methods are also transferable and accessible today.

On the other hand, the methods of survey sampling are highly transferable, but are not highly accessible. The basic methods of multistage sampling, clustering and stratification are transferable, but the specific procedures needed differ greatly both between fields (agriculture, schools, households, etc.) and between countries. The application of sampling methods requires intimate knowledge of the local situations, of available sampling frames, and local resources. The local workers must make the local resources and limitations understood by the outside expert; also how the survey

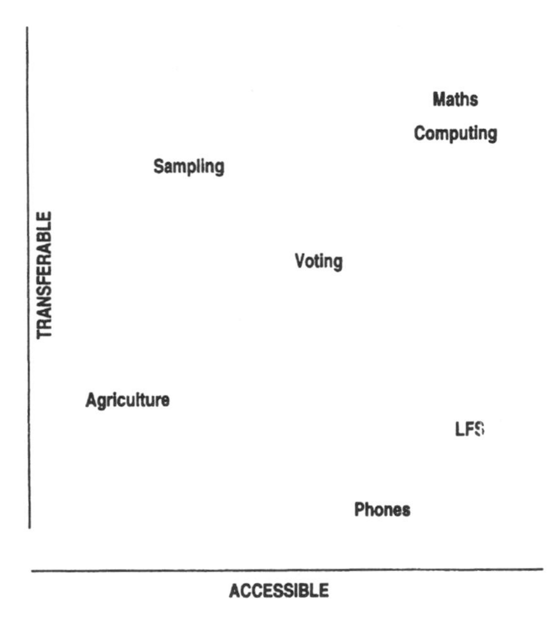

Figure 1. Transferability and Access

data can serve the policies and developments in the LDC. Therefore sampling experts must make personal visits to the LDCs to develop useful, practical, and specific procedures. Alternatively, the local statisticians must come to the expert, or both.

However, for survey methods transferability may be more difficult than for sampling, and that transferability may also differ greatly between various survey methods. For example, telephone surveys for household populations are not transferable to LDCs now. But the situation may be changing rapidly and this judgment may become obsolete before that same fate overtakes most of this paper. Political interviews may also need to be changed drastically. Consumer surveys must cover different items in different countries. Even on "standardized" surveys of the labor force (LFS) the meaning of employed/unemployed suffers from great differences of interpretation of "similar" questions (Kish 1994).

5. CRITERIA FOR EXPERTS' VISITS TO LDCs

Figure 2 represents an attempt at a *rational* order for reasons to invite, to send, and to go on experts' visits to LDCs. You may smile at this attempt, knowing that visits often depend on personal pleasures, curiosity, vanity, and friendships. Furthermore, even my list of rational motives is highly subjective. In any case, frequently the visits are prompted by a mixture of several reasons and motives, and several are better than only one.

For example, I believe that collaborating or consulting on research projects provides the best reasons for inviting an expert, but if the expert is also a good lecturer, she/he may also be invited to lecture, especially if he/she is a good methodologist. Incidentally, I give here even higher rating to longer term collaboration with joint responsibility for the project, than to short term consulting on only one aspect of the project. In either case I have been urging experts and their agencies to combine some lectures and teaching with their consulting visits. These combinations should be explored, and they make the visits both more useful and more enjoyable. Sometimes outside experts can serve two agencies simultaneously, perhaps as a catalyst to bring together the central statistical office with a university. My first and most exciting visit

a. COLLABORATOR on research projects
b. CONSULTANT on research projects, successful PRACTITIONER
c. Great LECTURER
d. Famous PERSONAGE with contacts
e. Important ADMINISTRATOR or STATESMAN
f. Famous THEORIST or METHODOLOGIST
g. Research author and authority
h. Good or famous textbook author
i. Good friend

Figure 2. Criteria for Inviting Visiting Experts

to the People's Republic of China for three weeks in 1982 had me lecturing on sampling at the Academy of Social Sciences every morning, and in the afternoon I worked with the Family Planning Commission designing their giant (1:1000) sample of 350,000 women. Every time I lectured in some country I also found some useful consulting tasks.

6. SURVEY METHODS

The separation between sampling methods and survey methods is not sharp and clean, and as several point out (and I agree with Verma's excellent five points[1]) there is a great overlap. Nevertheless the distinction is useful—like the separation of plants from animals in spite of rare examples of contradictions to all criteria. All samplers should know something about survey methods, and some can even be good in both areas, yet the distinction is useful for the reasons below.

Some of my best colleagues have questioned my concentration on survey sampling and neglecting other and broader aspects of survey methods. For example, my title concerns developing *samplers*. Also at the University of Michigan I organized and conducted for decades a *Sampling* Program for Foreign Statisticians. In my teaching and writing I also concentrated on survey sampling.

I willingly admit the importance of the nonsampling aspects of survey work and acknowledge the difficulties of working on nonsampling errors. I honor my colleagues who work on these problems, whether in Washington, in Ann Arbor, or in the less developed areas of the world. Nonsampling errors are often larger than sampling errors, especially for large samples. We also agree that sampling errors and nonsampling biases are related to each other through mean-square-errors. But sampling errors tend to predominate in small samples and also for the small subclasses even of large samples.

Let me also point to three outstanding examples of efforts to integrate sampling with survey methods. First, the Summer Institute in Survey Research Techniques has been going on at the University of Michigan since 1948. Second, the Joint Program in Survey Methodology started in 1993 as a joint tripartite effort between the Universities of Michigan and Maryland and Westat. Third, the American Statistical Association's Section on Survey Research Methods and the International Association of Survey Statisticians also attempt to combine both fields.

Nevertheless, the two fields are separate and distinct in several important ways, all of which affect the manner in which they can be taught. First, survey methods differ between countries, because of the different cultures, situations, developments, and resources. Survey methods are much less transferable between countries than are sampling methods.

Second, even more important, are differences between fields and disciplines of the survey methods needed to cover them adequately. Survey sampling belongs clearly to the field of statistics, and can be applied to any kind of quantitative research and

---

[1] Verma's discussion of Kish's lecture is given in Verma (1996).

investigation. On the contrary, the methods of observation and measurement are specific to the subject matter and are taught mostly in separate substantive departments. The distinct tools of demography, economics, sociology, marketing, health, medicine, biology, forestry, agriculture, physics, chemistry can only be learned as separate concentrations. I cannot imagine an investigator who would claim to be an expert in all those fields. Furthermore, many fields, like agriculture and sociology, vary from country to country. On the contrary, the methods of survey sampling, like statistics and mathematics, can be applied to obtain and treat quantitative data in any field; also in any country. Nevertheless two strong currents will narrow in the future this gap in transferability of survey methods. One comes from developments of resources in the LDCs. The second comes from improvements being wrought by scores of devoted researchers in survey methods, in cognitive science, and with new theories and techniques.

7. CURRICULA FOR SURVEY SAMPLING

The core of a sampling program for statisticians from LDCs should be a set of about three to six courses in survey sampling. These should include both theory and practical methods, plus workshops and laboratories in data processing and analysis procedures, as needed and available. These should be preceded and accompanied by 2 to 10 other courses in statistics, and I made this range wide deliberately for different individuals and situations. The statistics courses in turn need an adequate base in mathematics, probability, and computing. Hopefully most of these background courses would be part of the prerequisites for admitting statisticians to our sampling programs.

Courses in experimental design would strengthen sample design, especially as part of a master's program. Courses in observational studies would also be welcome, as would courses in other aspects of survey methods. Courses in econometrics, epidemiology, and other related sciences could also be useful additions.

Unfortunately, students of survey sampling face a wide, deep, and lamentable chasm between survey sampling and the mathematical statistical analysis that dominates the statistical curricula. This chasm separates statistical design, both survey sampling and experimental design, from the statistical analysis that occupies the curricula. This separation is confusing for all students, but doubly so for students from LDCs who must face this wide gap between the two branches of statistics they must learn and respect. This deplorable separation prevails in most universities and most countries, not only in the USA.

John Nelder and N.J.R. Jeffers (1994) complain "In many departments in the USA the subject (of experimental design) has almost disappeared from view..." and that "statistical methods seem currently to be focused on analysis... we should give more attention to the design of data collection... Proper attention to the fundamental principles of sampling and experimental design before any data are collected would save a great deal of anguish later...." These complaints come from Rothamsted in

England, the ancestral home of experimental design and of sampling, and the home of the union of statistical practice with theory.

My central complaint is that over 95 percent of statistical attention in curricula, textbooks, and publications is devoted to mathematical statistical analysis and only 5 percent to design. This neglect has deep and wide effects because statistical design is a statistical subject and is best taught in connection with other statistics, with mathematical statistics and analysis.

Why do statistical theory and analysis drive out design? I suggest that it is because theory and analysis can be made mathematical, so that with mathematics it is easier to teach and to set exams; also to write publishable articles. It is easier for faculties fluent in mathematics, and mathematics is a necessary condition to join the faculties of statistics departments. Necessary, I say, but not sufficient. At least some of the faculties should know enough design to be able to teach it. Usually they do not.

I admire mathematics, the noblest creation of the human mind. I also admire theory, which is needed for every instance of inference, decision, and action. But statistical methods are not synonymous with theory, and statistical theory is more than statistical mathematics. Statistical curricula should also have more statistical inference and philosophy of science.

There are several theoretical differences between mainline, classical, mathematical statistics (MS) and survey sampling (SS), and I merely summarize the most basic differences.

1. MS assumes a universe of IID (identically and independently distributed random variables), whereas the populations and samples of SS are stratified, clustered, and often non *epsem* (equal probability selection method), hence weighted. The confusion is even worse when the IID assumptions are not stated. These may be "obvious" to the authors, but they fool most readers. The IID assumptions allow such powerful results, they are so beautiful mathematically, that they are seductive. They seem "natural" like the circular path of planets and the spherical shape of the Earth seemed to be to the ancients, but just as wrong. The path from "If we only had IID, then..." to "Assume that we have IID..." is too smooth, and it should be rougher.
2. The real world—the physical, biological, and social worlds—are **never** IID in my experience and my philosophy. Furthermore, it is seldom convenient or efficient to select a simple random sample. I admit that the violations, either in the population or in the sample, especially for small samples, are not sometimes as violent as other violations (e.g., non-normality). But for large samples, departures from IID and SRS often and typically result in severe problems for statistical inference. (These differences pose obstacles to all statisticians and all students. But the obstacles become even more confusing for statisticians from LDCs studying sampling in the USA.)
3. There is no *clean* boundary between small samples and large samples, and MS asks us to disregard the boundary (ever since R.A. Fisher and likelihood). But SS says that we can't forget it; that n of 1 or 2 is too small for probability

samples, but n of 1000 or 100 is large enough, often n of 30 or even 12. For large samples we depend on ratio means, the Central Limit Theorem, etc. In the old books, Yule and Kendall (1937) for example, there are separate treatments for small and large samples, and it was a mistake to drop it.

4. It is not true, though often said or assumed, that "The object of most surveys is to produce one sample mean $\bar{y}$ or sample total Y." But samples are typically multipurpose, often multipopulation, multidomain, etc.
5. Samples do not consist of random variables but of population elements, each with a vector of measurements that become the survey variables. These differences signify that survey sampling is really based on a different paradigm and these differences pose formidable obstacles for degree students. We must look for ways to bridge them and overcome them.

8. RECRUITING AND PREREQUISITES

Unlike some other fields, perhaps literature and drama, survey sampling has one advantage: it does not attract playboys and idlers. But it has a disadvantage because we are recruiting into a small and obscure field, compared to physics, mathematics, and even to mathematical statistics.

For admissions to sampling programs I emphasize three prerequisites.

1. A reasonable background in mathematics and statistics is needed, but a great deal of judgment is also needed for comparisons of what that means for different LDCs. If the candidates lack enough statistics, these can be compensated by substituting sciences that use applied mathematics: economics, physics, engineering, agronomy, etc.
2. A working knowledge of English is needed. This is no problem in Malaysia, Nigeria, or India. But it is still a problem in Latin America, a monolingual (Spanish-Portuguese) continent, like North America. Also in "francophone" North Africa. These days English is rapidly penetrating, but even now we should reach those who do the real work in statistical offices, but many of whom do not yet speak good English.
3. To help the LDCs we must find and develop statisticians who are and will remain motivated and able to return to apply their sampling skills for and in their LDCs. This need is often in practical conflict with their knowledge of English, because learning English may indicate outbound intentions. Their return may be also in conflict with those professors who measure success by the number of successor PhDs they turn out for US universities.

Our graduates are the best recruiters into our programs, because they know our programs and we can trust their judgment. Also colleagues and experts visiting the LDCs can be helpful in discovering and recruiting talent. University departments can also find them if they are so motivated.

Here I add four actions I used for my Sampling Program at Michigan.

1. In 1963 I went on a four-week trip to statistical offices of South America on a personal, successful trip of recruiting.
2. I wrote sometimes to statistical offices that if they sent us statisticians with enough statistics, English, and motivation, we could turn them into survey samplers in the two summer months of our Sampling Program.
3. I wrote recruiting letters to selected departments of statistics and biostatistics in the USA, with similar promises. We would use their idle summer months to teach the survey sampling that their universities failed to offer.
4. These three sources were all cost effective and made good use of the three modest five-year tracking grants I received from the Ford Foundation. I used these funds for "pump-priming": I would accept good candidates in the fall and then urge and help them to apply for financial support from some international funding agencies. Thus our funds would have to pay for only 25 percent of our Fellows.

9. CRITERIA OF SUCCESS

I would consider our training a failure, from our aims of developing talent for the LDCs, when the trainee remained in the USA either as employee, or as professor, or as a spouse. Successes of training can take several forms. First, the trainee may become chief of the methods division or even the director of the central statistical office in her/his LDC. Or they may be statisticians in the ministry for agriculture, health, etc., or statisticians for the Central Bank or other institutions. Second, they may become university professors, teaching statistics, and, hopefully also some survey sampling courses. Sometimes this teaching may be combined fruitfully with office work.

Third, they may work for international agencies like the UN, FAO, UNFPA, or the World Bank. This activity should be for and with the LDCs and may be a highly useful application of the sampling skills acquired in training.

Fourth, in a few cases our graduates have become successful and useful international consultants, either full time or part time, engaged in interesting and useful roles.

10. MULTINATIONAL SURVEYS

In the past three decades multinational surveys have grown in numbers and size (Kish 1994); and they help the development of samplers for the LDCs. Much of the training in the LDCs has been done in connection with multinational surveys. The World Fertility Survey did a tremendous amount of recruiting and training in the 42 LDCs they covered in the 1970s, and many of their recruits play active roles now; some of them on the Demographic and Health Surveys in 56 countries by Macro International. The United Nations Statistical Office and its National Household Survey Capability Program had a multinational character and influence. The activities of the FAO's

Statistical Division has some of these aspects. More recently UNESCO and the International Education Association have begun to improve the sampling aspects of international comparisons.

11. LANGUAGE BARRIERS AND TRANSLATIONS

Most students and visitors to the USA are assumed, selected, and screened to possess sufficient English for sampling lectures. However, for lectures in the LDCs I have some advice based on trials and errors. It is sometimes said that "today every scientist knows English," especially in statistics, and especially in sampling, which has been developed and written mostly in English. But I doubt those claims: if you want to reach the women and men who do most of the real work, you must reach beyond those who are fluent in English, even if some of them can read sampling in English. Let me share with you two remedies I have used.

First, my lecture is divided into paragraphs, and while the translator presents paragraph n from blackboard 1, I put the outline, figures and formulas for paragraph $n + 1$ on blackboard 2. Then the translator and those fluent in English hear it from me the first time; and later again the second time from the translator, while I prepare paragraph $n + 2$ on blackboard 1.

Second, all questions and discussions from the participants must be in the local language and addressed to the audience, loudly. Then my answers in English are repeated by the translator in the local language. Although some participants can ask and receive in English, local language is essential to foster participation by others who cannot, or who would be shy, and they must be prodded. We announce that all questions must be in the local language, and that "all questions are good," and it is possible to turn any question to a good answer.

In Latin America I lecture in Spanish (with some help), but sending good experts who can talk the local languages is difficult in most countries. The situation may be easier in India, where most participants understand English, than in the Slavic countries where they don't.

French North Africa and the Francophone countries are closed to us, but they get help from INSEE in Paris in French. The 15 Arabic speaking countries pose special problems and opportunities. China and the Chinese culture are another complex and rich world.

12. HISTORY OF SAMPLE SURVEYS IN LDCs

It was only in 1926 that the ISI Assembly admitted the validity of random sampling and only in 1934 that Neyman gave his foundation paper, though this dealt with Census returns. The first national probability samples of households in the USA were the Unemployment Surveys of the Works Progress Administration (WPA) in 1939, the second was conducted by our Division of Program Surveys of the US Department of Agriculture in 1942, and in 1943 the US Census Bureau started its Labor Force Surveys, which became the famous Current Population Surveys.

So it is not surprising that in my trip to South America in 1963 I could not find a single national probability sample of households south of Mexico. I doubt there were any in any LDC, with the conspicuous counter example of the famous National Sample Survey of India. Today, only a generation later, the majority of LDCs have conducted several sample surveys. Often these are stimulated and aided by international agencies, like the UN, FAO, WHO, WFS, and the AID. Therefore I fail to share the pessimism and impatience of those who deplore the poor sampling methods still abounding in many LDCs, or the lack of sample surveys in them. We have come a long way in one generation and we shall go a long way in the next.

This progress was not even in all countries and continents, is much faster in China and fast developing East Asia than in Africa. But lest you become complacent with your generalization, share with me the amazing experience of the World Fertility Surveys. Not only were we able to introduce good sampling and survey methods into 42 LDCs, many of these the first good surveys for the country. In addition, each of these surveys was made "measurable"; that is, sampling errors, variances, and design effects, were also computed and published (Verma, Scott, and O'Muircheartaigh 1980). In contrast, for the 22 European developed countries, beyond the strict control of the London office of the WFS, not even one computed proper sampling errors. This illustrates the powerful effects of international control.

Of course, the development of sampling took place in the context of the many other technical developments of these decades. Perhaps the most important of these is the great spread of primary literacy and numeracy, and especially of secondary education. This is true not only of the industrialized world of Europe, but also of China and East Asia, much of Latin America, and some of Africa. This is most important for hiring good enumerators and good clerks for samples and for censuses.

The spread of telephones has changed many things for the better. It also took over most household surveys in some industrialized countries. This, alas, has introduced a gap in the transferability of survey methods. It will be some time before telephone or mail surveys will be practical for household surveys in most LDCs.

We can finish on a happy note by noting the amazing spread of computing technology. It offers great transferability and portability. The hardware has become rapidly, year by year, both much simpler and much cheaper, and thereby more available to LDCs. The software is also transferable and portable and the number of computing specialists is growing rapidly in the LDCs. The development of samplers certainly includes some computing skills these days and that will help a great deal.

Let me end on a doubly happy note: we have come a long way in one generation, but we have left a lot more work for the next one.

References

Abramowitz, M., and Stegun, I. A. (1965), *Handbook of Mathematical Functions*, New York: Dover Publications.

Aigner, D. J., Goldberger, A. S., and Kalton, G. (1975), "On the Explanatory Power of Dummy Variable Regressions," *International Economic Review*, 16, 503–510.

Alexander, C. H. (1999), "A Rolling Sample Survey for Yearly and Decennial Uses," *Bulletin of the International Statistical Institute*, Contributed Papers, 58, 1, 29–30.

Alexander, C. H. (2002), "Still Rolling: Leslie Kish's 'Rolling Samples' and the American Community Survey," *Survey Methodology*, 28, 35–41.

Anderson, D. W. (1976), "Gains From Multivariate Stratification," unpublished Ph.D. thesis, Ann Arbor, MI: University of Michigan.

Anderson, D., and Aitkin, M. (1985), "Variance Component Models With Binary Response," *Journal of the Royal Statistical Society*, B, Vol. 47, 203–210.

Anderson, D. W., Kish, L., and Cornell, R. G. (1976), "Quantifying Gains From Stratification for Optimum and Approximately Optimum Strata Using a Bivariate Normal Model," *Journal of the American Statistical Association*, 71, 887–892.

Anderson, D. W., Kish, L., and Cornell, R. G. (1980), "On Stratification, Grouping, and Matching," *Scandinavian Journal of Statistics*, 7, 61–66.

Anderson, R. L., and Bancroft, T. A. (1952), *Statistical Theory in Research*, New York: McGraw-Hill.

Aubert, V. (1959), "Chance in Social Affairs," *Inquiry*, 2, 1–24.

Australian Bureau of Statistics (1993), *The Australian Population Monitor*, Canberra: Australian Bureau of Statistics.

Bailar, B. A., Bailey, L., and Corby, C. (1978), "A Comparison of Some Adjustment and Weighting Procedures for Survey Data," in N. K. Namboodiri (ed.), *Survey Sampling and Measurement*, New York: Academic Press.

Bean, J. A., and Burmeister, L. F. (1978), "A Review of Optimal Sample Allocation for Multipurpose Surveys," *Biometrika*, 20, 3–14.

Berry, B. J. L. (1961), "A Method for Deriving Multi-Factor Uniform Regions," *Prezeglad Geograficzny*, 33, 263–279.

Biemer, P. P., Groves, R. M., Lyberg, L. E., Mathiowetz, N. A., and Sudman, S. (eds.) (1991), *Measurement Errors in Surveys*, New York: John Wiley and Sons.

Bjerve, P. J. (1975), "Presidential Address," *Bulletin of the International Statistical Institute*, Invited Papers, 46, 1, 41–48.

Bloom, D., and Idson, T. (1991), "The Practical Importance of Sample Weights," *Proceedings of the Section on Survey Research Methods*, American Statistical Association, 620–624.

Bogue, D. J. (1950), "A Technique for Making Extensive Population Estimates," *Journal of the American Statistical Association*, 45, 149–163.

Bogue, D. J., and Duncan, B. D. (1959), "A Composite Method for Estimating Postcensal Population of Small Areas by Age, Sex and Color," *Vital Statistics—Special Reports*, 47, No. 6, National Office of Vital Statistics.

Bousfield, M. V. (1977), "Intercensal Estimation Using a Current Sample and Census Data," *Review of Public Data Use*, 5, No. 6, 6–15.

Bowley, A. L. (1913), "Working Class Households in Reading," *Journal of the Royal Statistical Society*, 76, 672–691.

Bowley, A. L. (1926), "Measurement of the Precision Attained in Sampling," *Bulletin of the International Statistical Institute*, 22, 1, I 1–62.

Box, G. E. P. (1976), "Science and Statistics," *Journal of the American Statistical Association*, 71, 791–799.

Box, G. E. P., and Tiao, G. C. (1973), *Bayesian Inference in Statistical Analysis*, Cambridge, MA: Addison Wesley.

Brewer, K. R. W., and Mellor, R. W. (1973), "The Effect of Sample Structure on Analytical Surveys," *Australian Journal of Statistics*, 15, 145–152.

Brillinger, D. R. (1964), "The Asymptotic Behavior of Tukey's General Method of Setting Approximate Confidence Limits (the Jack-Knife) When Applied to Maximum Likelihood Estimates," *Review of the International Statistical Institute*, 32, 202–206.

Brillinger, D. R., and Tukey, J. W. (1964), *Asymptotic Variances, Moments, Cumulants, and Other Average Values*, Princeton: Memorandum.

Bryant, E. C., Hartley, H. O., and Jessen, R. J. (1960), "Design and Estimation in Two-Way Stratification," *Journal of the American Statistical Association*, 55, 105–124.

Campbell, D. T. (1957), "Factors Relevant to the Validity of Experiments in Social Settings," *Psychological Bulletin*, 54, 297–312.

Cannell, C. F. (1954), "A Study of the Effects of Interviewers' Expectation Upon Interviewing Results," Ph.D. thesis, Columbus, OH: Ohio State University.

Caplan, D., Haworth, M., and Steel, D. (1999), "UK Labour Market Statistics: Combining Continuous Survey Data Into Monthly Reports," *Bulletin of the International Statistical Institute*, Contributed Papers, 58, 2, 25–26.

Causey, B. D., Cox, L. H., and Ernst, L. R. (1985), "Applications of Transportation Theory to Statistical Problems," *Journal of the American Statistical Association*, 80, 903–909.

Chambers, R. L., and Feeney, G. A. (1977), "Log Linear Models for Small Area Estimation," paper presented at the Joint Conference of the CSIRO Division of Mathematics and Statistics and the Australian Region of the Biometrics Society, Newcastle, Australia, 29 August–2 September, *Biometrics*, Abstract No. 2655.

Chatterjee, S. (1967), "A Note on Optimum Stratification," *Skandinavisk Actuarietidskrift*, 50, 40–44.

Cicchitelli, G., Herzel, A., and Montanari, G. E. (1992), *Il Campionamento Statistico*, Bologna: Il Molino.

Cleland, J., and Scott, C. (1987), *The World Fertility Survey*, Oxford: Oxford University Press.

Cochran, W. G. (1937), "Problems Arising in the Analysis of a Series of Similar Experiments," *Journal of the Royal Statistical Society*, Ser. B, 4, 102–118.

Cochran, W. G. (1946), "Relative Accuracy of Systematic and Stratified Random Samples for a Certain Class of Populations," *Annals of Mathematical Statistics*, 17, 164–177.

Cochran, W. G. (1953), *Sampling Techniques*, New York: John Wiley and Sons.

Cochran, W. G. (1954), "The Combination of Estimates From Different Experiments," *Biometrics*, 10, 101–129.

Cochran, W. G. (1961), "Comparison of Methods for Determining Stratum Boundaries," *Bulletin of the International Statistical Institute*, 38, 345–358.

Cochran, W. G. (1963), *Sampling Techniques*, 2nd edition, New York: John Wiley and Sons.

Cochran, W. G. (1977), *Sampling Techniques*, 3rd edition, New York: John Wiley and Sons.

Cochran, W. G. (1983), *Planning and Analysis of Observational Studies*, New York: John Wiley and Sons.

Cohen, S. B. (1978), "A Modified Approach to Small Area Estimation," unpublished Ph.D. thesis, Chapel Hill, NC: University of North Carolina.

Cohen, S. B., Burt, V. L., and Jones, G. K. (1986), "Efficiencies in Variance Estimation for Complex Survey Data," *American Statistician*, 40, 157–164.

Cohen, S. B., Xanthopoulos, J. A., and Jones, G. K. (1988), "An Evaluation of Statistical Software Procedures Appropriate for the Regression Analysis of Complex Survey Data," *Journal of Official Statistics*, 4, 17–34.

Cohen, S. E. (1969), "A Device for Sampling From Finite Populations on Successive Occasions," *American Statistician*, 23, 4, 26–27.

Cornfield, J. (1954), "Statistical Relationships and Proof in Medicine," *American Statistician*, 8, 19–21.

Cornfield, J., and Tukey, J. W. (1956), "Average Values of Mean Squares in Factorials," *Annals of Mathematical Statistics*, 27, 907–949.

Cox, D. R. (1958), "Some Problems Connected With Statistical Inference," *Annals of Mathematical Statistics*, 29, 357–372.

Curnow, R. N., and Dunnett, C. W. (1962), "The Numerical Evaluation of Certain Multivariate Normal Integrals," *Annals of Mathematical Statistics*, 33, 571–579.

Dalenius, T. (1957), *Sampling in Sweden*, Stockholm: Almquist and Wicksell.

Davis, J. A., Spaeth, J. L., and Hudson, C. (1961), "Analyzing Effects of Group Composition," *American Sociological Review*, 26, 215–225.

Deming, W. E. (1944), "On Errors in Surveys," *American Sociological Review*, 9, 359–369.

Deming, W. E. (1950), *Some Theory of Sampling*, New York: John Wiley and Sons.

Deming, W. E. (1956), "On Simplification of Sampling Design Through Replication With Equal Probabilities and Without Stages," *Journal of the American Statistical Association*, 51, 24–53.

Deming, W. E. (1960), *Sample Design in Business Research*, New York: John Wiley and Sons.

Deming, W. E., and Stephan, F. F. (1940), "On a Least Squares Adjustment of a Sampled Frequency Table When the Expected Marginal Totals are Known," *Annals of Mathematical Statistics*, 11, 427–444.

DuMouchel, W. H., and Duncan, G. S. (1983), "Using Sample Survey Weights in Multiple Regression Analyses of Stratified Samples," *Journal of the American Statistical Association*, 78, 535–543.

DuMouchel, W. H., Govindarajulu, Z., and Rothman, E. (1973), "A Note on Estimating the Variance of a Sample Mean in Stratified Sampling," *Canadian Journal of Statistics*, 1, 267–274.

Duncan, D. B. (1955), "Multiple Range and Multiple *F* Tests," *Biometrics*, 11, 1–42.

Duncan, G. J., and Kalton, G. (1987), "Issues of Design and Analysis of Surveys Across Time," *International Statistical Review*, 55, 97–117.

Durbin, J. (1958), "Sampling Theory for Estimates Based on Fewer Individuals Than the Number Selected," *Bulletin of the International Statistical Institute*, 36, 113–119.

Efron, B., and Morris, C. (1973), "Stein's Estimation Rule and Its Competitors—An Empirical Bayes Approach," *Journal of the American Statistical Association*, 68, 117–130.

Ericksen, E. P. (1971), "A Method for Combining Sample Survey Data and Symptomatic Indicators to Obtain Estimates for Local Areas," unpublished Ph.D. thesis, Ann Arbor, MI: University of Michigan.

Ericksen, E. P. (1973), "A Method for Combining Sample Survey Data and Symptomatic Indicators to Obtain Population Estimates for Local Areas," *Demography*, 10, 137–160.

Ericksen, E. P. (1974), "A Regression Method for Estimating Population Changes of Local Areas," *Journal of the American Statistical Association*, 69, 867–875.

Fay, R. E. (1979), "Some Recent Census Bureau Applications of Regression Techniques to Estimation," in J. Steinberg (ed.), *Synthetic Estimates for Small Areas: Statistical Working Papers and Discussion*, 155–184, NIDA Research Monograph 24, Washington, DC: U.S. Government Printing Office.

Fay, R. E., and Herriot, R. (1977), "Estimates of Income for Small Places: An Application of James-Stein Procedures to Census Data," *Journal of the American Statistical Association*, 74, 269–277.

Fellegi, I. P. (1964), "Response Variance and its Estimation," *Journal of the American Statistical Association*, 59, 1016–1041.

Fellegi, I. P. (1966), "Changing the Probabilities of Selection When Two Units are Selected With PPS Without Replacement," *Proceedings of the Social Statistics Section*, American Statistical Association, 434–442.

Fellegi, I. P. (2000), "Leslie Kish—A Life of Giving," *Survey Methodology*, 26, 119–120.

Fellegi, I. P., and Sunter, A. B. (1974), "Balance Between Different Sources of Survey Errors," *Sankhyā*, 36, 119–142.

Fisher, R. A. (1925), *Statistical Methods for Research Workers*, Edinburgh: Oliver and Boyd.

Fisher, R. A. (1953), *The Design of Experiments*, 6th ed., London: Oliver and Boyd.

Flyer, P., Rust, K., and Morganstein, D. (1989), "Complex Survey Variance Estimation and Contingency Table Analysis Using Replication," *Proceedings of the Section on Survey Research Methods*, American Statistical Association, 110–119.

Folsom, R. E., and Judkins, D. R. (1997), *Substance Abuse in States and Metropolitan Areas: Model Based Estimates From the 1991–1993 NHDSA – Methodology Report*, Rockville, MD: Substance Abuse and Mental Health Services Administration.

Foreman, E. K. (1983), "Integrated Programmes of Household Surveys: Design Aspects," *Bulletin of the International Statistical Institute*, Invited Papers, 50, 2, 1344–1362.

Francis, I. (1981), *Statistical Software: A Comparative Review*, New York: North Holland.

Frankel, L. R., and Stock, J. S. (1942), "On the Sample Survey of Unemployment," *Journal of the American Statistical Association*, 37, 77–80.

Frankel, M. R. (1971), *Inference From Survey Samples*, Ann Arbor, MI: Institute for Social Research, University of Michigan.

Frankel, M. R., and King, B. (1996), "A Conversation With Leslie Kish," *Statistical Science*, 11, 65–87.

Frechtling, J. A., Lorie, J. H., and Schweiger, I. (1950), "1950 Survey of Consumer Finances, Part V, The Distribution of Assets, Liabilities and Net Worth of Consumers, Early 1950," *Federal Reserve Bulletin*, 36, 1595–1597.

Gales, K., and Kendall, M. G. (1957), "An Inquiry Concerning Interviewer Variability," *Journal of the Royal Statistical Society*, Ser. A, 120, 121–147.

Ghangurde, P. D., and Singh, M. P. (1977), "Synthetic Estimation in Periodic Household Surveys," *Survey Methodology*, 3, 152–181.

Gini, C., and Galvani, L. (1929), "Di una Applicazione del Metodo Representativo," *Annali di Statistica*, 6, 4, 1–107.

Glass, G. V. (1976), "Primary, Secondary and Meta-Analysis of Research," *Educational Researcher*, 5, 3–8.

Golder, P. A., and Yeomans, K. A. (1973), "The Use of Cluster Analysis for Stratification," *Applied Statistics*, 22, 213–219.

Gonzalez, M. E. (1973), "Use and Evaluation of Synthetic Estimates," *Proceedings of the Social Statistics Section*, American Statistical Association, 33–36.

Gonzalez, M. E., and Hoza, C. (1978), "Small Area Estimation With Application to Unemployment and Housing Estimates," *Journal of the American Statistical Association*, 73, 7–15.

Gonzalez, M. E., and Waksberg, J. (1973), "Estimation of the Error of Synthetic Estimates," paper presented at the first meeting of the International Association of Survey Statisticians, Vienna, Austria, 18–25 August.

Goodman, L. A. (1959), "Some Alternatives to Ecological Correlation," *American Journal of Sociology*, 64, 610–625.

Goodman, R. (1947), "Sampling for the 1947 Survey of Consumer Finances," *Journal of the American Statistical Association*, 42, 439–448.

Goodman, R., and Kish, L. (1950), "Controlled Selection—a Technique in Probability Sampling," *Journal of the American Statistical Association*, 45, 350–372. Reproduced as Chapter 4 in this volume.

Graubard, B. I., and Korn, E. L. (1991), "The Use of Classical Quadratic Test Statistics for Testing Hypotheses With Complex Survey Data: An Application to Testing Informativeness of Sampling Weights in Multiple Linear Regression Analysis," *Proceedings of the Section on Survey Research Methods*, American Statistical Association, 631–636.

Gray, P. G. (1956), "Examples of Interviewer Variability Taken From Two Sample Surveys," *Applied Statistics*, 5, 73–85.

Greenberg, B. G. (1929), *The Story of Evolution*, New York: Garden City.

Groves, R. M., and Couper, M. P. (1998), *Nonresponse in Household Interview Surveys*, New York: John Wiley and Sons.

Groves, R. M., and Hess, I. (1975), "An Algorithm for Controlled Selection," Chapter 7 in I. Hess, D. C. Riedel, and T. B. Fitzpatrick (eds.), *Probability Sampling of Hospitals and Patients*, 2nd edition, Ann Arbor, MI: Health Administration Press.

Groves, R. M., and Magilavy, L. J. (1986), "Measuring and Explaining Interviewer Effects in Centralized Telephone Surveys," *Public Opinion Quarterly*, 50, 251–256.

Groves, R. M., Dillman, D. A., Eltinge, J. L., and Little, R. J. A. (eds.) (2002), *Survey Nonresponse*, New York: John Wiley and Sons.

Hagood, M. J., and Bernert, E. H. (1945), "Component Indexes as a Basis for Stratification in Sampling," *Journal of the American Statistical Association*, 40, 330–341.

Hansen, M. H., and Hauser, P. M. (1945), "Area Sampling—Some Principles of Sample Design," *Public Opinion Quarterly*, 9, 183–193.

Hansen, M. H., and Hurwitz, W. N. (1943), "On the Theory of Sampling From Finite Populations," *Annals of Mathematical Statistics*, 14, 333–362.

Hansen, M. H., and Hurwitz, W. N. (1949), "On the Determination of Optimum Probabilities in Sampling," *Annals of Mathematical Statistics*, 20, 426–432.

Hansen, M. H., Dalenius, T., and Tepping, B. J. (1985), "The Development of Sample Surveys of Finite Populations," chapter 13 in *A Celebration of Statistics*, The ISI Centenary Volume, Berlin: Springer-Verlag.

Hansen, M. H., Hurwitz, W. N., and Bershad, M. A. (1960), "Measurement Errors in Censuses and Surveys," *Bulletin of the International Statistical Institute*, 38, 2, 359–374.

Hansen, M. H., Hurwitz, W. N., and Madow, W. G. (1953), *Sample Survey Methods and Theory, Volumes I and II*, New York: John Wiley and Sons.

Hansen, M. H., Madow, W. G., and Tepping, B. J. (1978), "On Inference and Estimation From Sample Surveys (With Discussion)," *Proceedings of the Section on Survey Research Methods*, American Statistical Association, 82–91.

Hansen, M. H., Madow, W. G., and Tepping, B. J. (1983), "An Evaluation of Model-Dependent and Probability-Sampling Inferences in Sample Surveys," *Journal of the American Statistical Association*, 78, 776–793.

Hanson, R. H., and Marks, E. S. (1958), "Influence of the Interviewer on the Accuracy of Survey Results," *Journal of the American Statistical Association*, 53, 635–655.

Hartley, H. O. (1959), "Analytic Studies of Survey Data," in *Volume in Onora di Corrado Gini*, Rome: Instituto di Statistica.

Hauser, P. M., and Hansen, M. H. (1944), "On Sampling in Market Surveys," *Journal of Marketing*, 9, 26–31.

Hedges, L. V., and Olkin, I. (1985), *Statistical Methods for Meta-Analysis*, Orlando, FL: Academic Press.

Heeler, R. M., and Day, G. S. (1975), "A Supplementary Note on the Use of Cluster Analysis for Stratification," *Applied Statistics*, 24, 342–344.

Heeringa, S. G., and Liu, J. (1999), "Complex Sample Design Effects and Inference for Mental Health Survey Data," *International Journal of Methods in Psychiatric Research*, 7, 56–65.

Hess, I., Riedel, D. C., and Fitzpatrick, T. B. (eds.) (1975), *Probability Sampling of Hospitals and Patients*, 2nd edition, Ann Arbor (MI): Health Administration Press.

Hess, I., Sethi, V. K., and Balakrishnan, T. R. (1966), "Stratification: A Practical Investigation," *Journal of the American Statistical Association*, 61, 74–90.

Hidiroglou, M. A., and Srinath, K. P. (1981), "Some Estimators of a Population Total From Simple Random Samples Containing Large Units," *Journal of the American Statistical Association*, 76, 690–695.

Hoffman, B. (1972), *Albert Einstein, Creator and Rebel*, New York: Viking Press.

Holt, D., Smith, T. M. F., and Winter, P. D. (1980), "Regression Analysis of Data From Complex Surveys," *Journal of the Royal Statistical Society*, Ser. A, 143, 474–487.

Horvitz, D. G. (1952), "Sampling and Field Procedures of the Pittsburgh Morbidity Survey," *Public Health Reports*, 67, 1003–1012.

Hox, J. J., de Leeuw, E. D., and Kreft, I. G. G. (1991), "The Effect of Interviewer and Respondent Characteristics on the Quality of Survey Data: A Multilevel Model," in *Measurement Errors in Surveys,* P. P. Biemer, R. M. Groves, L. E. Lyberg, N. A. Mathiowetz, and S. Sudman (eds), New York: John Wiley and Sons.

Houseman, E. E. (1947), "The Sample Design for a National Farm Survey by the Bureau of Agricultural Economics," *Journal of Farm Economics*, 24, 241–245.

Hyman, H. H., Cobb, W. J., Feldman, J. J., Hart, C. W., and Stember, C. H. (1954), *Interviewing in Social Research*, Chicago: University of Chicago Press.

Iannacchione, V. G., Milne, J. G., and Folsom, R. E. (1991), "Response Probability Weight Adjustments Using Logistic Regression," *Proceedings of the Section on Survey Research Methods*, American Statistical Association, 637–642.

Isnard, M. (1999), *Alternatives to Traditional Census Taking: The French Experience*, Paris, France: INSEE.

Jensen, A. (1926), "The Representative Method in Practicc," *Bulletin of the International Statistical Institute*, 22, 1, 381–439.

Jensen, A. (1928), "Purposive Selection," *Journal of the Royal Statistical Society*, 91, 541–547.

Johnson, E. G., and King, B. V. (1987), "Generalized Variance Functions for Complex Sample Surveys," *Journal of Official Statistics*, 3, 235–250.

Kalsbeek, W. D. (1973), "A Method for Obtaining Local Postcensal Estimates for Several Types of Variables," unpublished Ph.D. thesis, Ann Arbor, MI: University of Michigan.

Kalton, G. (1968), "Standardization: A Technique to Control for Extraneous Variables," *Applied Statistics*, 17, 118–136.

Kalton, G. (1979), "Ultimate Cluster Sampling," *Journal of the Royal Statistical Society*, Ser. A, 142, 210–222.

Kalton, G. (1983a), *Compensating for Missing Survey Data*, Ann Arbor (MI): Institute for Social Research.

Kalton, G. (1983b), "Models in the Practice of Survey Sampling," *International Statistical Review*, 51, 175–188.

Kalton, G., and Blunden, R. M. (1973), "Sampling Errors in the British General Household Survey," *Bulletin of the International Statistical Institute*, 45, 3, 83–97.

Kalton, G. (2002), "Leslie Kish's Impact on Survey Statistics," *Survey Methodology*, 28, 25–29.

Kaplan, W. (1952), *Advanced Calculus*, Reading, MA: Addison-Wesley.

Katona, G., Kish, L., Lansing, J. B., and Dent, J. D. (1950), "Methods of the Surveys of Consumer Finances," *Federal Reserve Bulletin*, 36, 685–809.

Kempthorne, O., et al. (1954), *Statistics and Mathematics in Biology*, Ames, IA: The Iowa State College Press.

Kendall, M. G. (1948), *The Advanced Theory of Statistics, Volume 1*, London: Griffin & Co.

Kendall, M. G. (1951), "Regression, Structure and Functional Relationship, Part I," *Biometrika*, 38, 12–25.

Kendall, M. G. (1952), "Regression, Structure and Functional Relationship, Part II," *Biometrika*, 39, 96–108.

Kendall, M. G. (1968), "Chance," in *Dictionary of History of Ideas*, New York: Charles Scribner & Sons, 335–340.

Kendall, M. G., and Buckland, W. R. (1957), *A Dictionary of Statistical Terms*, London: Oliver and Boyd.

Kendall, M. G., and Buckland, W. R. (1982), *A Dictionary of Statistical Terms*, London: Longman Group.

Kendall, P. L., and Lazarsfeld, P. F. (1950), "Problems of Survey Analysis," in R. K. Merton and P. F. Lazarsfeld (eds.), *Continuities in Social Research*, Glencoe, IL: Free Press.

Keyfitz, N. (1951), "Sampling With Probabilities Proportional to Size," *Journal of the American Statistical Association*, 58, 183–201.

Keyfitz, N. (1953), "A Factorial Arrangement of Comparisons of Family Size," *American Journal of Sociology*, 58, 470–480.

Kiaer, A. N. (1895a), "Observations et Expériences Concernant les Dénombrements Représentetifs," *Bulletin of the International Statistical Institute*, 9, I, 176–183.

Kiaer, A. N. (1895b), *The Representative Method of Statistical Surveys*, English translation 1976, Oslo: Statistik Centralbyro.

King, A. J., and Jessen, R. J. (1945), "The Master Sample of Agriculture: I, Development and Use; II, Design," *Journal of the American Statistical Association*, 40, 38–56.

Kish, L. (1949), "A Procedure for Objective Respondent Selection Within the Household," *Journal of the American Statistical Association*, Vol. 44, 380–387. Reproduced as Chapter 3 in this volume.

Kish, L. (1957), "Confidence Intervals for Clustered Samples," *American Sociological Review*, 22, 154–165.

Kish, L. (1959), "Some Statistical Problems in Research Design," *American Sociological Review*, 24, 328–338. Reproduced as Chapter 2 in this volume.

Kish, L. (1961a), "Efficient Allocation of a Multipurpose Sample," *Econometrica*, 29, 363–385.

Kish, L. (1961b), "A Measurement of Homogeneity in Areal Units," *Bulletin of the International Statistical Institute*, 38, 4, 201–209.

Kish, L. (1962), "Studies of Interviewer Variance for Attitudinal Variables," *Journal of the American Statistical Association*, 57, 92–115. Reproduced as Chapter 12 in this volume.

Kish, L. (1963), "Changing Strata and Selection Probabilities," *Proceedings of the Social Statistics Section*, American Statistical Association, 139–143.

Kish, L. (1965a), *Survey Sampling*, New York: John Wiley and Sons.

Kish, L. (1965b), "Sampling Organizations and Groups of Unequal Sizes," *American Sociological Review*, 30, 564–572. Reproduced as Chapter 5 in this volume.

Kish, L. (1968), "Standard Errors for Indexes From Complex Samples," *Journal of the American Statistical Association*, 63, 512–529.

Kish, L. (1969), "Design and Estimation for Subclasses, Comparisons, and Analytical Statistics," in N. L. Johnson and H. Smith (eds.), *New Developments in Survey Sampling*, New York: John Wiley and Sons.

Kish, L. (1971), "Multipurpose Programs for Sampling Errors," *Bulletin of the International Statistical Institute*, Contributed Papers, 44, 216–221.

Kish, L. (1975), "Representation, Randomization and Control," chapter 9 in H. M. Blalock, A. Aganbegian, F. M. Borodkin, R. Boudon, and V. Capecchi (eds.) *Quantitative Sociology*, New York: Academic Press.

Kish, L. (1976), "Optima and Proxima in Linear Sample Designs," *Journal of the Royal Statistical Society*, Ser. A, 139, 80–95.

Kish, L. (1977), "Robustness in Survey Sampling," *Bulletin of the International Statistical Institute*, 47, 515–528.

Kish, L. (1978), "Chance, Statistics, and Statisticians," *Journal of the American Statistical Association*, 73, 1–6. Reproduced as Chapter 16 in this volume.

Kish, L. (1979a), "Samples and Censuses," *International Statistical Review*, 47, 99–109.

Kish, L. (1979b), "Rotating Samples Instead of Censuses," *Asian and Pacific Census Forum* (East-West Center, Honolulu), 6, 1–13.

Kish, L. (1980), "Design and Estimation for Domains," *Statistician*, 29, 209–222.

Kish, L. (1981), *Using Cumulated Rolling Samples*, Washington: Library of Congress.

Kish, L. (1984), "Analytical Statistics From Complex Samples," *Survey Methodology*, 10, 1–7.

Kish, L. (1985), "Sample Surveys Versus Experiments, Controlled Observations, Censuses, Registers, and Local Studies," *Australian Journal of Statistics*, 27, 111–122.

Kish, L. (1987a), *Statistical Design for Research*, New York: John Wiley and Sons.

Kish, L. (1987b), "Discussion" in R. Platek, J. N. K. Rao, C.-E. Särndal, and M. P. Singh (eds.) *Small Area Statistics*, New York: John Wiley and Sons.

Kish, L. (1988a), "Multipurpose Sample Designs," *Survey Methodology*, 14, 19–32. Reproduced as Chapter 13 in this volume.

Kish, L. (1988b), "A Taxonomy of Elusive Populations," *Proceedings of the Section on Survey Research Methods*, American Statistical Association, 44–46.

Kish, L. (1989), *Sampling Methods for Agricultural Surveys*, Rome: FAO.

Kish, L. (1990a), "Rolling Samples and Censuses," *Survey Methodology*, 16, 63–71.

Kish, L. (1990b), "Weighting: Why, When, and How," *Proceedings of the Section on Survey Research Methods*, American Statistical Association, 121–130.

Kish, L. (1992), "Weighting for Unequal P_i," *Journal of Official Statistics*, 8, 183–200. Reproduced as Chapter 10 in this volume.

Kish, L. (1994), "Multipopulation Survey Designs," *International Statistical Review*, 62, 167–186.

Kish, L. (1995a), *Questions/Answers (1978–1994)*, Paris: International Association of Survey Statisticians.

Kish, L. (1995b), "Methods for Design Effects," *Journal of Official Statistics*, 11, 55–77. Reproduced as Chapter 9 in this volume.

Kish, L. (1995c), "The Hundred Years' Wars of Survey Sampling," *Statistics in Transition*, 2, 813–830. Reproduced as Chapter 16 in this volume.

Kish, L. (1996), "Developing Samplers for Developing Countries," *International Statistical Review*, 64, 143–162. Reproduced as Chapter 17 in this volume.

Kish, L. (1998), "Space/Time Variations and Rolling Samples," *Journal of Official Statistics*, 14, 31–46.

Kish, L. (1999), "Cumulating/Combining Population Surveys," *Survey Methodology*, 25, 129–138. Reproduced as Chapter 15 in this volume.

Kish, L., and Anderson, D. W. (1978), "Multivariate and Multipurpose Stratification," *Journal of the American Statistical Association*, 73, 24–34. Reproduced as Chapter 7 in this volume.

Kish, L., and Frankel, M. (1970), "Balanced Repeated Replications for Standard Errors," *Journal of the American Statistical Association*, 65, 1071–1094.

Kish, L., and Frankel, M. R. (1974), "Inference From Complex Samples," *Journal of the Royal Statistical Society*, Ser. B, 36, 1–37. Reproduced as Chapter 8 in this volume.

Kish, L., and Hess, I. (1958), "On Noncoverage of Sample Dwellings," *Journal of the American Statistical Association*, 53, 509–524.

Kish, L., and Hess, I. (1959a), "Some Sampling Techniques for Continuing Survey Operations," *Proceedings of the Social Statistics Section*, American Statistical Association, 139–143.

Kish, L., and Hess, I. (1959b), "A 'Replacement' Procedure for Reducing Bias of Nonresponse," *American Statistician*, 13, 17–19.

Kish, L., and Lansing, J. B. (1954), "Response Errors in Estimating the Value of Homes," *Journal of the American Statistical Association*, 49, 520–538. Reproduced as Chapter 11 in this volume.

Kish, L., and Scott, A. J. (1971), "Retaining Units After Changing Strata and Probabilities," *Journal of the American Statistical Association*, 66, 461–470. Reproduced as Chapter 6 in this volume.

Kish, L., and Slater, C. W. (1960), "Two Studies of Interviewer Variance of Socio-Psychological Variables," *Proceedings of the Social Statistics Section*, American Statistical Association, 66–70.

Kish, L., Frankel, M. R., and Van Eck, N. (1972), "SEPP: Sampling Errors Program Package," Ann Arbor, MI: Institute for Social Research, University of Michigan.

Kish, L., Frankel, M. R., Verma, V., and Kaciroti, N. (1995), "Design Effects for Correlated $(P_i - P_j)$," *Survey Methodology*, 21, 117–124.

Kish, L., Groves, R. M., and Krotki, K. (1976), "Sampling Errors for Fertility Surveys," Occasional Paper No. 17, World Fertility Survey, Voorburg, Netherlands: International Statistical Institute.

Kish, L., Lovejoy, W., and Rackow, P. (1961), "A Multi-Stage Probability Sample for Traffic Surveys," *Proceedings of the Social Statistics Section*, American Statistical Association, 227–230.

Kish, L., Namboodiri, N. K., and Pillai, R. K. (1962), "The Ratio Bias in Surveys," *Journal of the American Statistical Association*, 57, 863–876.

Korn, E. L., and Graubard, B. I. (1999), *Analysis of Health Surveys*, New York: John Wiley and Sons.

Kott, P. S. (1991a), "A Model-Based Look at Linear Regression in the Survey Data," *American Statistician*, 45, 107–112.

Kott, P. S. (1991b), "Hypothesis Testing of Linear Regression Coefficients With Survey Data," *Proceedings of the Section on Survey Research Methods*, American Statistical Association, 24–33.

Kovalevsky, A. G. (1924), *Basic Theory of Sampling Methods*, Saratov USSR: Vestnik Statistiki, 20.

Kruskal, W. H., and Mosteller, F. (1979a), "Representative Sampling, I: Non-Scientific Literature," *International Statistical Review*, 47, 13–24.

Kruskal, W. H., and Mosteller, F. (1979b), "Representative Sampling, II: Scientific Literature," *International Statistical Review*, 47, 111–127.

Kruskal, W. H., and Mosteller, F. (1979c), "Representative Sampling, III: The Current Literature," *International Statistical Review*, 47, 245–265.

Kruskal, W. H., and Mosteller, F. (1980), "Representative Sampling, IV: The History of the Concept in Statistics, 1895–1939," *International Statistical Review*, 48, 169–195.

Laake, P. (1978), "An Evaluation of Synthetic Estimates of Employment," *Scandinavian Journal of Statistics*, 5, 57–60.

Laake, P., and Langva, H. K. (1976), "The Estimation of Employment Within Geographical Regions: On the Bias, Variance and Mean Square Error of the Estimates" (in Norwegian with English summary), Central Bureau of Statistics of Norway, Article 88.

Lahiri, D. B. (1963), "Multi-Subject Sample Survey System—Some Thoughts Based on the Indian Experience," in *Contributions to Statistics*, presented to Mahalanobis on his 70th birthday, London: Pergamon Press, and Calcutta: Statistical Publishing Society.

Lê, T. N., and Verma, V. K. (1997), *An Analysis of Sample Designs and Sampling Errors of the Demographic and Health Surveys*, Calverton, MD: Macro International.

Lee, H. (1991), "Model Based Estimators That are Robust to Outliers," *Proceedings of the Bureau of the Census Annual Research Conference*, 178–196, Washington: U.S. Bureau of the Census.

Lepkowski, J. M. (1980), "Design Effects for Multivariate Categorical Interactions," Ph.D. thesis, Ann Arbor, MI: University of Michigan.

Levy, P. S. (1971), "The Use of Mortality Data in Evaluating Synthetic Estimates," *Proceedings of the Social Statistics Section*, American Statistical Association, 328–331.

Li, C. C. (1956), "The Concept of Path Coefficient and its Impact on Population Genetics," *Biometrics*, 12, 190–209.

Lin, T. K. (1992), "Some Improvements on an Algorithm for Controlled Selection," *Proceedings of the Section on Survey Research Methods*, American Statistical Association, 407–410.

Little, R. J. A., and Rubin, D. B. (1987), *Statistical Analysis With Missing Data,* New York: John Wiley and Sons.

Lyberg, L. E., Biemer, P. P., Collins, M., de Leeuw, E., Dippo, C., Schwarz, N., and Trewin, D. (eds.) (1997), *Survey Measurement and Process Quality*, New York: John Wiley and Sons.

Macro International Inc. (1994), "An Assessment of the Quality of Health Data in the DHS-I Surveys," DHS Methodological Reports No. 2, Calverton, Maryland: Macro International Inc.

Macura, M., and Cleland, J. (1985), *A Celebration of Statistics: the ISI Centenary Volume*, A. C. Atkinson, and S. E. Fienberg (eds.), New York: Springer Verlag.

Madow, W. G. (1949), "On the Theory of Systematic Sampling II," *Annals of Mathematical Statistics*, 20, 333–354.

Madow, W. G., and Madow, L. (1944), "On the Theory of Systematic Sampling I," *Annals of Mathematical Statistics*, 15, 1–24.

Mahalanobis, P. C. (1944), "On Large-Scale Sample Surveys," *Philosophical Transactions of the Royal Society of London*, Ser. B, 231, 329–451.

Mahalanobis, P. C. (1946), "Recent Experiments in Statistical Sampling in the Indian Statistical Institute," *Journal of the Royal Statistical Society*, 109, 325–370.

Mahalanobis, P. C., and Sen, S. B. (1954), "On Some Aspects of the Indian NSS," *Bulletin of the International Statistical Institute*, 34, 2, 5–14.

Martin, J. H., and Serow, W. J. (1978), "Estimating Demographic Characteristics Using the Ratio-Correlation Method," *Demography*, 15, 223–233.

McCarthy, P. J. (1966), *Replication: An Approach to the Analysis of Data From Complex Surveys*, Vital and Health Statistics, Series 2, No. 14, U.S. Department of Health, Education, and Welfare, Washington: U.S. Government Printing Office.

McGinnis, R. (1958), "Randomization and Inference in Sociological Research," *American Sociological Review*, 23, 408–414.

McNeill, W. H. (1976), *Plagues and Peoples*, Garden City, NY: Anchor Press.

Molinari, G. (1993), "Design Effects and Ratio of Homogeneity in Complex Sampling Designs," *Statistica*, 53, 633–646.

Monod, J. (1971), *Chance and Necessity*, New York: Vintage Books.

Mooney, H. W. (1956), *Methodology in Two California Health Surveys, San Jose (1952) and Statewide (1954–55)*, U.S. Public Health Monograph No. 70.

Morrison, P. A., and Relles, D. A. (1975), "A Method for Monitoring Small Area Population Changes in Cities," *Review of Public Data Use*, 3, 2, 10–15.

Moser, C. A., and Stuart, A. (1953), "An Experimental Study of Quota Sampling," *Journal of the Royal Statistical Society*, Ser. A, 106, 315–383.

Murthy, M. N. (1967), *Sampling Theory and Methods*, Calcutta: Statistical Publishing Society.

Namboodiri, N. K. (1972), "On the Ratio-Correlation and Related Methods of Subnational Population Estimation," *Demography*, 9, 443–453.

Namekata, T. (1974), "Synthetic State Estimates of Work Disability," unpublished Ph.D. thesis, Champaign, IL: University of Illinois.

Namekata, T., Levy, P. S., and O'Rourke, T. W. (1975), "Synthetic Estimates of Work Loss Disability for Each State and the District of Columbia," *Public Health Reports*, 90, 532–538.

Nathan, G. (1972), "On Asymptotic Power of Tests for Independence in Contingency Tables From Complex Stratified Samples," *Journal of the American Statistical Association*, 67, 917–920.

Nathan, G. (1973), *Approximate Tests of Independence in Contingency Tables From Complex Stratified Cluster Samples*, Vital and Health Statistics, Ser. 2, No. 53, Washington, DC: National Center for Health Statistics,.

National Research Council (2000a), *Small-Area Income and Poverty Estimates: Priorities for 2000 and Beyond.* Panel on Estimates of Poverty for Small Geographic Areas, C. F. Citro and G. Kalton (eds.). Committee on National Statistics, Washington, DC: National Academy Press.

National Research Council (2000b), *Small-Area Estimates of School-Age Children in Poverty: ITAC Evaluation of Current Methodology*, Panel on Estimates of Poverty for Small Geographic Areas, C. F. Citro and G. Kalton (eds.), Committee on National Statistics, Washington, DC: National Academy Press.

Nelder, J., and Jeffers, J. N. R. (1994), *News and Notes*, February, London: Royal Statistical Society.

Neyman, J. (1933), *An Outline of the Theory and Practice of the Representative Method Applied in Social Research*, Institute for Social Problems, Warsaw (in Polish, English summary).

Neyman, J. (1934), "On the Two Different Aspects of the Representative Method: The Method of Stratified Sampling and the Method of Purposive Selection," *Journal of the Royal Statistical Society*, 97, 558–625.

Neyman, J. (1938/1952), *Lectures and Conferences on Mathematical Statistics and Probability*, Washington, DC: Graduate School of the U.S. Department of Agriculture. Mimeographed lectures were printed and bound in 1952.

Nicholls, A. (1977), "A Regression Approach to Small Area Estimation," Canberra, Australia: Australian Bureau of Statistics (mimeographed).

O'Hare, W. (1976), "Report on a Multiple Regression Method for Making Population Estimates," *Demography*, 13, 369–379.

O'Muircheartaigh, C. A., and Campanelli, P. C. (1998), "The Relative Impact of Sampling and Interviewer Effect on Survey Variance," *Journal of the Royal Statistical Society*, Series A, 161, 1, 63–78.

O'Muircheartaigh, C. A., and Campanelli, P. C. (1999), "A Multilevel Exploration of the Role of Interviewers in Survey Nonresponse," *Journal of the Royal Statistical Society*, Series A, 162, 3, 437–446.

O'Muircheartaigh, C. A., and Wiggins, R. D. (1981), "The Impact of Interviewer Variability in an Epidemiological Study," *Psychological Medicine*, 11, 817–824.

O'Muircheartaigh, C., and Wong, S. T. (1981), "The Impact of Sampling Theory on Survey Sampling Practice: A Review," *Bulletin of the International Statistical Institute*, Invited Papers, 49, 1, 465–493.

Pannekoek, J. (1991), "A Mixed Model for Analyzing Measurement Errors for Dichotomous Variables," in P. P. Biemer, R. M. Groves, L. E. Lyberg, N. A. Mathiowetz, and S. Sudman (eds.), *Measurement Errors in Surveys*, New York: John Wiley and Sons.

Pfeffermann, D., and La Vange, L. (1989), "Regression Models for Stratified Multi-Stage Cluster Samples," chapter 12 in C. J. Skinner, D. Holt, and T. M. F. Smith (eds.), *Analysis of Complex Surveys*, New York: John Wiley and Sons.

Piaget, J., and Inhelder, B. (1975), *The Origin of the Idea of Chance in Children*, New York: W. W. Horton.

Platek, R., Rao, J. N. K., Särndal, C.-E., and Singh, M. P. (eds.) (1987), *Small Area Statistics*, New York: John Wiley and Sons.

Pocock, D. C. D., and Wishart, D. (1969), "Methods of Deriving Multi-Factor Uniform Regions," *Transactions of the Institute of British Geographers*, 47, 73–98.

Popper, K. R. (1959), *The Logic of Scientific Discovery*, London: Hutchinson.

Porter, T. M. (1987), *The Rise of Statistical Thinking: 1820–1900*, Princeton: Princeton University Press.

Potter, F. J. (1988), "Survey of Procedures to Control Extreme Sampling Weights," *Proceedings of the Section on Survey Research Methods*, American Statistical Association, 453–458.

Potter, F. J. (1990), "A Study of Procedures to Identify and Trim Extreme Sampling Weights," *Proceedings of the Section on Survey Research Methods*, American Statistical Association, 225–230.

Purcell, N. J. (1979), "Efficient Estimation for Small Domains: A Categorical Data Analysis Approach," unpublished Ph.D. thesis, Ann Arbor, MI: University of Michigan.

Purcell, N. J., and Kish, L. (1979), "Estimation for Small Domains," *Biometrics*, 35, 365–384.

Purcell, N. J., and Kish, L. (1980), "Postcensal Estimates for Local Areas (Or Domains)," *International Statistical Review*, 48, 3–18. Reproduced as Chapter 14 in this volume.

Purcell, N. J., and Linacre, S. (1976), "Techniques for the Estimation of Small Area Characteristics," paper presented at the 3rd Australian Statistical Conference, Melbourne, Australia, 18–20 August.

Pursell, D. E. (1970), "Improving Population Estimates With the Use of Dummy Variables," *Demography*, 7, 87–91.

Rao, C. R. (1971), "Inference in Sampling From Finite Populations," in V. P. Godambe, and D. A. Sprott (eds.), *Foundations of Statistical Inference*, Toronto: Holt, Rinehart & Winston.

Rao, J. N. K. (1966), "Alternative Estimators in PPS Sampling for Multiple Characteristics," Sankhyā, A, 28, 47–60.

Rao, J. N. K. (2003), *Small Area Estimation*, New York: John Wiley and Sons.

Rao, J. N. K. and Scott, A. J. (1981), "The Analysis of Categorical Data From Complex Sample Surveys," *Journal of the American Statistical Association*, 76, 221–230.

Rao, J. N. K., and Scott, A. J. (1987), "On Simple Adjustments to Chi-Square Tests With Sample Survey Data," *Annals of Statistics*, 15, 385–397.

Rao, J. N. K., and Shao, J. (1999), "Modified Balanced Repeated Replication for Complex Survey Data," *Biometrika*, 86, 2, 403–415.

Rao, J. N. K., and Wu, C. F. J. (1985), "Inference From Stratified Samples," *Journal of the American Statistical Association*, 80, 620–630.

Rao, J. N. K., and Wu, C. F. J. (1987), "Resampling Inference From Complex Survey Data," *Journal of the American Statistical Association*, 83, 231–241.

Robinson, W. S. (1959), "Ecological Correlations and the Behavior of Individuals," *American Sociological Review*, 15, 351–357.

Rodriquez-Vera, A. (1982), *Multipurpose Optimal Sample Allocation Using Mathematical Programming*, Ph.D. thesis, Ann Arbor, MI: University of Michigan.

Roethlisberger, F. J., and Dickson, W. J. (1939), *Management and the Worker*, Cambridge: Harvard University Press.

Rosenbaum, P. R. (2002), *Observational Studies*, 2nd edition, New York: Springer-Verlag.

Rosenberg, H. (1968), "Improving Current Population Estimates Through Stratification," *Land Economics*, 44, 331–338.

Rubin, D. B. (1987), *Multiple Imputation for Nonresponse in Surveys*, New York: John Wiley and Sons.

Rust, K. F. (1984), "Techniques for Estimating Variances for Sample Surveys," Ph.D. thesis, Ann Arbor, MI: University of Michigan.

Salmon, W. C. (1967), *The Foundation of Scientific Inference*, Pittsburg: The University of Pittsburgh Press.

Savage, R. J. (1957), "Nonparametric Statistics," *Journal of the American Statistical Association*, 52, 332–333.

Scardovi, I. (1976), "Pagina Dimenticate di Storia della Scienzia," *Genus*, 32, 1–44 (English summary).

Schaible, W. L. (1996), *Indirect Estimators in U.S. Federal Programs*, New York: Springer-Verlag

Schaible, W. L., Brock, D. B., and Schnack, G. A. (1977), "An Empirical Comparison of the Simple Inflation, Synthetic and Composite Estimators for Small Area Statistics," *Proceedings of the Social Statistics Section*, American Statistical Association, 1017–1021.

Scheffé, H. (1953), "A Method for Judging All Contrasts in the Analysis of Variance," *Biometrika*, 40, 87–104.

Schmitt, R. C., and Crosetti, A. H. (1954), "Accuracy of the Ratio-Correlation Method of Estimating Postcensal Population," *Land Economics*, 30, 279–280.

Sedransk, J. (1965), "Designing Some Multi-Factor Studies," *Journal of the American Statistical Association*, 62, 1121–1139.

Selvin, H. C. (1957), "A Critique of Tests of Significance in Survey Research," *American Sociological Review*, 22, 519–527.

Sethi, V. K. (1963), "Contributions to Stratified Sampling and Some Related Problems," unpublished Ph.D. thesis, Institute of Social Sciences, Agra, India: Agra University.

Sewell, W. H. (1952), "Infant Training and the Personality of the Child," *American Journal of Sociology*, 58, 150–159.

Shadish, W. R., Cook, T. D., and Campbell, D. T. (2002), *Experimental and Quasi-Experimental Designs for Generalized Causal Inference*, Boston, MA: Houghton-Mifflin.

Sharot, T. (1986), "Weighting Survey Results," *Journal of the Market Research Society*, 28, 269–284.

Simon, H. A. (1954), "Spurious Correlation: A Causal Interpretation," *Journal of the American Statistical Association*, 49, 467–479; also in his *Models of Man*, New York: John Wiley and Sons, 1956.

Sitter, R. R., and Skinner, C. J. (1994), "Multi-Way Stratification by Linear Programming," *Survey Methodology*, 20, 65–73.

Skinner, C. J. (1989), "Domain Means, Regressions and Multivariate Analysis," chapter 3 in C. J. Skinner, D. Holt, and T. M. F. Smith (eds.), *Analysis of Complex Surveys*, New York: John Wiley and Sons.

Skinner, C. J., Holt, D., and Smith, T. M. F. (eds.) (1989), *Analysis of Complex Surveys*, New York: John Wiley and Sons.

Smith, T. M. F. (1976), "The Foundations of Survey Sampling: A Review," *Journal of the Royal Statistical Society*, Ser. A, 139, 183–204.

Spencer, B., and Cohen, T. (1991), "Shrinkage Weights for Unequal Probability Samples," *Proceedings of the Section on Survey Research Methods*, American Statistical Association, 625–630.

Starsinic, D. E. (1974), "Development of Population Estimates for Revenue Sharing Areas," *Census Tract Papers, Series GE* 40, No. 10, U.S. Bureau of the Census, Washington, DC: U.S. Government Printing Office.

Stephan, F. F. (1948), "History of the Uses of Modern Sampling Procedures," *Journal of the American Statistical Association*, 43, 12–39.

Stock, J. S., and Hochstim, J. R. (1951), "A Method of Measuring Interviewer Variability," *Public Opinion Quarterly*, 15, 322–334.

Strand, N. V., and Jessen, R. J. (1943), "Some Investigations of the Suitability of the Township as a Unit for Sampling Iowa Agriculture," Iowa State College Research Bulletin 315.

"Student" (1908), "The Probable Error Of The Mean," *Biometrika*, 6, 1–25.

"Student" (1909), "The Distribution of Means of Samples Which are not Drawn at Random," *Biometrika*, 7, 210–215.

Sukhatme, P. V. (1954), *Sampling Theory of Surveys With Applications*, Ames, IA: Iowa State College Press.

Swanson, D. A. (1978), "An Evaluation of 'Ratio' and 'Difference' Regression Methods for Estimating Small, Highly Concentrated Populations: The Case of Ethnic Groups," *Review of Public Data Use*, 6, 4, 18–27.

Szalai, A. (1972), *The Use of Time*, The Hague: Mouton.

Tambay, J.-L., and Catlin, G. (1995), "Sample Design of the National Population Health Survey, *Health Reports*, Catalogue No. 82-003, 7, 29–38.

Tchuprow, A. A. (1923), "On the Mathematical Expectation of the Moments of Frequency Distributions in the Case of Correlated Observations," *Metron*, 2, 646–680.

Tepping, B. J. (1968), "Variance Estimation in Complex Surveys," *Proceedings of the Social Statistics Section*, American Statistical Association, 11–18.

Tepping, B. J., Hurwitz, W. N., and Deming, W. E. (1943), "On the Efficiency of Deep Stratification in Block Sampling," *Journal of the American Statistical Association*, 38, 93–100.

Tharakan, T. C. (1969), *Inference Based on Complex Samples From Finite Populations*, Ph.D. thesis, Ann Arbor, MI: University of Michigan.

Thompson, J. R. (1968), "Some Shrinkage Techniques for Estimating the Mean," *Journal of the American Statistical Association*, 63, 113–122.

Thomsen, I. (1977), "On the Effect of Stratification When Two Stratifying Variables are Used," *Journal of the American Statistical Association*, 72, 149–153.

Townsend, P. (1962), *The Last Refuge*, London: Routledge and Kegan Paul.

Tukey, J. W. (1949), "Comparing Individual Means in the Analysis of Variance," *Biometrics*, 5, 99–114.

Tukey, J. W. (1954), "Causation, Regression and Path Analysis," chapter 3 in O. Kempthorne et al. (eds.), *Statistics and Mathematics in Biology*, Ames, IA: The Iowa State College Press.

Tukey, J. W. (1954), "Unsolved Problems of Experimental Statistics," *Journal of the American Statistical Association*, 49, 710–713.

Tukey, J. W. (1958), "Bias and Confidence in Not-Quite Large Samples," Abstract, *Annals of Mathematical Statistics*, 29, 614.

U.S. Bureau of the Census (1947), *A Chapter in Population Sampling*, Washington, DC: U.S. Government Printing Office.

U.S. Bureau of the Census (1949), "Illustrative Examples of Two Methods of Estimating the Current Population of Small Areas," *Current Population Reports*, Series P-25, No. 20, Washington, DC: U.S. Government Printing Office.

U.S. Bureau of the Census (1966), "Methods of Population Estimation: Part I, Illustrative Procedure of the Census Bureau's Component Method II," *Current Population Reports*, Series P-25, No. 339, Washington, DC: U.S. Government Printing Office.

U.S. Bureau of the Census (1967), "The Current Population Survey—A Report on Methodology," Technical Paper No. 7, Washington, DC: U.S. Government Printing Office.

U.S. Bureau of the Census (1969), "Estimates of the Population of Counties and Metropolitan Areas, July 1, 1966: A Summary Report," *Current Population Reports*, Series P-25, No. 427, Washington, DC: U.S. Government Printing Office.

U.S. Bureau of the Census (1973). *County and City Data Book 1972* (A Statistical Abstract Supplement), Washington: United States Government Printing Office.

U.S. Bureau of the Census (1975a), "Population Estimates and Projection," *Current Population Reports*, Series P-25, No. 580, Washington, DC: U.S. Government Printing Office.

U.S. Bureau of the Census (1975b), *The Methods and Materials of Demography*, by H. S. Shryock, J. S. Siegel, and associates, 3rd Printing, Washington, DC: U.S. Government Printing Office.

U.S. Bureau of the Census (1978), *The Current Population Survey: Design and Methodology*, Technical Report 40, Washington, DC: U.S. Government Printing Office.

U.S. National Bureau of Standards (1959), *Tables of the Bivariate Normal Distribution and Related Functions*, Applied Mathematics Series 50, Washington, DC: U.S. Government Printing Office.

U.S. National Center for Health Statistics (1958), *Statistical Designs of the Health Household Interview Surveys*, Public Health Series, 584-A2, Washington, DC: U.S. Government Printing Office.

U.S. National Center for Health Statistics (1968), *Synthetic State Estimates of Disability*, Public Health Service Publication No. 1759, Washington, DC: U.S. Government Printing Office.

U.S. National Center for Health Statistics (1977a), *State Estimates of Disability and Utilization of Medical Services, United States, 1969–1971*, D.H.E.W. Publication No. (HRA) 77-1241, Washington, DC: U.S. Government Printing Office.

U.S. National Center for Health Statistics (1977b), *Synthetic Estimation of State Health Characteristics Based on the Health Interview Survey*, D.H.E.W. Publication No. (P.H.S.) 78-1349, Washington, DC: U.S. Government Printing Office.

United Nations Statistical Office (1950), *The Preparation of Sampling Survey Reports*, Statistical Papers, Series C, No. 1, Lake Success, NY: United Nations (see also Revision 2 in 1964).

United Nations Statistical Office (1981), *The National Household Survey Capability Programme Prospectus*, New York: United Nations.

Verma, V. (1992), "Household Surveys in Europe: Some Issues in Comparative Methodologies," in *Seminar: International Comparisons of Survey Methodologies*, Athens.

Verma, V. (1996), "Discussion of Developing Samplers for Developing Countries by Leslie Kish," *International Statistical Review*, 64, 156–162.

Verma, V. (1999), "Combining National Surveys for the European Union," paper presented at the 52nd Session of the International Statistical Institute, Helsinki.

Verma, V., and Lê, T. (1996), "An Analysis of Sampling Errors for the Demographic and Health Surveys," *International Statistical Review*, 64, 265–294.

Verma, V., Scott, C., and O'Muircheartaigh, C. (1980), "Sample Designs and Sampling Errors for the World Fertility Survey," *Journal of the Royal Statistical Society*, Ser. A, 143, 431–473.

Veroff, J., Atkinson, J. W., Feld, S. C., and Gurin, G. (1960), "The Use of Thematic Apperception to Assess Motivation in a Nationwide Interview Study," *Psychological Monographs: General and Applied*, Fall, 1–32.

Wallis, W. A. (1966), "Economic Statistics and Economic Policy," *Journal of the American Statistical Association*, 61, 1–10.

Watson, A. N. (1947), "Respondent Pre-Selection Within Sample Areas," No. 2 of Technical Series on Statistical Methods in Market Research, Philadelphia: Curtis Publishing Company.

Wiggins, R. D., Longford, N., and O'Muircheartaigh, C. A. (1992), "A Variance Components Approach to Interviewer Effects," in A. Westlake, R. Banks, C. Payne, and T. Orchard (eds.), *Survey and Statistical Computing*, Amsterdam: North-Holland.

Wilk, M. B., and Kempthorne, O. (1955), "Fixed, Mixed and Random Models," *Journal of the American Statistical Association*, 50, 1144–1167.

Wilk, M. B., and Kempthorne, O. (1956), "Some Aspects of the Analysis of Factorial Experiments in a Completely Randomized Design," *Annals of Mathematical Statistics*, 27, 950–985.

Wold, H. (1956), "Causal Inference From Observational Data," *Journal of the Royal Statistical Society*, Ser. A, 119, 28–61.

Wolter, K. M. (1985), *Introduction to Variance Estimation*, New York: Springer Verlag.

Woodruff, R. S. (1952), "Confidence Intervals for Median and Other Positional Measures," *Journal of the American Statistical Association*, 58, 454–467.

Woodruff, R. S. (1966), "Use of a Regression Technique to Produce Area Breakdowns of the Monthly National Estimates of Retail Trade," *Journal of the American Statistical Association*, 61, 496–504.

Woodruff, R. S. (1971), "A Simple Method for Approximating the Variance of a Complicated Estimate," *Journal of the American Statistical Association*, 66, 411–414.

World Fertility Surveys (1984), *Major Findings and Implications*, The Hague: International Statistical Institute.

Wright, S. (1954), "The Interpretation of Multi-Variate Systems," chapter 2 in O. Kempthorne et al. (eds.), *Statistics and Mathematics in Biology*, Ames, IA: The Iowa State College Press.

Yates, F. (1946), "A Review of Recent Statistical Developments in Sampling and Sampling Surveys," *Journal of the Royal Statistical Society*, 109, 12–43.

Yates, F. (1949, 1953, 1960, 1981), *Sampling Methods for Censuses and Surveys*, 1st, 2nd, 3rd, and 4th editions, London: Griffin & Co.

Yates, F. (1951), "The Influence of *Statistical Methods for Research Workers* on the Development of the Science of Statistics," *Journal of the American Statistical Association*, 46, 32–33.

Yates, F., and Cochran, W. G. (1938), "The Analysis of Groups of Experiments," *Journal of Agricultural Science*, 28, 556–580.

Yeomans, K. A., and Golder, P. A. (1975), "Further Observations on the Stratification of Birmingham Wards by Clustering: A Riposte," *Applied Statistics*, 24, 345–346.

Yule, G. U., and Kendall, M. G. (1937), *An Introduction to the Theory of Statistics*, London: Griffin & Co.

Yule, G. U., and Kendall, M. G. (1965), *An Introduction to the Theory of Statistics*, 14th edition, London: Griffin & Co.

Zarkovich, S. S. (1956), "Note on the History of Sampling Methods in Russia," *Journal of the Royal Statistical Society*, Ser. A, 119, 336–338; addendum in 1962.

Index